WALDEN'S SHORE

WALDEN'S SHORE

Henry David Thoreau and Nineteenth-Century Science

Robert M. Thorson

Harvard University Press
Cambridge, Massachusetts
London, England
2014

Printed in the United States of America

Library of Congress Cataloging-in-Publication Data

Thorson, Robert M., 1951–
Walden's shore : Henry David Thoreau and nineteenth-century science / Robert M. Thorson.
pages cm
Includes bibliographical references and index.
ISBN: 978-0-674-72478-5
1. Thoreau, Henry David, 1817–1862. Walden. 2. Literature and science—United States—History—19th century.
3. Nature in literature. 4. Walden Woods (Mass.) I. Title.
PS3048.T55 2013 2013013531
818'.303—dc23

To those who share Thoreau's vision
that Nature is a sacred place

Erratic boulders at Nonesuch Pond by Herbert Gleason. From undated, hand-tinted lantern slide originally titled "Learning Rock." Thoreau described and sketched this scene in Natick, Massachusetts, in his *Journal* on November 7, 1851.
Courtesy Concord Free Public Library.

If you have built castles in the air, your work need not be lost;
that is where they should be. Now put the foundations under them.

—Henry David Thoreau, 1854

CONTENTS

ILLUSTRATIONS

PREFACE

We were sojourners in Thoreau Country: an odd assortment of birdwatchers, pilgrims, and the merely curious. Gathered in a loose huddle, we stood on a high cliff overlooking Fair Haven Bay, a sheet of blue water similar to nearby Walden Pond, but with a river running through it. The Sudbury River, a sluggish stream that wiggles its way northward to become the Concord River before flowing past the oldest inland town in New England, settled in 1635.

Figure 1. Thoreau's "observatory" at Fair Haven Hill by Herbert Gleason (1903). Original caption reads "Across face of Cliffs toward Conantum." Rock is locally resistant Andover Granite. View is across a broad glacial "notch."
Courtesy of Concord Free Public Library, Image #57 (October 15, 1903).

It was a sparkling Sunday morning, church-like among tall pines. Our "congregation" was a field trip, one of three sponsored by the 69th Annual Gathering of the Thoreau Society, the oldest and largest organization devoted to an American author. Our guide was the lanky Peter Alden, an indefatigable naturalist whose family roots go back to the Ur-settlement of New England at Plymouth Rock. At each stop, his modus operandi was to point out some species of flora or fauna, and then say something about it. With his encouragement, and from our cliff-top observatory, I digressed to more earthly matters, giving a spontaneous discourse on ice-quakes, aquifers, granite, tundra times, channel hydraulics, tectonics, acid soils, and sundry other geo-topics. That field trip digression precipitated this book.

Above us we heard the screech of a spiraling hawk, perhaps the progeny of those seen almost two centuries ago by the Thoreau children during one of their family outings. All eyes went upward. I had to redirect them back down to the Andover Granite, which poked through the forest floor here and there. Craggy and jagged in some places. Rounded and smooth in others. Sometimes glaciers yank rock loose. Sometimes they grind rock down. But always they melt away, releasing and washing whatever residues they were transporting at the time. The net result over Walden Woods about 16,000 years ago was an irregular drape of silt, sand, gravel, and stone let down on the older and more substantial crust.

My field trip companions were especially intrigued by the idea that two similar kettles—Walden Pond and Fairhaven Bay—could have such different literary outcomes. Walden's elevation, isolation, and depth made it a "deep green well" in the forest, one that Thoreau claimed as a symbol of eternal purity. In contrast, Fairhaven Bay became a wide spot in a sometimes-murky river carrying runoff from agricultural meadows and fields to the south. Similar landforms. Different floras, faunas, and fates. Loons in one place. Gulls in another. Walden a timeless "forest mirror," and Fair Haven a quiver for time's arrow, "the stream I go a-fishing in." Indeed, the contingencies of pre-history govern the outcomes of landscapes just as surely as history governs the outcomes of nations, and biography the paths of our own lives. What if Thoreau had never met Ralph Waldo Emerson? What if Thoreau had left Walden Pond early and gone west to help survey the mineral lands of Michigan, a job he was qualified for and interested in doing?

While walking past the Andromeda Ponds on the way back to our parked cars, one of the field trip participants—I forget which one—

suggested I write a summary of my cliff-top comments for the *Thoreau Society Bulletin.* I toyed with the idea later that morning by outlining what I remembered saying. To that list I added several points from my talk the previous afternoon titled "Walden as the Fourth Derivative of Geology." That evening, during a reception at the Thoreau Institute for our keynote speaker, Megan Marshall, I broached my plans for an article to a former executive director of the Thoreau Society, Jayne Gordon. She was the first to encourage me to carry on.

Then came two surprises that transformed my loose outline into the book you're now reading. In July 2010, during my initial background research, I pulled out my dog-eared copy of Barksdale Maynard's *Walden Pond: A History,* which I had found to be a very useful compendium of the local social intercourse. On prior readings, I had glossed over its single paragraph on geology because I was already familiar with the material. On this more careful reading, however, I was stunned to find several misleading statements that the author unwittingly passed on from his sources: errors that I've since traced back to an un-reviewed field trip guidebook. In that moment, it struck me that a short, plain-language, peer-reviewed book on the creation story of Walden Pond might be appreciated by those with an interest in this famous place. By August, I was ensconced in a cabin overlooking a lake in Maine and working on such a manuscript. Beginning at sunrise each morning, I wrote like a man possessed until I had a draft in hand. Within the year, I had an editor, my agent had a contract, and my geo-narrative was nearly complete.

The second surprise came in May 2011, and more than doubled the scope of what I originally had in mind. As a self-reward for finishing a preliminary draft and turning in my grades for spring semester, I took some time off to read Thoreau's *Journal* for the first time. Its combination of chaos and delight blew my *Walden* project far off course, adding nearly two years to the schedule. Two passages were especially critical. On February 3, 1852, Thoreau narrated an astonishingly accurate vision for the ice-sheet glaciation of Concord, a subject I thought he knew nothing about. Two months later on April 11, 1852, he revealed his genius for river-channel hydraulics, something readers of *Walden* would never suspect. In that passage, he described the three-dimensional helicoidal flow responsible for shaping the meandering channel of Nut

Meadow Brook, putting him ahead of the state geologist in his understanding. Struck by his prescient insights, and with my editor's encouragement, I expanded my narrative to showcase Thoreau's achievements as a physical landscape scientist, and to put his ideas in the context of nineteenth-century American geology.

The first glimmer of this project dates back to 2003, when I developed an experimental course titled "Geoscience through American Studies," now the oldest one in the Honors Core Curriculum at the University of Connecticut. Within a year, the New England historian Robert A. Gross and I—along with the landscape photographer Janet Pritchard—were collaborating on a second interdisciplinary course titled "Walden and the American Landscape." This became my portal into what Leo Marx called the "sedate precincts of Thoreau scholarship," but which I found quite exciting. Since 2004 we've taught this course with additional colleagues from the departments of History, English, and Art & Art History, namely Christopher C. Clark, Wayne Franklin, Matthew MacKenzie, and Sydney Landon Plum. All helped lead me through the maze of ideas in *Walden,* and encouraged me to keep working on the physical science informing Thoreau's masterpiece. In a parallel scholarly seduction, I became involved in a campus-wide reading group on environmental history.

In 2009 I shared some of my emerging Thoreauvian ideas in *Beyond Walden: The Hidden History of America's Kettle Lakes and Ponds.* This book featured the galaxy of sandy, glacial lakes extending from the cranberry bogs of Nantucket to the prairie potholes of Great Falls, Montana, the most famous of which is Walden Pond. During that project, the tug of *Walden*'s intellectual gravity kept pulling me back toward Concord. Eventually, I yielded to its attraction and joined the Thoreau Society. One thing led to another, and pretty soon I was reading my way through Robert Hudspeth's bibliography as serialized in the society's *Bulletin.* Especially helpful works were Robert Sattelmeyer's *Thoreau's Reading,* Walter Harding's *The Days of Henry Thoreau,* and Jeffrey Cramer's *Walden: A Fully Annotated Edition.* My clearest introductions to Thoreau's scientific thinking were Donald Worster's *The Economy of Nature,* Robert Richardson's *Thoreau, A Life,* and Laura Dassow Walls's *Seeing New Worlds.* For ecocriticism I turned to Lawrence Buell's *The Environmental Imagination* and Leo Marx's *The Machine in the Garden.* For the history of geology, I depended on Frank Dawson Adams's *The Birth and Development of the Geological Sciences,* and Stephen Jay Gould's *Time's Arrow—Time's Cycle.* The data-rich *Geo-*

hydrology and Limnology of Walden Pond by John Colman and Paul Friesz of the U.S. Geological Survey became my essential reference for how Lake Walden works as a natural system.

Completion of *Walden's Shore* was supported by a sabbatical leave from the Department of Ecology & Evolutionary Biology and the Department of Anthropology at the University of Connecticut. A visiting appointment in American Studies at Harvard University during its finishing stages helped me complete the work, thanks to the help of Arthur Patton-Hock after a referral from Lawrence Buell.

I gratefully acknowledge research support from the library staff at the University of Connecticut, especially its Interlibrary Loan Department and the archivists at the Dodd Center, namely Melissa Watterworth Batt and Terri Goldich. Others who helped me access resources include Jeffrey Cramer, Curator of the Thoreau Institute collections; Michael Frederick, Executive Director of the Thoreau Society; Susan Halpert of Harvard's Houghton Library; William Hosley of Terra Firma Northeast; Joseph Kopera of the Massachusetts Geological Survey; Mary Sears of Harvard's Museum of Comparative Zoology; staff at the A.C. Clark Library at Bemidji State University in Minnesota; Mike Volmar, Head Curator of Fruitlands Museum; Leslie Perrin Wilson, Special Collections Librarian at the Concord Free Public Library; and Violetta Wolf of the Boston Museum of Science. Astronomer Cynthia Peterson of the University of Connecticut and atmospheric scientist Don Hampton of the Geophysical Institute at the University of Alaska helped me with solar zenith calculations and atmospheric phenomenology, respectively.

This project benefited from many helpful reviews. My greatest debt is to Wayne Franklin, who—while chairing the Department of English at UConn—found the time to read the first draft of my geology manuscript in January 2011 and advise me what to do next. Sherman Clebnik and Jon Inners provided helpful technical corrections on large blocks of the manuscript. The insights of two anonymous readers for Harvard University Press had significant influence. Collegial reviews of discipline-based portions were offered by Steve Grant, Michael Frederick, John Inners, Joseph Kopera, Edward Mooney, Will Ouimet, and Byron Stone. Conversations with emeritus glacial geologist Carl Koteff and other participants of a Geological Society of America field trip to Walden Pond in March 2012 helped guide my interpretations of that section.

At Harvard University Press, I wish to thank my editor, John Kulka, for his insight, steerage, and support, and Heather Hughes for her able administrative skills. Heidi Allgair and Sarah Perillo were as fastidious with their manuscript editing as they were patient with my many corrections. My agent, Lisa Adams of the Garamond Agency, was always supportive. And finally, I thank my wife, Kristine, for too many things to mention.

INTRODUCTION

The historic steeple of First Parish Church pokes the sky just south of the Concord town green. There in 1817, a healthy infant named David Henry was baptized with water by the town's spiritual patriarch, Unitarian minister Rev. Ezra Ripley. There in 1862, the coffin of a man who called himself Henry David was strewn with wildflowers while his memory was eulogized by Ralph Waldo Emerson, whose fall from Unitarian grace had made him world famous. "I know not any genius," Emerson said of his best friend, "who so swiftly inferred universal law from the single fact."[1]

Thoreau was indeed a genius. And what they do best is a mental maneuver called induction: the taking of disparate facts and linking them into more general ideas. Novelists do it when they employ archetypes, biographers when they sum up a life, and scientists when they derive fundamental laws. Sir Isaac Newton's seventeenth-century law of gravity, for example, unified the disparate facts of planetary orbits, the fall of an apple, and the buoyancy of a ship. His intellectual heir at Cambridge, physicist Stephen Hawking, traced this maneuver back to Pythagoras, one of Thoreau's heroes from the fifth century BCE.[2]

Using his powerful induction, Henry Thoreau came within an inch of discovering a scientific law that would have explained why the landscape he saw from his doorway looked the way it did. A general explanation that would have unified the "natural facts" of Walden's middling size, roughly rounded shape, deep bottom, steep banks, acid soils, pine-oak forest, stony shore, and the astonishing clarity of its water. Geologists have since learned that all these attributes derive from a single unnamed "thaw" law governing the collapse of loose sediment near stagnant glacial ice to create a landform known as a kettle. This law is so general that it can be applied anywhere in any epoch, and yet specific

Figure 2. Slope of inward collapse above Walden Pond. View to east shows southeastern portion of Walden Pond below Heywood's Peak.
January 2013.

enough to be seen as a special case of conical collapse within granular materials. Examples familiar to Thoreau included the dimple of sand in a flowing hourglass, the death trap of an ant lion, and the perfectly "circular" hollows he saw on outer Cape Cod, whose "sands had run out" millennia before he encountered them in October 1849.[3]

Henry's close call with discovering a local "theory of everything" was inspired by his legendary fathoming project, which he undertook "before the ice broke up, early in '46." Using a surveying compass and Gunter's chain to lay out transects on the frozen surface, something to

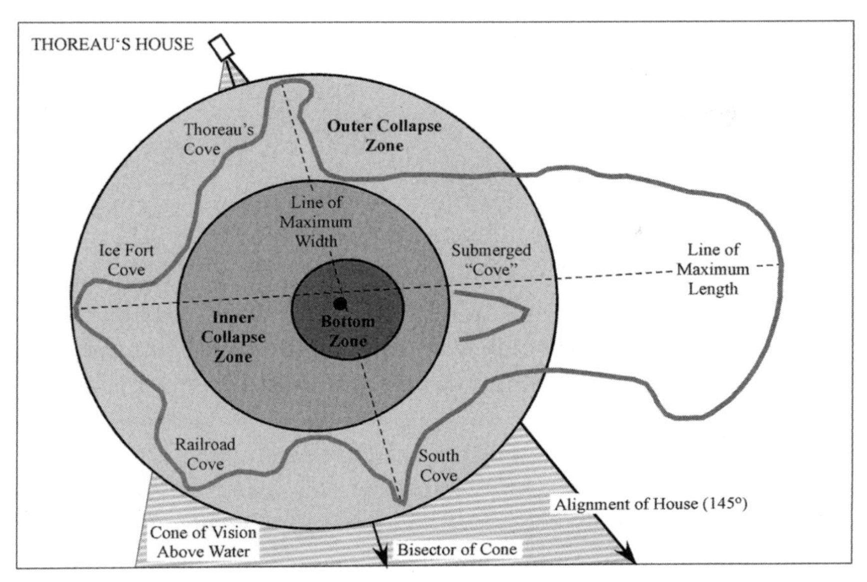

Figure 3. Thoreau's cone of vision over the bull's-eye symmetry of Walden's western basin. From his doorway, his horizontal arc of vision coincided roughly with the diameter of the basin. Progressively shaded circles show outer and inner limits of its zone of collapse and its flat bottom. All are concentric around the nadir at −102 to −107 feet depth, shown by black dot. The bisector of the cone lies very near the nadir and line of maximum width.

chop more than a hundred holes through sixteen inches of ice, a fist-sized stone tied to a cod line to serve as a fathometer, and a home-made plumbago pencil, he discovered that the main part of the lake in front of his house—its star-shaped, cove-tipped western basin—exhibited a radial symmetry from its deepest hole to its forested rim.[4]

Eureka!

So important was this discovery that, before publication of *Walden,* he converted his desk-sized bathymetric survey map into a page-sized "Reduced Plan," and insisted that it be included in his otherwise literary text, creating what one leading scholar called an "unexpected and slightly bizarre interruption." Thoreau's publishers probably had a similar reaction because they omitted this technical drawing from some copies of *Walden*'s first edition, and removed it entirely from those shortly after his death. One nineteenth-century reviewer wondered if the bathymetry was a satire on then-popular, but costly government coastal surveys. A more recent critic saw the map as parody: "a wonderful, fanciful conceit that critiques strictly empirical accounts of nature as woefully inadequate."[5]

More than a century after *Walden*'s publication, however, critic Leo Marx saw what Henry Thoreau wanted us to see: that the crosshairs of length and width for the lake coincided with its maximum depth at the bull's-eye of Thoreau's local universe. Only from this "mystic center" did the *lake*-scape ripple outward from the deep blue of its open water to the "vivid green" of its shallows, and the *land*-scape radiate upward from its white stony shore to the steep forested bank, flaring kettle brink, and "blue cauldron of the air." Only from the geometric origin of that circle could the Fates send his drifting rowboat to whatever summer shore they willed. Only from that sweet spot does the word "environment" make literal sense, given that its Latin root *environs* means "to surround." In a nutshell, the radial symmetry of Walden's western basin gave rise to what Marx called the "unitary coherence and power" of *Walden* the book.[6]

Beneath the Epicenter

In seismology, the epicenter of an earthquake is the geographic point on Earth's surface directly above the place where fault rupture began at depth, its hypocenter or focus. Though the epicenter gets practically all the media attention, geoscientists are far more interested in the focus, because that's where fundamental causes and explanations are to be found. The same principle holds true for *Walden* scholarship. Though deep thinkers from many disciplines have waxed eloquently about Henry's *epiphenomenal* experiences on and near the water, precious few have plunged downward below the world of diving loons and blundering bullheads in search of the pond's *focus,* where deeper causes and more general explanations are to be found. "I cannot come nearer to God and Heaven," Henry wrote of Walden's "deepest resort," his closest approach to the focus. Only there did the "magic circle" of its shore surround the line of the gravitational vector that—during deglaciation—sucked everything downward toward the crystalline iron core of our rapidly spinning and highly magnetic planet.[7]

Thoreau moved into his drafty house overlooking the lake on July 4, 1845. Within two days, he'd noticed that the immediacy and volatility of life at the surface quickly gave way to the greater stability of gentle currents within the upper layer of the lake, its epilimnion. "Even time has a depth," he wrote in his *Journal*, "and below its surface the waves do not lapse and roar." At greater depth lay nearly motionless water of

Walden's deep hole, its hypolimnion, which is cupped by clay and mixes only twice per year. And below that tranquil liquid lay the slushy muck Thoreau detected with his stone-on-a-string, his "soft" bottom. That is as far as he got. We now know that below his soft bottom are eleven inches of semisolid organic sediment dating back to Concord's colonial days, and holding an independent archive of its history. And below that are another twelve feet of solid organic sediment spanning the entire postglacial epoch. And below that is a thin sequence of silty mud, muddy sand, glacial gravel, and rough-handled boulders dating to about 16,000 years ago; all of which is younger than the magnificent European Paleolithic cave paintings of Lascaux in France and Altamira in Spain. And below everything is Walden's "hard bottom and rocks in place," a mass of smoothed granite older than the oldest dinosaurs, a small portion of North America's stable tectonic basement that's been billions of years in the making.

At this point of contact between the deepest unconsolidated sediments and solid rock, time has reached a depth where nothing lapses and nothing roars. This is Walden's ground zero. The place where the vector of meltdown subsidence pulled most strongly on what had been the flat gravel plain of a dusty glacial delta. The initial result was a shallow sag, perhaps with a few random ruptures. They coalesced to outline a shallow crater, and then a growing hollow, and then an unstable lake whose flanking slopes were being scalloped by slumps, tumbled by stones, choked by mudflows, and gashed by gullies. Growing both above the water and below it, this enlarging depression eventually reached bedrock more than a hundred feet farther down. Meltdown ended. The final result was a flat-bottomed kettle lake "walled in" by steep collapse slopes and fluted by triangular coves. Tundra crept in to paint the land velvet green. Turbid water clarified to paint the lake azure blue. Between them was the white "magic circle" of *Walden's Shore*.[8]

Just as surely as one of Thoreau's *Wild Fruits* emerged in stages from seed to flower to food, so too did Walden Pond emerge in stages from buried ice, to sterile lake, to an aquatic ecosystem. More parsimoniously, a stagnant block of subarctic ice blossomed into a nineteenth-century pastoral retreat.

Enter Thoreau. A childhood visitor in the 1820s, his eyes filling with wonder. An adolescent wanderer in the 1830s, falling in love with this scarcely used "outback." A Harvard alumnus and Emersonian mystic during the 1840s, building a snug house near its shore. A self-taught sojourning scientist during the early 1850s, using his skills and experience to leverage detailed field observations into incomparable prose-poetry. Quite correctly, he compared Walden's crosshairs symmetry to that of a mountain, especially Mount Monadnock, whose central dome and buttressing spurs were topographic reciprocals to Walden's deep hole and flanking coves, respectively. The geologists of his day correctly understood that such isolated mountains were erosional residuals, places anomalously resistant to long-term weathering. What they lacked was a collateral scientific theory to explain how deep, isolated basins like Walden could open up beneath a flat "shrub oak plateau" composed mostly of sandy river gravel.[9]

Henry got stuck.

Lamentably, his career as a nature writer coincided with one of the greatest goofs in the history of American science: the official rejection of the glacial theory between 1842 and 1862. These dates exactly bracket Thoreau's "Natural History of Massachusetts," "his first published essay in the genre that was to make him famous," and his deathbed dictations to sister Sophia for posthumous publication. Under the leadership of Dr. Charles T. Jackson (Ralph Waldo Emerson's brother in law) and Reverend Edward Hitchcock (New England's leading natural theologian), the American Association of Geologists and Naturalists erred as cultural conservatives, resistant to the novel idea that a sluggishly creeping ice sheet of continental proportions had once buried their landscape. They erred as liberal patriots, manifesting Emerson's plea to turn away from the "courtly muses" of Europe: in this case the figure of Jean Louis Rodolphe Agassiz, who "epitomized Continental science" and whose famous work *Etudes sur les Glaciers* had been published in French. They erred as descriptive natural historians, preoccupied with classification and unwilling to embrace the more mathematically rigorous glaciology of Scottish physicist James D. Forbes, who proved that snowflakes could metamorphose into a quasi-plastic solid capable of slow viscous motion. And finally, they erred because of bad luck, what historian of science Thomas Kuhn called the "apparently arbitrary element" of scientific progress "compounded of personal and historical accident."[10]

So, instead of adopting the now-obvious concept of ice-sheet glaciation, these self-appointed New England "Men of Science" clung steadfast to idea that the salient features of New England's physical landscape

Figure 4. *The Iceberg* by Frederic E. Church (1891). During Thoreau's era, the conservative scientific establishment preferred the iceberg drift theory as a mechanism for creating the sediments and landforms of Concord.
Frederic Edwin Church, American (1826–1900), oil on canvas, Height 28⅓ × Width 39¼ inches. Courtesy Carnegie Museum of Art, Pittsburg. Howard N. Eavenson Americana Collection, 72.7.3.

were created by debris-laden icebergs drifting from higher latitudes via powerful oceanic currents. Historians of geology conclude that this blunder "set back glacial theory in America by almost a generation." Thoreau's generation![11]

Midway through this lacuna of confusion, and two years before the submission of his *Walden* manuscript, Thoreau left unambiguous evidence in his *Journal* that he accurately envisioned a great continental ice sheet invading from Canada to engulf Concord. By the summer of 1852 he had adopted the glacial lexicon—moraine, erratic, glacier, iceberg—to help explain his sojourning terrain, and had read compelling descriptions of existing ice sheets from Greenland by explorer Elisha Kent Kane, and from Patagonia by Charles Darwin. By the time of *Walden's* publication in 1854 he had dabbled with the ice physics of James Forbes and had at least skimmed portions of Agassiz's catastrophic vision.[12]

Yet not one tweet of all this highly relevant information shows up in *Walden*. In place of cold, hard theories, we get quasi-comic creation myths and obscure allegories that are as accurate as they are delightful to read. The "old settler and first proprietor" who arrived during snowy winters to divine, dig, and stone the pond. The "new views" from frozen lakes, which resembled those of Baffin Bay. The demigod "Thaw with his gentle persuasion," who "sunk" Walden below its prairie plateau. The Indian fable of downward collapse, tumbling stones, and seismic adjustment borrowed from a kettle in Connecticut. The "lusty old dame," who gave birth to the postglacial ecosystem. The "Cold Fridays" and "Great Snows" alerting his readers to the precariousness of human existence on what must surely have been—in the not-too-distant-past—a very frigid planet. And worst of all, Thoreau's deadpan vision of human extinction due to a return performance of Agassiz's *Eizeit*, his global ice age.[13]

My Thoreau

Richard Schneider's *Thoreau's Sense of Place* is a broad, multi-authored review of recent *Walden* scholarship from the humanities that I found very helpful, especially as a goad to keep working on the geoscience. In the book's preface, Lawrence Buell asks this question: "To what extent was Thoreau biocentric, to what extent anthropocentric?" This was rhetorical, because he'd answered this question in his earlier treatise, *The Environmental Imagination:* "The development of Thoreau's thinking about nature seems pretty clearly to move along a path from homocentrism toward biocentrism."[14]

This statement is certainly true for the Alpha and Omega of Thoreau's intellectual biography, and for its general trajectory from his poetic neoclassical transcendentalism of the late 1830s to his data-boggled phenological Kalendar of the early 1860s. But the mid-career masterpiece of his writing career—the *Walden* upgrade between 1852 and 1854—is thematically geocentric: "I am its stony shore." And its strongest metaphors are earthly ones: The clay of the universal potter at the Deep Cut. The "golden mean" of the "rule of the two diameters." The clean sand of his cellar as the most "solid and honest" part of his house. And most important of all, Thoreau's conviction that bedrock and truth are one in the same. "Let us settle ourselves, and work and wedge our feet downward," he invites his readers, "till we come to a hard bottom and rocks in place, which we can call reality."[15]

This book explores that rock reality, beginning with the Andover Granite, the official geological "formation" mapped beneath Walden Pond. Decades of outcrop mapping by the U.S. Geological Survey, combined with aeromagnetic data from the 1970s and seismic reflection from the 1990s, prove that this granite forms the "hard bottom and rocks in place" beneath Walden's aqueous epicenter. Working upward from this igneous simplicity, I review the physical science—geology, hydrology, limnology, meteorology, pedology, optics, acoustics, chemistry, and physics—of Thoreau's nature writing in *Walden*. My treatment includes the mechanical and chemical roles of living organisms but ignores their biological aspects—botany, zoology, physiology, and phenology—because they are treated so well elsewhere. Ecology, being part of the pond's energy flow, and evolution, being part of paleontology and archaeology, are touched on here and there.[16]

Focusing on the physical science returns me to what, in 1939, the biographer Henry Canby called his "new Thoreau," the man who carried out "an exacting study of the depth, flows, colors, seasons, ice and fishy denizens of Walden Pond, to the last exactitude." Importantly, wrote Canby, "Walden Pond was put through as elaborate a physical test as a recruit for the army. No more could be said of it when he was through, except what a chemist, biologist, or geologist might have added. Here was a 'new Thoreau.'"[17]

That same year, a young Connecticut zoologist named Edward Deevey made a holiday visit to Walden Pond to explore this unusually deep lake. Because Thoreau's bathymetric map had been excised by the publisher from Deevey's edition of *Walden,* he had no choice but to resurvey the pond to re-locate its deepest hole. When he later found a copy of Thoreau's bathymetry, he was taken aback by both its measurement accuracy and the scientific quality of Thoreau's insights. The Yale community, to which both Canby and Deevey were connected, soon claimed Thoreau as a pioneering limnologist, an idea that gained less traction closer to Boston.[18]

Their "new Thoreau" is my Thoreau. The pioneering field scientist who, working alone with primitive instruments and what the latter called a "pathetic" research program, nevertheless made hundreds of astute observations and prescient inductions about the lake and its surrounding landscape. He clearly meets the legal definition of a scientist being used by the U.S. Supreme Court: one who posits "theoretical explanations about the world that are subject to further testing and refinement."

Though my Thoreau was undoubtedly one of Edgar Allan Poe's transcendental "frogpondians," he was the one more interested in why the pond was blue than why the frog was green. My Thoreau was curious for curiosity's sake about the physics of ice, air, water, light, sound, and heat, especially during his scientific sojourning stage (1850–1854), the results of which he translated into the lion's share of the second half of *Walden*. He was the one who "worked diligently to find nature's meaning within the physicality of nature itself," as Rochelle Johnson concluded, rather than somewhere "out there" in Emersonian hyperspace. He was the one who wrote what Philip Gura called "crystalline" prose.[19]

To conclude, as William Howarth did, that "mineral elements did not intrigue him" is patently false. Quoting Shakespeare, my Thoreau sought "sermons in stones," eventually building a mineral and rock collection as a distinct part of his natural history program that is virtually unknown. Some specimens he gathered for the beauty of their crystals, some as natural curiosities, and some as mundane voucher specimens of local rock. My Thoreau kept rocks on his writing desk. He sought wealth from golden spangles of Concord mica, rather than nuggets of California ore. He invoked obscure technical names from mineral taxonomy as veiled teases: essonite (garnet) against Reverend Hitchcock's supernaturalism; and selenite (gypsum) against himself for the empty "white spots" in his summer journal, meaning its blank pages. His specimen of selenite is indeed both white and clear, linking it to his text. Finally, Thoreau lovingly celebrated his favorite mineral—H_2O ice—for its optical properties on bright days and its acoustic properties on dark nights. Metaphorically, it was "azure tinted marble" when quarried from the lake and stacked up like a fortress in the sun. Literally, it was the butt of his jokes at night, a restless bedfellow prone to "flatulency" beneath a blanket of snow.[20]

My Thoreau needs resuscitation, having been nearly drowned beneath breaking intellectual waves of the last century. A wave of recognition during the late 1930s for his literary contributions to the American Renaissance. A wave of hero-worship during the mid-1940s that culminated in the desecration of his house site with a "hammered" stone monument of the sort he would have loathed. "To what end, pray, is so much stone hammered?" he wrote of such architecture. A wave of interpretation during the "'myth and symbol' school of American Studies dating to the 1950s and 1960s." A wave of social reform during the civil rights foment of the 1960s and 1970s. A wave of ecological thinking, wilderness preservation, and pollution-control during the 1970s and 1980s when he spoke "a word for nature." And in this millennium, a wave of marketing Thoreau as the symbolic "green" man—whether urban or rural—complete with Walden video games and folk ballads.[21]

All true. All well deserved. All important. All shoaling the transcendental shores. But gasping for air beneath these successive waves was a pretty good non-transcendental scientist, especially during his Apollonian moods. A "regular hairshirt" of a field worker whose compass and measuring stick were as important as his "old book" of a plant press. The late Ed Schofield was hot on his trail twenty years ago when he wrote: "There is a direct cause-and-effect link between the surficial geology of the Walden Woods area and both the substance and character of Thoreau's writings."[22]

More generally, I seek to counterbalance what strikes me as a recent trend in ecocriticism that refracts science through literature without being scientific. Tom Calderwood's "An Astronomer Reads Thoreau" provides a notable exception to an otherwise almost purely descriptive literature. Finally, we must not forget that in the climax of *Walden,* Thoreau asks us to imagine a "living earth; compared with whose great central life all animal and vegetable life is merely parasitic." This book explores the geodynamics of that "living earth" beneath the more congenial domain of ecology.[23]

This book—perhaps because of my archaeological and paleontological background—follows a downward path to Thoreauvian simplicity. I excavate more than I explicate. "Down to the pond" for reasons galore, if only to get drunk on the water, as E. B. White quipped. Down for spiritual *katabasis* to obtain a reverential, rather than a magisterial, view. Down to scientific law through induction. Down to the holism of radial geometry. Down to the simplicity of the cone and hemisphere. Down to the thermodynamic origin of life at the Deep Cut. Down from the swaying branches of history to the solid trunk of myth, and from modern languages to the "father tongue." Down to actual "Contact!" with rock, whether on the summit of Katahdin or in the bowels of a quarry. Down to the sands of Sleepy Hollow, where Thoreau's bones remain today. To my geological mind, Thoreau's lifelong search for simplicity was more *de*scendental than it was *trans*cendental.[24]

Geo-*Walden*

"For the first week, whenever I looked out on the pond it impressed me like a tarn high up on the side of a mountain." Robert Sattelmeyer cites this passage from *Walden* as evidence that Thoreau was transforming a "busy commercial and agricultural site with a long and complex history of human settlement into a remote forest lake." Agreed. An additional, and more parsimonious explanation is suggested by Thoreau's use of the phrase "reminded me of" in *Walden* Version I (1847), rather than "impressed me like" in Version VII (1854). Though kettle and tarn lakes are indeed very different in terms of setting, material, and origin, they are surprisingly similar in terms of size, shape, curvature, and purity. Perhaps these were the shared attributes that Thoreau wanted to call our attention to, not their differential elevation and degree of remoteness.[25]

When describing his house interior, Thoreau jolts us with this: "I had three pieces of limestone on my desk," which he threw "out the window in disgust," ostensibly because he did not want to dust them. But why limestone? Why three pieces? Was it really limestone? Why "disgust" at something that, in other circumstances, he enjoyed like the "bloom on fruits, not to be swept off." Could something other than housekeeping or a veiled critique of marriage have been involved? I suggest that Thoreau was having a "fit" of frustration brought on by his failure to burn his own lime for mortar? This forced him to buy two casks from the factory, which devalued his house as a symbol of self-reliance and inflated its construction cost, adding nearly $75 in equivalent modern dollars.[26]

Half a century ago, Melvin Lyon concluded that the "theme of Walden and the chief factor in determining its inner structure is the desire for redemption, or rebirth." If so, then Thoreauvians of the twenty-first century might appreciate knowing that the cyclical frequency of rebirth can now be extended to the glacial-interglacial metronome. Just as surely as "frost coming out of the ground . . . precedes the green and flowery spring," so too does the physical withdrawal of an ice sheet precede the biological re-colonization of the land, a transition also driven by orbital planetary astronomy. Given Walden's setting in an ancient bedrock valley, the lake we know today may be the third or fourth incarnation of nearly the same thing in nearly the same place. Just as surely as every spring thaw erases the evidence of earlier thaws, so too might the last ice sheet have erased evidence of previous Waldens. Thoreau's *Journal,* which links climate change to orbital variations, suggests he was on this seemingly modern train of thought by February 1854, only one month shy of submitting his *Walden* manuscript.[27]

In an oft-quoted and oft-misunderstood passage about drifting on Walden Pond in his hand-made rowboat, "The Rover," Henry reminisced about spending "many an hour when I was younger, floating over its surface as the zephyr willed, . . . and I arose to see what shore my fates had impelled me to." H. Daniel Peck asserts that Thoreau sought the middle to "gain a perspective of centrality and serenity." Agreed. To this I would add Thoreau's sophisticated knowledge of lake micrometeorology. In the full passage of *Walden,* and even more explicitly in the *Journal*, he left several clues as to why he sought the middle for physical reasons alone. Can you find them?[28]

These geocentric tidbits—tarn, limestone, Waldens, and zephyrs—illustrate how Thoreau used his knowledge of physical science to set up countless *Walden* passages. But after setting them up, he threshed the technical details and rigorous explanations away, leaving behind kernels of ultra-dense prose. Luckily for literary scholars, the informing scientific content for most *Walden* passages survives in the *Journal* as fragments that can be pieced together if one knows what to look for. In fact, the concordance spreadsheet I created to compare these texts proves that Thoreau thought long and hard about landscape science while composing his masterpiece. That he continued this habit to the end of his life is proven by the very last sentence of his *Journal*, which has absolutely nothing to do with society, philosophy, poetry, or proto-ecology. Instead it has to do with geology: the asymmetric "stratification" created by rain-splash erosion on the exposed bank of the railroad causeway. From this analytically refined observation, he drew his last induction: "Thus each wind is self-registering."[29]

Finally, the creation narrative for Walden Pond remains blurry in popular culture, even though it's straightforward and uncontroversial. Consider the official "geology" page of the Walden Pond State Reservation Web site, which reads as follows: "A large block of ice broke off into glacial Lake Sudbury," was "surrounded by sediments . . . flowing from the glacier," and "left behind an indentation that eventually filled with water." This "easy-speak" is fine for the 99 percent who want to skim the surface of understanding. But for more sophisticated readers, it's misleading, especially for writers because their metaphors magnify misunderstandings. More accurately, the glacial lake expanded northward as debris-rich stagnant ice melted downward beneath a fixed shoreline elevation, leaving submerged residual masses. Little, if any, breakage was involved. Multiple blocks, rather than one, were present. All were buried, rather than merely "surrounded." "Indentation" connotes a force that never was. Finally, the water didn't "eventually" arrive. It had been there all along, actually dropping down to the present level after the ice-dam failed.[30]

Purpose and Structure

I write for anyone who cares about Thoreau and his works. I assume some previous knowledge of *Walden* and the *Journal,* though certainly not expertise, which is why I provide a glossary of specialized words at the back. Within academia, I target the disciplines of American Studies and Environmental Studies at the level of a typical undergraduate survey course. The literary wing of these fields has long recognized *Walden* as "Thoreau's masterpiece," "one of America's indispensible books," "one of the first of the vitamin-enriched American dishes," a work of "monumental preeminence," the "most central text in the ecocritical canon," the "summit of Thoreau's literary career," as "important today as on the day of publication," and "arguably the most widely translated and available book by an American author." "Certainly no American," wrote Joyce Carol Oates, "has ever written more beautiful, vigorous, supple prose." The environmental wing has claimed Thoreau as "*the* environmental prophet," "the founding father of environmental thought in America," and the "patron saint of American Environmental writing." They have turned the 1856 Maxham daguerreotype—with neck beard *sin* moustache—into a global icon. A few card-carrying scientists like David Foster, Richard Primack, and E. O. Wilson have leveraged the popularity of ecology to a wider appreciation of *Walden,* and vice versa. I follow their example, but on their road not taken, the one leading toward geoscience.[31]

Consider this a work of "cultural geology," an embryonic field that seeks to understand American culture through this traditional disciplinary lens. My thesis is simple: *Walden* scholarship will benefit by adding geoscience to the mix. My interest is inclusive: the hard rock and deep time of classical geology; the physics and chemistry of soils, the erosion and sediment transport of geomorphology; and the rapid flux of air and water claimed by hydrology, limnology, and meteorology. As I put it a decade ago, "If physical geography is the house in which regional culture makes its home, then geology is its foundation, plumbing, and wiring." Because I lack formal training in "the basic analytical premises used by 'trained' readers of literature," scholars in the humanities may find my work naïve and over-reaching, deficiencies I hope will be offset by its original contributions. And to help my spoonful of medicine go down, I drench this book with Thoreau's quotes, and with those from scholars who helped me understand them. If I step on a few expert toes during the process, I do so either unknowingly or unavoidably.[32]

And now, a few caveats.

This book is restricted to aspects of *Walden* that deal with nature. When I refer to *Walden,* please read it as "that part of *Walden* dealing with material Nature," or, more succinctly, geo-*Walden,* a phrase I sometimes insert as a reminder. This includes Thoreau's spiritual and philosophical reflections about the physical world, but excludes social, economic, and political history, and the bulk of personal psychology, content nicely subsumed by the title of Robert Gross's extended essay: "That Terrible Thoreau: Concord and Its Hermit." Though I weigh in strongly on published literary criticism, sometimes I'm so perplexed that I can only gasp. For example, I've nothing to say about Perry Miller's interpretation of the Deep Cut's sand foliage as a "slime of sand" constituting a "filthy afterbirth." Or Paul Friedrich's phallic rhythm of Thoreau's lost axe relative to Patrick Morgan's "geo-feminism" of the shapely curves of local meanders. Or Sharon Cameron's claim that the *Journal* is seductive, sadistic, and voyeuristic. Or Patrick Chura's contention that *Walden* began as "a materialization of the earth's magnetic field."[33]

This book is limited to Thoreau's works and biography before April 27, 1854. That's when he added the last known content to *Walden,* in this case two sentences transferred in three steps from jottings on an unrelated letter to the *Journal,* and then to the page proofs, skipping the early manuscript stage altogether. Statements such as "Thoreau never . . ." or "Thoreau understood that . . ." refer only to events before that date. Within this thirty-seven-year span, I zoom in on the "Walden years," the nine-year interval beginning in late March 1845 when he borrowed an ax, walked into the woods, and began framing his house amidst snow flurries. I further zoom in on Henry's "scientific sojourning stage," which began near the end of 1850, accelerated in mid-1851 after reading Charles Darwin's *Journal,* peaked in mid-1852 during his "Year of Observation," and continued strongly to the submission of *Walden,* probably on March 13, 1854.[34]

This book is heavily biased toward presenting Thoreau as a competent, pioneering geoscientist. With few exceptions, I emphasize what he got right and overlook what he got wrong or didn't notice. Mine is not a fair and balanced treatment.

One of my primary interests is the biography of Thoreau's self-taught scientific education between childhood and the submission of *Walden.* For this part of my investigation, my beloved 1962 Dover edition of Houghton Mifflin's 1906 *Journal* was more than sufficient. Hence, my citations to Thoreau's *Journal* are by day and year, rather than to hard-copy pages to any particular edition. And because textual analysis was irrelevant to my conclusions, I used whatever edition of Thoreau's work came into my possession first. This strikes me as both practical and egalitarian because it does not presume that all readers are literary scholars with hardbound access to the expensive Princeton standard, regardless of how good it is. Besides, it is only a matter of time before fully searchable, standardized electronic texts will be freely available.

Using language from Stanley Cavell's *The Senses of Walden,* understanding Thoreau's "book of the place" (*Walden*) requires that we first understand his "place of the book" (Walden). Henry worked hard to understand Walden's geological narrative and its timeless limnology prior to composition of his masterpiece. To help modern readers explore this same material, I treat these challenging subjects separately, slowly and incrementally, and ask that they stay with me.[35]

Finally, "Walden" meant two things to Thoreau. As a familiar geographic place, it was Walden Pond, a proper noun. As a natural phenomenon, it was a "lake," usually in lowercase, as in this sentence: "A lake is the landscape's most beautiful and expressive feature." To the Puritan colonists who named Walden before 1652, "lakes" were freshwater lochs, generally large, rugged, rocky, and elongated bodies of water like those of the English Lake District. Anything smaller, shallower, or softer, especially if it could be used to water livestock or turn the wheel of a mill, was considered a "pond." Based on this vernacular, "Walden Pond" became an official geographic place name, abbreviated to "Pond" or "pond" as necessary. In contrast "Lake Walden" is the physical place where limnology happens. It is not only the deepest lake in Massachusetts, but is six times larger than necessary to qualify as such. Throughout this book, I adopt Thoreau's protocol of using "pond" for place and "lake" for the attributes of that place. The name "Walden" used alone could mean either, depending on context. The name *Walden* always refers to the published book whose full title is *Walden; or, Life in the Woods.*[36]

Thoreau's book pivots on "The Ponds," its central and most descriptive chapter. This book pivots on "The Walden System," its central and most scientific chapter, which forms an interlude between two narratives. "Part I - The Place of the Book" narrates the story of how Walden came to be. Each of its five chapters has its own internal narrative of downward creation, for example of the bedrock by tectonic under-thrusting and of the lake by meltdown subsidence. In this first half of the book, the goal of showcasing Thoreau's understanding of the geology is the first order of business, with necessary side-glances to fill out the rest. "Part II - The Book of the Place," narrates how Thoreau's scientific understanding of Walden informed *Walden*. Its five chapters move through the quality of his observations, the role they played in his composition, my interpretations of key passages, thoughts about his main mythological purpose, and finally, Thoreau's downward path toward simplicity. My ultimate goal is to show that *Walden* is as strongly grounded in physical reality as it is in language, literature, and mythology.

Binding the two halves of my book together is the slow accretion of Thoreau's scientific vitae, drawn largely from overlooked *Journal* passages. This third narrative moves from the "Nature-boy" instincts of his carefree youth to his highly systematic work as a watershed scientist and local consultant. Along the way we examine Thoreau the matriculated student, poet-naturalist, amateur geologist, surveyor, quantifier, and theorizer. Bits and pieces of my own scientific life show up here and there as a consequence of writing in the first person.

And now, onward and downward.

I

THE PLACE OF THE BOOK

1

ROCK REALITY

The "cinnamon stone," wrote Thoreau, is the only "precious stone" found in Concord. "Large as a brick and as thick, and yet you could distinguish a pin through it, it was so transparent." Though this reads like a typical *Journal* entry, it was actually a fictional setup for the wisecrack to follow. "If not a mountain of light," he continued, "it was a brickbatful, at any rate." By this, he meant something to throw at someone, almost certainly the Reverend Edward Hitchcock, who irked Thoreau for being a wolf in sheep's clothing: in this case, a "sacred historian" posing as a "Man of Science."[1]

The wolf was a Christian divine who used his presidency of Amherst College as a bully pulpit to evangelize "natural theology" across the region, a program Thoreau wanted no part of. The sheep's clothing was Hitchcock's otherwise excellent *Final Report on the Geology of Massachusetts in Four Parts,* complete with six hundred hand-tinted copies of a beautiful foldout map of the state. When published in 1841, five hundred copies of the full report were delivered by government order to various institutions, including "one copy each to each incorporated Athenaeum, Lyceum, and Academy." This gave Thoreau free and convenient public access to the best such work in the nation at a time when geology was the most fashionable science of the day, and when Hitchcock was one of its leading fashionistas. Seizing his historic moment nearly a decade earlier, Hitchcock had persuaded the Commonwealth to appoint him as "geologist of the state," fund a long-term mapping project, and publish its massive report at taxpayer expense. Thus it was that Thoreau's most valuable scholarly reference for the most exciting new field of science had been written by someone whose lifelong purpose was to "defend and illustrate" the truth of "Christian religion" by aligning the facts of geology to it.[2]

Thoreau preferred the pagan delight and practical utility of ordinary minerals. Plain "white quartz" from local outcrops and "hornstone" from Maine's Mount Kineo were more precious to him than gems because they could be quickly shaped into the useful tools of archaeology. Muscovite, a bronze mica, glittered the "golden sands" of his Concord dreamscape. Limonite, a natural rust, stained them to their yellow-brown hue. Magnetite—a shiny black mineral from his "cabinet" of specimens—is my pick for Thoreau's favorite mineral. Being widely distributed in nearby outcrops, this iron oxide was the quasi-magical component of his surveying compass, the "lodestone" or "leading stone" that kept him gainfully employed.[3]

In his *Journal,* Thoreau traced the historic usefulness of magnetite from the ancient dynasties of China to the year 1427, when explorer Vasco da Gama used it to open up the spice trade for Portugal. Mariners during this age of exploration had no idea why their compasses worked because theirs was a pre-scientific era, rife with superstition, alchemy, and magic. A scientific, albeit qualitative understanding of Earth's magnetism would await the year 1600, when William Gilbert, private physician to Queen Elizabeth I, correctly theorized that our planet acted like a giant bar magnet, to which crystals of lodestone aligned.[4]

The Puritans who colonized New England within the next few decades—in 1607 at Popham Point, Maine, 1620 at Plymouth, 1630 at Massachusetts Bay, and 1635 at Concord—paid scant attention to such secular ideas from the royal court. For them, answers to "life's persistent questions" were to be found in the traditions and scripture of the Protestant Reformation. Today, it is hard for us moderns to appreciate the power the Christian Bible held over the lives of Concord's English pioneers and their descendants. Most read it as literal truth. And beginning in 1650, this truth included the calendar dates of Genesis printed in the margins of the King James Bible. These had been calculated by James Ussher, Archbishop of Armaugh, and published in his *Annales Vertas Testamenti.* Improving on earlier calculations by Martin Luther, he summed the chain of Abrahamic begats within the Mosaic chronology to determine the precise moment of Creation: the nightfall before Sunday, October

23, 4004 BCE. To question such authority in late seventeenth-century New England might get you burned at the stake. Herein lies the taproot of scientific conservatism in early America, and Thoreau's gripe with Reverend Hitchcock's doublespeak.[5]

Secular readers of this twenty-first century who lack an introduction to scientific cosmology and nuclear chemistry must also accept on faith the chronologies of others. But thanks to the triumph of Enlightenment science, most college-educated Americans today have little doubt about the almost inconceivable span of deep time. Astronomers have used the radiation from distant galaxies to prove that the universe was created approximately 13.75 billion years ago, plus or minus statistical error. Geologists have used the uranium-lead isotope decay series (U^{238}-Pb^{206}) to show that Earth was created about 4.57 billion years ago, making it approximately one-third the age of the cosmos. This latter date comes quite close to the roughly four billion years suggested by the *Bhagavad Gita*, a sacred Hindu text that Thoreau treasured his entire adult life. Both the *Gita*'s inspired guess and the lab-calculated ages for Earth's age are sufficiently old to imagine "no vestige of a beginning,—no prospect of an end." This famous quote from Scotsman James Hutton—the father of modern geology—was known to Thoreau five years before he moved to Walden Pond to begin his experiment in deliberate living.[6]

Bedrock

Truth is bedrock. "Let us settle ourselves, and work and wedge our feet downward through the mud and slush of opinion, and prejudice, and tradition, and delusion, and appearance, that alluvion which covers the globe . . . till we come to a hard bottom and rocks in place, which we can call reality." Note that Thoreau does not invite us downward to a place *inside* the bedrock. Rather, he invites us down to its *upper* surface, to a *point d'appui* where we can stand firmly on our deepest personal convictions, beneath the mud and slush of sociopolitical life.[7]

This famous metaphor helps explain Thoreau's lifelong love affair with lichens. They hug the earth tighter than any other multicellular life form, living in the nick of space between the gaseous veil of Earth's atmosphere and the mineral chemistry of its crust. Lichens—whether on bedrock ledges, erratic boulders, or Concord's famed fieldstone walls—symbolized more plainly to him than anything else the dependence of organic life on inorganic mineral substance. His cliff-hanging columbine

Figure 5. Glacially smoothed outcrop of the Andover Granite. Complex and highly deformed textures include veins, folds, pinch-outs, and blobs (xenoliths) of host rock. Vegetated low spots are associated with veins. Brunton Compass for scale. From Hudson, Massachusetts, courtesy of Joseph P. Kopera.

in bloom told the same story more beautifully and complexly. Even more beautiful and more complex was the version told by Concord's rich ecosystems: the forest above its "vegetable mould," the swamp above its peat, the ponds above their muck, and the meadows above their annual "slime" deposited by river freshets. As with the Nile slime of antiquity, this *fluvial* "alluvion," now called "alluvium," was the economic base of Thoreau's Euro-American agrarian culture, the Algonquin peoples they replaced, and those from the Archaic Period who lived several thousand years earlier at the time of the Pharaohs.[8]

Beneath Concord's ubiquitous veneer of organic residues lay a much thicker mélange of unconsolidated clay, silt, sand, gravel, and boulders whose origins were then being worked out. Thoreau called this material "earth" with a lowercase "e." And beneath his "earth" lay the solidity and certainty of the Andover Granite, the strong, but heavily fractured crystalline rock of nearby outcrops. It was named for the town of Andover, Massachusetts, to the northeast, which was named after a village in Hampshire, England, much farther to the northeast. This official map unit is one small portion of a larger tectonic amalgam known as the Nashoba Terrane, which, is one small portion of the much larger amalgam informally referred to as New England bedrock, meaning everything east of the Hudson River–Lake Champlain Lowland. This, in turn, is one small portion of the much, much larger amalgam known as the continental crust of North America. This type of crust—which underlies everything between Portland, Maine, and Portland, Oregon—floats higher than the younger oceanic crust on either side because its minerals are richer in silicon and aluminum, and therefore less dense than the warmer, softer, darker, and more iron- and magnesium-rich minerals of the mantle, Earth's underbelly.[9]

In common parlance, granite is a grainy, light-colored rock common to New England's ledges and boulders, and which is not conspicuously layered at the scale of an individual stone. Its mineral crystals are large enough to see, and visibly interlock with one another to create a three-dimensional mosaic having negligible pore space. By definition, it must contain two kinds of feldspar, generally with milky pastel tones, and a significant fraction of quartz, generally the greasy gray infillings. Speckles of mica and black accessory minerals are also present. As with granites everywhere, the Andover crystallized slowly from a molten mass called magma that froze into a homogenous mass. As with the crystalline "ice" of liquid water, the crystalline "ice" called granite is the colder, solid phase of a previously warmer liquid of the same chemical composition. This is self-evident for the black Hawaiian basalt that cools from orange-hot lava right in front of our eyes, and for the glass in a glass factory that cools from its syrupy yellow-hot antecedent. The early naturalists who extended this "hot-molten to cold-solid" origin to granite and similar rocks became known as plutonists, named for the Roman god of the underworld.[10]

When sun-dried and exposed to the natural acidity of Concord's rainfall, Andover Granite bleaches and stains to a dull, but pleasant buff-colored patina. Bright reflections usually come from broken crystals of feldspar, which fracture along cleavage planes. This rock unit is so highly variable that it is sometimes hard to map. In some outcrops it is broken to the point of crumbling. In some ledges are veins so coarse that

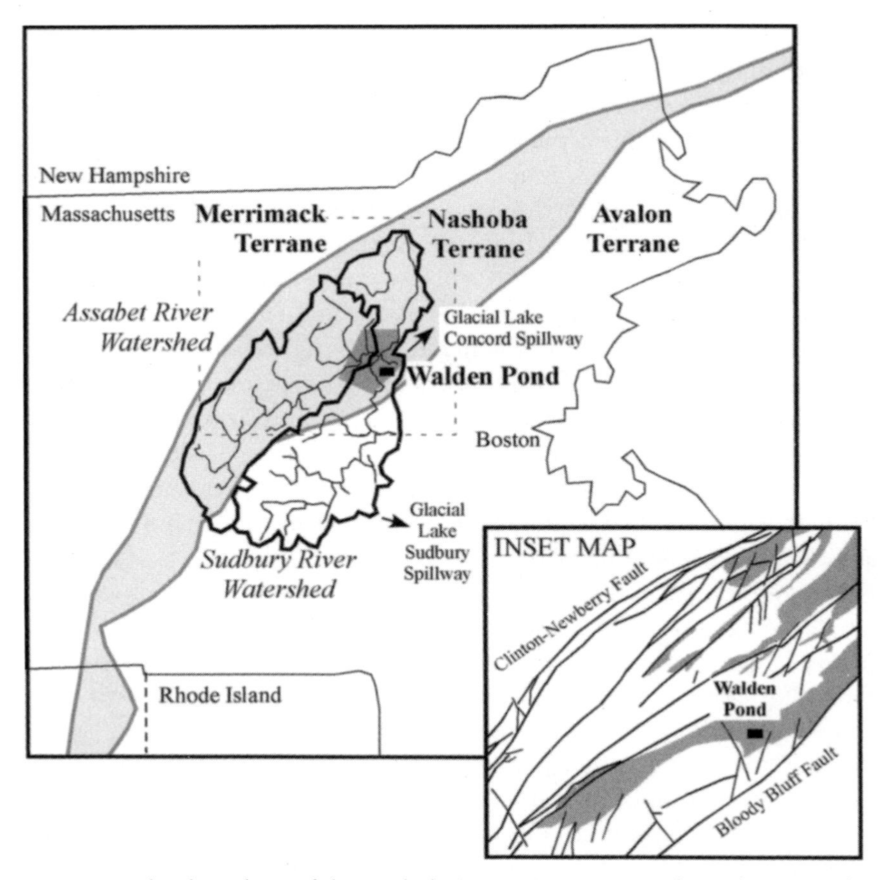

Figure 6. Bedrock geology of the Nashoba Terrane. Main map shows location of Walden Pond (black rectangle) within the town of Concord (dark gray shading) relative to the watershed of the Concord River (heavy black line) and the Nashoba Terrane (light shading). Inset map shows extent of the Andover Granite (dark stipple) relative to thrust faults of the Acadian orogeny (heavy black lines) within and bounding the Nashoba Terrane. Younger block faults (thin lines) occur in northwesterly and northeasterly sets.

Bedrock modified from Barosh, 1993, Figures 1 (p. 213) and 2 (p. 214) and Zen et al., 1983. Watersheds from U.S. Geological Survey. Glacial lake spillways from Koteff, *Glacial Lakes*.

the crystals are several inches across. Others contain dikes and sills so fine-grained that you can barely see the minerals, even with a magnifying lens. Also commonly present are residual blobs of the preexisting formations that were engulfed into the stiff liquid as it was forced up from below, but which didn't completely melt. Resembling plums in a pudding, these inclusions are called "xenoliths" because they seem strangely out of place.[11]

Thoreau's townsmen complained of a dearth of granite in Concord. Actually, there is plenty of it around, though not of a quality high enough to be profitably quarried. That sort of quality comes from larger, more uniform vats of magma such as those of the Presidential Range of New Hampshire and the quarries of Quincy, Massachusetts, whose granite was used to build the Bunker Hill Monument. Because Concord's native granite was sheared and broken, "bowlders" of high-quality granite carried southward from the Granite State by glacial ice were highly prized. Many disappeared into the house sills, barn foundations, and posts of colonial Middlesex towns.[12]

Happily, the rough breakage and poor quality of Andover granite made it a source of abundant boulders, especially the large glacial erratics that Thoreau enjoyed so much for their titanic suggestions. The smaller, rounder boulders found throughout Walden Woods were typically more far-traveled. Regardless of where each had come from, all had originally been yanked up from some distant outcrop as jagged fragments before being milled into sub-rounded shapes by transport within the glacier's basal shear zone. These were the stones that paved Walden's shore. When hauled up to his house site and fitted together, these were the stones that supported his house and chimney. When dropped one at a time by visiting pilgrims, these were the stones that grew into the disheveled cairn marking his house site. This secular shrine was inaugurated in 1872 when Bronson Alcott, Henry's philosophical friend, cast the first proverbial stone, probably a cobble of granite.[13]

Not until the 1970s did scientists finally prove that Andover Granite extends beneath Walden Pond. They did so using the differential magnetic signatures within rock detected by airborne sensors. More recently, they have towed an array of seismic instruments across the pond; tracing intact bedrock from its western end near the railroad all the way east to the headquarters for Walden Pond State Reservation. That is where roughnecks drilling boreholes through the glacial overburden encountered intact rock meeting the unit's identification criteria. As their drill bit passed through the overlying sand and gravel to enter in situ bedrock, it crossed what geologists call an "unconformity." This is a physical boundary so sharp that one can put a finger on it, a contact so abrupt that the remains of one ancient "dynasty" are directly overlain by the remains of a younger one.[14]

By definition, time is always missing at such contacts: usually lots of it. The most famous one in the United States lies near the base of Arizona's Grand Canyon. There, millions of tourists have seen where the younger, flat-lying and cliff-forming strata of the upper canyon—yellow sandstone, creamy limestone, and red shale—abruptly overlie the much older, steeply tipped, and highly contorted Vishnu schist of the lower canyon. The lacuna of time missing on this unconformity exceeds a billion years, which is greater than the grand sum of all strata rising to the canyon rim. Using a familiar analogy, the canyon's famous "book of time" overlies a beveled mass of crumpled and partially melted cuneiform tablets.

A major unconformity is also ubiquitously exposed throughout New England above the crystalline rock of its sea cliffs, river banks, and road cuts. As with the Vishnu schist, it is also highly contorted, partially melted, and crosscut by veins and dikes, because both masses formed deep within the roots of their respective ancient mountain systems at a time when the rock was pliant, compressed, and steaming with geothermal fluids. In New England, however, there is no well-layered polychrome sequence of younger sedimentary rocks above the eroded mountain root because this region has not since been submerged beneath the sea and lifted back up again, as was the case for the Grand Canyon. Instead there is a nearly continuous sequence of glacial strata pasted, washed, and gently let down on the bones of an older world.

Neptunism and Plutonism

Though it defies common sense, most New England geologists of the early nineteenth century actually believed that its rock had precipitated from a primeval, universal ocean. They belonged to the so-called "neptunist" school of thought, named for the Roman god of the sea and tacitly linked to Christian scripture through Noah's mythical deluge. Leader of this school was Abraham G. Werner, father of Germany's chemical-mineralogical-mining school of geology. Born in 1749 in Prussian Silesia, his family's fortunes had long been tied to the mining industry. As a popular professor during the late eighteenth century, he taught his students that shortly after its cosmic beginnings, Earth had been completely covered by an aqueous brine, from which all of its ancient rocks crystallized in sequence, beginning with granite. One of his contemporaries, the Swiss aristocrat Horace-Benedict de Saussure, likened the folded limestone strata of the nearby Alps to the precipitated laminations of alabaster in caverns. The concentric layers within an oyster's pearl provide another water-based analogy.[15]

Though almost comical today, Werner's scientific theory was a genuine improvement over the scripturally driven cosmogonies that preceded it. One example of such "biblical geology" was Thomas Burnet's four-volume *The Sacred Theory of the Earth, Containing an Account of the Original of the Earth, and of All the General Changes Which It Hath Already Undergone, or is to Undergo, till the Consummation of all Things* (1681–1689). He posited an egg-like early earth with liquid water beneath its hard shell which, when cracked, flowed out to submerge the surface. John Woodward's *Essay Toward a Natural History of the Earth* (1695) and William Whiston's *New Theory of the Earth* (1696) were similarly fanciful late seventeenth-century attempts to fit the square peg of proto-geology into the round hole of scripture.[16]

Two works from the mid-eighteenth century gave neptunism its pre-Wernerian scientific flair. Gottfried Leibniz, the co-inventor of calculus, suggested in his *Protogea* (1749) that Earth began with a universal ocean that gradually disappeared, leaving continents in its wake. Georges-Louis Leclerc, Comte de Buffon, in his thirty-four-volume *Histore Naturalle* (1749), emphasized protracted floods and deluges, treating the seven biblical days of creation as surrogates for longer epochs culminated by major upheavals. To historian of geology Frank Dawson Adams, this "fairyland of science," and "these early fables of geological science," "should be read by all who are in need of mental recreation and who possess the required leisure and a certain sense of humor." To historian of science Thomas Kuhn, however, such seemingly screwball explanations are perfectly normal prior to the consolidation of scholarly endeavors into "normal science."[17]

Three of Thoreau's most important intellectual heroes hailed from the neptunist-Wernerian school. Carl Linnaeus was of "the opinion that sea sand" on the world's beaches "was chemically precipitated out of the waters of the sea," something like pellets of sleet crystallizing out of the air. German poet-scientist Wolfgang Goethe was a Werner devotee long before the young Thoreau adopted him as an role model for his "exact descriptions of things as they appear to him." But as a theorizer, Goethe came down on the wrong side of the debate, being "bitterly hostile" to the plutonists. In non-fiction, he lambasted them with *Naturwissenschaftliche Schriften.* In fiction, he used Faust as a neptunian protagonist to uphold the unmarred beauty of a progressively developing and increasingly beautiful earth as it precipitated, layer by layer, from water. His antagonist was the evil Mephistopheles, who represented the hell-bent plutonist view. Another one of Werner's students was Alexander von Humboldt. According to his biographer, Laura Dassow Walls, his early work "advanced Werner's own Neptunian theories and he was duly enrolled and warmly received by the great professor."[18]

When the Geological Society of London was formed in 1807, many of its early members were devotees of Werner's so-called "continental" theory. Neptunism was carried to Scotland by a Werner protégé named Roderick Jamiesen, who was Regius Professor of Natural History at the University of Edinburgh. In 1808, he published *Elements of Geognosy,* the "most scholarly and refined presentation of Abraham Werner's system." To honor his mentor, Jamiesen created a Wernerian Natural History Society in Edinburg, where he locked horns with John Playfair, one of the world's most Prominent Plutonist. One of Jamiesen's most lackluster students was the young Charles Darwin, who was struck by the absurdity of his professor's neptunian views, and who was greatly put off by his dull and deadening lectures. "The sole effect they produced on me," Darwin wrote in his autobiographical *Recollections* many years later, "was the determination never as long as I lived to read a book on Geology or in any way to study the science."[19]

American geology was founded within the Jamiesen tradition. Leading the way was a fellow Scotsman William Maclure, who emigrated to Virginia, and in 1809, published America's first geological map. Returning to Edinburg, he "used a Wernerian classification of rocks" in his career-summing *Observations on the Geology of the United States* (1817). Bringing this tradition to New England from Philadelphia was Benjamin Silliman, who went on to become the region's most luminous physical scientist. In 1829, he created for his students at Yale College a modified version of the world's first geology textbook, Robert Bakewell's *Introduction to Geology* (1815). To it Silliman adding an appendix where "the tone is catastrophic, Wernerian, and even favorable to the aqueous origin of granite." When supplemented by his own neptunist-leaning *Outline for the Course of Geological Lectures Given in Yale College,* Silliman's "course-pack" became the standard geology text in America for more than a decade.[20]

One of his post-graduate students named Amos Eaton, went on to co-found the famed Rensselaer school in Albany. There he remained "an avowed Wernerian," going beyond what Silliman had done by publishing America's first truly indigenous *Geological Textbook* (1832). Thoreau was almost certainly exposed to Eaton's theories when he was a student at Harvard, from 1833 to 1837, and when he returned to Concord. There he read Lemuel Shattuck's *A History of the Town of Concord* (1835), which contained a geology section loosely based on Werner's classification. Thoreau's vision that it may have once "rained rocks" on Maine's Mount Katahdin, and the mystical "water-world" imagery of *Walden* derive in part from this historical legacy.[21]

By the late 1830s, however, neptunism lay in ruins, thanks to Charles Lyell's upgrade (1830) of James Playfair's upgrade (1802) of James Hutton's upgrade (1795) of his earlier published lectures (1783–1788) gathered as the *Theory of the Earth with Proofs & Illustrations.* Hutton's theory was based on original field observations made at the scale of cliffs and mountains, rather than on inherited theories involving the origin of crystals. It held that landscapes are endlessly created and destroyed during episodes of vigorous mountain building that alternated with more protracted and passive episodes of erosion. During times of landscape *construction*, the sedimentary residues of the previous age are re-crystallized and re-melted by geothermal heat and re-raised as highlands. This made Hutton an obligate plutonist, even though this meta-process covered only the first half of his theory. The second half came during *deconstruction*, when these crystalline highlands were laid low by weathering processes within the soil, and by the flushing of the residues to the sea by rivers. As Playfair summed up, "the decay of one part" is "subservient to the restoration of another, and gives stability to the whole." Unconformities were key to the system, being the boundaries between the rock-generating and rock-destroying hemicycles of the full cycle, referred to by Hutton as a "revolution." Using Hutton's term in its proper context, Thoreau described two unconformities in his *Journal,* both of which became covertly significant in the final version of *Walden.*[22]

Hutton's *Theory* did not come out of nowhere. Rather, it was one of many manifestations of the eighteenth-century Scottish Enlightenment that also brought David Hume's philosophy, Adam Smith's economics, Robert Burn's poetry, and Robert Black's chemistry. According to Hutton's biographer Jack Repcheck, Hutton's Earth was an unchanging machine analogous to Isaac Newton's unchanging solar system. His cyclical thinking—which came to be called uniformitarianism—can be traced directly back to Newton himself through Professor Colin Maclurin (1648–1746), who had learned his science at the feet of the aging master in Cambridge before moving to the University of Edinburgh and teaching it to Hutton in a course on natural philosophy. Both Newton's clockwork universe and Hutton's clockwork planet became important symbols of late eighteenth-century deism because they were run by natural laws, rather than intelligent design and oversight.[23]

Hutton wisely avoided speculating about Earth's origin and ultimate end to avoid drawing parallels to biblical Creation and the eschatology of Armageddon. He also made sure to include the caveat that "geology is silent" on the "system of chronology which rests on the authority of the Sacred Writings," and that "in no way related to the morality of human actions." In spite of this respectful deference to religious teachings, the orthodox Christian backlash to Hutton's work was as intense as it was disingenuous. By loading "his theory with the reproach of atheism and impiety," theologians won the battle for the hearts and minds of science-minded thinkers well in the early decades of the nineteenth century when Thoreau's future professors were being trained. "Uniformitarianism was unpopular," wrote historian Herbert Hovenkamp, "not because it eliminated God from the picture, but because it forgot to mention Him." Oxford-trained geologist-turned-writer Simon Winchester put the matter more broadly. Geology in general and Hutton in particular "suffered more than the usual share of growing pains," due to "bigotry, intolerance, churchly disapproval, and the fundamental assumptions it sought to challenge."[24]

An equally serious hurdle for the acceptance of Hutton's *Theory* was his turgid prose style. Almost unreadable, it fell dead from the press shortly before his death. Luckily, his protégé, John Playfair, chair of Natural Philosophy at the University of Edinburgh, understood it well enough from conversations and lectures to write his own book-length summary, titled *Illustrations of the Huttonian Theory with Proofs and Illustrations* (1802), adding to it his own excellent observations and commentary. To wit, Playfair became Hutton's Boswell.

Enter Charles Lyell, a young man whose entire professional life was changed by a book he read during the year of Thoreau's birth (1817).

From his father's library, he borrowed the same book Silliman had modified to launch academic geology in New England: Bakewell's *Introduction.* His interest piqued, Lyell enrolled in lectures being given at Exeter College, Oxford, by the world's first academic geologist, the Reverend William Buckland. Drawn in by the double-scoop of Bakewell's book and Buckland's charisma, Lyell—like both Hutton and Silliman before him—abandoned a promising law career to study geology. Next he declined a lectureship at King's College in order to write one of the most underrated books of the nineteenth century: *Principles of Geology,* published in three volumes between 1830 and 1835. Lyell followed Hutton by treating Earth as a colossal recycling machine, but toned down the drama of his "revolutions" to emphasize even more strongly that slow, gradual changes achieved great results through eons of deep time. Central to Lyell's theory was the cyclical rise and fall of the sea in different regions at different times.[25]

To emphasize this last point, Lyell chose for his frontispiece an engraving of the ruins of the Temple of Serapis in the seaport of Pozzuoli, Italy. Its marble columns had acted like dip sticks to record clear evidence for historical changes in sea level, a fall and then a rise. And to bookshelf his text with Newton's *Principia Mathematica,* Lyell cloned its title as an anglicized version of *"Principia Geologica."* And to ensure that even the most casual reader took home its central idea, he chose for his epigraph a quote from Playfair paraphrasing Hutton: "Amid all the revolutions of the globe the economy of Nature has been uniform, and her laws are the only thing that have resisted the general movement. The rivers and the rocks, the seas and the continents have been changed in all their parts; but the laws which direct those changes, and the rules to which they are subject, have remained invariably the same."[26]

Reprinted in nine editions before Thoreau's *Walden,* Lyell's *Principles* almost single-handedly created modern geology by bringing it to an eager world audience. It established the discipline's first true paradigm, last on the list in Thomas Kuhn's *Structure of Scientific Revolutions:* "Aristotle's *Physica,* Ptolemy's *Almagest,* Newton's *Principia* and *Opticks,* Franklin's *Electricity,* Lavoisier's *Chemistry,* and Lyell's *Geology.*" It stimulated rapid growth in knowledge in both Europe and America, making it the "science of the day." In Boston, it prepared the way for Reverend Hitchcock's public funding. In Concord, it brought cyclical uniformitarianism to the library of Ralph Waldo Emerson. Intrigued, he recommended the book to his twenty-three-year-old novice, Henry Thoreau, who read it with gusto in the fall of 1840.[27]

I often imagine the young Thoreau opening the *Principles* for the first time. After defining his subject carefully, Lyell reviews the history of geology beginning with this passage: "The earliest doctrines of the Indian and Egyptian schools of philosophy, agreed in ascribing the first creation of the world to an omnipotent and infinite Being. They concurred also in representing this Being, who had existed from all eternity, as having repeatedly destroyed and reproduced the world and all its inhabitants." Lyell's cited source was the *Institutes of Menu,* the very same book Thoreau was then reading as his introduction to Eastern spirituality. In it, the word "*menu,*" *or* "*manu,*" "comes from the ultimate Indo-European root for our words *mind* and *man,* more generally, human consciousness," explained biographer Robert Richardson. Thoreau was "deeply, repeatedly, and lastingly moved by the book; his response was that of a strongly religious nature to a great revelation." Speaking of the *Institutes of Menu* in June 1841, Thoreau wrote: "I know of no book which comes to us with grander pretensions."[28]

Thus, Emerson the philosopher—following Lyell, Playfair, and Hutton—introduced his novice not only to the fashionable science of geology, but also to its compelling link with Hindu spirituality. Fourteen years later, Thoreau would combine this science and this spirit to underwrite *Walden.*

Katahdin

Being a big fan of Vulcan made Thoreau an easy mark for the plutonist school of thought, despite his formal education during the neptunian era. He sensed beneath Concord the presence of a great "subterranean fire," one responsible for creating gemlike crystals of quartz within the veins of the Andover Granite. These, he collected for his specimen cabinet. These he interpreted as the "frost-work of a longer night." This six-word snippet of prose-poetry densely abstracts three Huttonian verities. Crystals of water-ice and silica-ice were indeed both hexagonal "frosts" originating from fluids, whether vapor or liquid. Freezing quartz requires a "longer night" than freezing water. And these respective nights have different causes. In Playfair's words, the "revolutions within the earth are independent of revolutions within the celestial spheres."[29]

Thoreau correctly envisioned planet Earth emerging from an initially molten state under darkened skies. "Mornings of creation, I call them . . . A morning which carries us back beyond the Mosaic creation, where crystallizations are fresh and unmelted. It is the poet's hour." This passage was inspired by a Promethean scene coming from a Concord field on a moonless night. From a distance, Henry saw a burning "heap of stumps half covered with earth," a "dim light . . . which appeared to come from the earth," a "phosphorescence . . . a strange, Titanic thing this Fire, this Vulcan. . . . Within are fiery caverns, incrusted with fire as a cave with saltpetre . . . the glass men are nearer the truth than the men of science." This last clause offered playful support for the plutonists who, as "glass men," invoked a molten origin for local rock. Conversely, it was a dig at the neptunist thrall for their aqueous version of creation.[30]

Thoreau's closest encounter with Earth's origin came during his cosmic experience on the summit plateau of Maine's Mount Katahdin in early September 1846. Alone in a "cloud factory," his view was mostly blocked except for the coarse granite on which he walked. In that primitive scene, one hostile even to lichens, he envisioned Earth as a brand new planet, its primitive rocky crust being "some star's surface, some hard matter in his home!" Every rock on Earth is indeed cosmic, being an aggregate of minerals, which are themselves aggregates of elements created during stellar explosions called supernovae. Approximately five billion years before Thoreau's Katahdin experience, this "stardust" was being gathered by gravity into planetesimals, which accreted to form the Earth about 4.57 ± 0.02 billion years ago. "All the earth" he wrote, was but "the outside of the planet bordering on the hard-eyed sky . . . solitary and cool-aired."[31]

Shortly after its initial melting and differentiation into layers, Earth was struck by a Mars-sized object with great force. The incoming kinetic energy blasted so much debris into low orbit that it created a ring that congealed to form the moon. It also gave Earth its rapid spin, tilted its axis, and set in motion the orbital wobbles and harmonic oscillations ultimately responsible for the glacial-interglacial cycles giving rise to Walden. The remaining share of that kinetic energy was converted to heat, which re-melted much of the planet. When cooled, Earth re-crystallized into a spherical dark shell of frozen lava resembling a glossy coat of silicate paint, making it "as celestial as that sky." A final bombardment came about 3.9 billion years ago, a flurry of asteroids so intense that the moon remains riddled with its impact craters today. Earth was similarly struck, but its craters have long since been healed on our "living planet."

Earth's first eon, the Hadean, was over. The volatiles steamed out by whole-scale melting had condensed and fallen back to Earth during a seemingly endless rain far longer than forty days and forty nights. The skies had cleared. What had been the black surface of a hot orb beneath a pitch black sky became the bottom of a bright blue universal ocean, through which a few massive shield volcanoes likely protruded.[32]

After the hellish Hadean came the strange Archean. Microbial life arose before 3.8 billion years, likely underwater on some volcanic vent. We suspect this because the oldest living things on Earth are thermophyllic chemo-autotrophic cells without a nucleus. Translation? Primitive microbes that thrive in scalding water and use metal brines for food. At the time, Earth's landmasses were small, austere, and volcanic. Green things did not exist. The atmosphere was without oxygen. And the prototypes of tectonic plates were thinner, smaller, and more transient. Earth was a Venus-like world of "flake tectonics."

"Are not all lands islands?" This question from Thoreau's *Journal* quotes Sir John Richardson, who was quoting the "esquimaux," who had jolted the British explorer by asking a question so obvious it hardly needs an answer. On a planet that is still nearly three-quarters ocean, it seems self-evident that even the most stable and most interior parts of our largest continents had once been islands. This was, in fact, one of Lyell's major *Principles*, which Thoreau quoted in a passage from *Cape Cod:* "In short, the dry land itself came through and out of the water in its way to the heavens, for 'in going back through the geological ages, we come to an epoch when, according to all appearances, the dry land did not exist, and when the surface of our globe was entirely covered with water.'" Though this makes both Thoreau and Lyell sound like a neptunists, this interpretation works equally well from the plutonist perspective. "Dry land" did, in fact, come "through and out of the water" by the same mechanism creating new volcanic islands near Iceland today. It was progressively refracted by differential melting of Earth's most primitive surface rocks, komatites. These had crystallized from Earth's most primitive lavas, which had flowed more freely than those of hottest Hawaiian lavas today. Microbial weathering also played an important role in converting the greenish-black colors of primitive rocks to the blanched shades of continental crust, because it liberated the lighter elements for future igneous concentration.[33]

Next came the seemingly endless Proterozoic, which marked the halftime of Earth history beginning about 2.5 billion years ago. By then, photosynthesis had begun to pollute the atmosphere with oxygen. Masses of newly refracted and recycled rock—granite, gneiss, greenstone, schist, and quartzite—had gathered like clumps of flotsam to create even larger and more stable masses of crust called cratons, which constitute the bulky cores of our continents today. One each lies beneath rain-drenched Amazonia, the sandblasted Australian outback, and glacier-scraped Hudson Bay. Africa has two. Asia has four. In each case, they were resistant enough to survive what Arctic explorer Elisha Kent Kane called "all the disguises of climate," a comment Thoreau borrowed from western Greenland for use in his *Journal*. There, Kane had just seen some of the oldest and toughest rock on the planet, so resistant that an expanded version of the Greenland Ice Sheet had merely scratched the surface when gliding over them.[34]

Once formed, cratons are too strong, thick, and buoyant to be dragged back down into the mantle by tectonic forces. Instead, new continental crust accretes to their edges like new tread on an old tire. All of New England east of the Hudson River constitutes the latest series of re-treads. All were added during the last half-billion years, the last one-ninth of the planet's history. Earlier re-treads include the circa billion-year-old Grenville rocks of the Adirondacks, the circa two billion-year-old Penokian rocks of Lake Michigan, and even older re-treads northwest of Lake Superior circa three billion years in age. Between each episode of accretion, the highlands rumpled up by collision were dissolved and washed away, and their elements were incorporated into whatever life forms had evolved by that time.[35]

Before two billion years ago, the only life forms on Earth were microbes lacking a nucleus (prokaryotes). By one billion years ago, there were single-celled protists with a nucleus (eukaryotes). By a half billion years ago, there were hordes of complex creatures, one for every phylum, including our own (Chordata). By a quarter billion, the evolutionary line leading toward mammals (Synapsida) had branched away from other reptiles. By a twentieth of a billion, Earth had monkeys screaming in the jungles of African (Primata). Finally, about 0.0002 billion years ago, the neurochemistry of human consciousness erupted into "man thinking" *(Homo sapiens)*.

"I stand in awe of my body," Thoreau wrote at the climax his Katahdin soliloquy, "this matter to which I am bound . . . Think of our life in nature,—daily to be shown matter, to come in contact with it, rocks, trees, wind our cheeks! The solid earth!" manifest as the granite of Katahdin, "The actual world!" created from recycled elements; "The common sense!" of our primitive minds. "Contact! Contact! Who are we? Where are we?"[36]

These were rhetorical questions for Thoreau because he had already answered them a decade earlier. On September 15, 1838, when he was a recent college graduate uncertain where life might take him, he cuddled down against the gray stones of a farmer's fieldstone wall, knowing that rock was the mother of all flesh, his "Alma Natura." That affinity was still with him sixteen years later, when he confided to H. G. O. Blake: "Rocks—earth—brute beasts comparatively are not so strange to me" as is "contact" with "humanity." And in the final version of *Walden* he draws the reader's attention to the myth of "Deucalion and Pyrrha," who "created men by throwing stones over their heads behind them." More succinctly, "as Raleigh rhymes it in his sonorous way . . . our bodies of a stony nature are."[37]

Making Mountains

"My instinct" Thoreau mused, "tells me that my head is an organ for burrowing as some creatures use their snout and fore paws." And with it," he continued, "I would mine and burrow my way through these hills." He knew that it was rough country down there: far rougher than the thrice-smoothed surface of Walden Woods. Smoothed most recently by fifteen millennia of downhill soil creep beneath the forest canopy. Smoothed before that by the thick drape of "alluvion" let down on rocky terrain. And smoothed before that by the grinding and smearing of the ice sheet, rubbing down the high spots and spackling the low spots with till. If we could clear-cut Walden Woods, roll up its forest sod, power wash the sand and gravel, cart away the boulders, and chisel off the veneer of glacial till, we would unearth a hard, corrugated landscape of gutter-like valleys trending southwest to northeast.

This is the bone-work of the northern Appalachians. Running opposite this bedrock grain would be narrower, southeast-to-northwest trending notches and defiles. In miniature, this landscape would have the feel of weather-beaten petrified wood, perhaps like that of an old barn door or a fence post, cracked across the grain in a few places. Rub your fingers with the grain (southwest to northeast) and it will feel smooth. Run them against the grain (northwest to southeast) and it will feel bumpy. In such a landscape, Thoreau could have burrowed a beeline from Sudbury to Billerica. Burrowing in the other direction, however, from Littleton to Lincoln, he would either have to move up and down all the way, or a zigzag from one notch to another, as with a labyrinth.[38]

Today, Concord bedrock is so extensively mantled with glacial sediment that only the largest landscape elements reveal this tectonic grain. One of the most prominent is an ancient bedrock valley running straight northeast from Fairhaven Bay to beyond Little Goose Pond. Within this "paleo-valley" is a "chain of ponds," the largest and deepest of which is Walden. From a moraine hilltop near his house known as Heywood Peak, Thoreau described it as a "wide indentation in the hills . . . where their opposite sides slope to each other." Based on bedrock mapping by geologist Patrick Barosh and others, the bottom of this buried valley is aligned by some kind of ductile thrust fault or shear zone passing very near the center of Walden's largest and deepest basin. This is not coincidence. As Thoreau knew full well, bedrock ruled his "little world" from below.[39]

By the first decade of the nineteenth century, John Playfair had clarified Hutton's theory about how mountains were made. By its last decade, this "big" idea had been given a proper technical name. Orogeny: a spasm of intense tectonic deformation, uplift, regional metamorphism, and crustal re-melting spanning at least several million years, but brief by the standards of deep time. Unfortunately, naming this phenomenon didn't explain it. A proper understanding of the mountain building mechanism had to wait for the theory of plate tectonics, which would revolutionize geoscience during my own undergraduate years of the late 1960s and early 1970s.[40]

One of the first geniuses to take a crack at the problem of making mountains was the famous polymath, Leonardo da Vinci (1452–1519). While working as a civil engineer in the Alps, he noticed that torrential mountain rivers did far more than destroy communities, bridges, and roads. They also transported enormous masses of rock and earth to lower elevations and sluiced it away. From this, he correctly inferred that the mountains of the Alps were, in fact, erosional remnants: that which remained after everything else had been carried away. This was equivalent to saying that the marble statues he carved from raw rock were also erosional remnants: that which remained after the rest had been chiseled, scraped, and rubbed away. Though true, this interpretation of mountain formation was only a partial explanation, because it left out the creation of the rock, its quarrying, and its transport to the studio.

Shortly thereafter, and following some of Leonardo's ideas about crustal uplift, the broader context of mountain building was explored by Georgius Agricola (1494–1555), a mining *Burgermeister* in Bohemia. To the obvious external agencies, he added three internal ones responsible for creating the raw material, and for raising it to the studio of Earth's surface: subterranean winds, earthquakes, and volcanic fires. In his *Ne Natura Eorum quae Effluent ex Terra* (1546)—which predated Hutton and Lyell by more than two centuries—he wrote in plain uniformitarian language that "all these varied and wonderful processes . . . have been in operation since the most remote antiquity, so far back in the dim distance of the past beyond the memory of man that none can tell when they had their beginning."[41]

Next was Steen Pederson (1636–1686) from Copenhagen. He followed the example of Agricola (originally a German named Georg Bauer) by moving to Italy, exploring mountain scenery, and Latinizing his name: in his case to Nicolas Steno. To this very day, all beginning geology students are taught Steno's laws of stratigraphy, which state that all "secondary" rocks such as limestone and sandstone were originally laid down horizontally, sequentially, and extensively, and that any departure from this initial condition must be due to deformation of the earth's crust. These laws were published in what is widely considered the founding work of historical geology, *De Solido intra Solidum Naturaliter Content Dissertationis Prodromus* printed in Florence in 1669. Following Agricola again, it concludes that the tilting and folding within mountain belts was due largely to the venting of fiery gasses from inside the earth and the subsequent collapse after support was removed. Italian Abbe Anton Lazzaro Moro (1687–1740) carried Steno's ideas deeper into the crust to the geothermal fire itself, making him a true plutonist. He correctly argued that volcanoes, volcanic islands, larger islands like Britain, and even the crust of continents were ultimately the result of heated uplift, and ocean basins the result of adjacent subsidence. To him, a mountain range was little more than a chain of volcanoes associated with older rocks.[42]

The main problem with all these early theories is that none could account for the strong horizontal compressional folding and thrusting shown by rocks of the Alps and adjacent Mediterranean mountain ranges. This compelling evidence for crustal shortening had been a nagging problem, even to Lyell, whose theories depended mostly on vertical motions. During Thoreau's era, the favored explanation for compressing mountains was wrinkling of the Earth's crust due to shrinkage of the planet during progressive cooling. This theory, published by French geologist Leonce Elie de Beaumont in 1828 (and updated in 1852), got the local mechanics right, but the driving process wrong. That process, we now know, is the horizontal compression caused by the tectonic convergence of rigid plates drifting above the softer, warmer rock of the mantle. Elsewhere on Earth, these large plates of lithosphere diverge and slide to produce other effects, such as ocean basins and enormous linear valleys, respectively.[43]

Taken as one coherent package New England's bedrock can be crudely visualized as the final result of an intermittent, ultra-slow-motion collision of three super-sized tectonic barges against the pier of the North American craton. Each impact created its own separate orogeny—Taconic, Acadian, and Alleghenian—culminating about 440, 380, and 280 million years ago, respectively. But unlike conventional barges, these masses were soft enough at depth to squish into each other like moist clay, and hot enough to locally re-melt, welding a unified landmass together.

Prior to the Taconic orogeny, ancestral North America was a large, strong, and buoyant mass overlain by flat limy marine strata not unlike the Bahamas today. Rammed into it from what is now the east was a volcanic island arc not unlike modern Japan. Strong compression forced great masses of volcanic strata over the craton to create a linear mountain chain, the stubble of which are the Taconic Mountains near the New York state line. It also deformed and metamorphosed the ocean mud and carbonate bank into the slate-marble landscape of the Green Mountains in western Vermont, the Berkshires of Massachusetts, and the Litchfield Hills of Connecticut.

On the back side—or seaboard side—of this first collision was the muddy bottom of an ancient sea called the Iapetos Ocean, named after the father of Atlas, for whom the present Atlantic is named. Accumulating in it was sediment of volcanic, terrestrial, and planktonic origin. Incoming for the second collision was an elongated mass of far-traveled, very old, and very strong continental crust known as the Avalon Terrane, the remains of which now dominate the New England shore between southern Connecticut and the Maine coast. The collision of Avalon with post-Taconic North America is known as the Acadian orogeny. Caught up within it were the oceanic sediments of Iapetos, which were squeezed, thickened, and heated in the relentless vice-grip of tectonic closure to create the main mountain root of central and eastern New England.[44]

Also caught up within the squeeze was the Nashoba Terrane, a discrete sliver of crust about thirty miles wide and a hundred miles long. Originally composed of island arc volcanic rocks, it was an exotic fragment that had, millions of years earlier, been sliced away from somewhere else, and had since been behaving like Baja California today. After docking against the growing edge of North America somewhere to the north, the Nashoba Terrane (hereafter Nashoba) slid southeasterly into place along a giant strike slip fault called the Clinton-Newbury Fault, coming to rest in the vicinity of modern Concord.[45]

During the Acadian culmination, all of Nashoba was compressed so strongly that it was wedged downward in a northwesterly direction beneath a growing mountain root, whose axis lay to the northwest. In that direction, vast underground pools of locally re-melted crust were cooling and crystallizing into a host of giant plutons spanning a broad arc from southwestern Connecticut to beyond Katahdin. Rising toward it from the southeast were a series of lower ranges, the roots of which Thoreau called "vergers, crests of the waves of the earth, which in the highest break at the summit into granitic rocks over which the air beats." This remarkable episode created what Thoreau called the "spine" of "granite to be the backbone of the world." Following one of his westward-looking mountain meditations, he concluded: "I have seen how the foundations of the world are laid, and I have not the least doubt that it will stand a good while."[46]

Meanwhile, back to the southeast where Concord is today, the original sediments of Nashoba were being metamorphosed almost beyond recognition. Heated to temperatures just below the melting point. Softened to the consistency of warm taffy. Squeezed and thickened. Smear-sheared against one another along dipping thrust faults far too hot to generate earthquakes. Intruding these faults from below, and forcing the softened layers apart, were injections of truly molten rock being squeezed up like pressurized caulk by the weight of the overlying rocks. This liquid magma, which had the consistency of stiff porridge, solidified about 350 million years ago to become the Andover Granite. Walden Pond, Concord, and the bulk of Thoreau's sojourning country lie above this discrete volume of crust.[47]

The eastern edge of Nashoba is defined by the Bloody Bluff Fault, named for a hematite-stained cliff along the old Lexington Road where, on April 19, 1775, colonial militia inflicted heavy casualties on the British redcoats during their retreat back to Boston. This boundary was recognized and correctly mapped during Thoreau's era by state geologist Hitchcock. What he called "sienite," and dated to his "First Series," is the older Avalon Terrane to the southeast of the fault. These were the beautiful outcroppings that so intrigued Thoreau at Cohasset. What Hitchcock called "schist" and "horneblende slate associated with Gneiss" and dated to his "Second Series" are the high-grade metamorphic rocks of the Nashoba block, which Thoreau explored along the old Marlborough Road.[48]

This boundary also parallels the prominent topographic highland extending from Wayland northeast through Lincoln, and then Lexington. It forms the watershed divide between the north-draining Concord River, which empties into the Merrimack River at Lowell, and the south-draining Charles River, which empties into Boston Harbor. Without the Bloody Bluff Fault, there would have been no watershed divide. And with no divide, no glacial lake. And with no lake, no Walden Pond. And with no pond, no *Walden.*

Young Naturalist

Thoreau was the iconic barefoot boy of rural New England, driving the family cows to pasture. He was curious, independent-minded, inventive, physically able, and above all, highly imaginative. Drawn to nature by instinct, this tendency was reinforced by his family's devotion to outdoor rambles and to his sister Sophia's fashionable interest in botany.[49]

His formal education laid the groundwork for the self-education to follow. When attending the Concord Academy, now an elite college preparatory school, his main purpose was to acquire background knowledge and basic skills. There he was exposed to astronomy, botany, natural philosophy, natural history, and mathematics. And there, when eleven or twelve years old, he wrote his oldest extant essay, *The Seasons*. Naturally, it involved nature.[50]

After graduating he left his family home as a teenager to attend Harvard College in Cambridge between 1833 and 1837. There he studied math through calculus, and the physical sciences of mechanics, optics, electricity, mineralogy, geology, and astronomy, using a sequence of texts called the *Cambridge Natural Philosophy.* His only curriculum exposure to nature was through a course in "natural history taken at the end of his senior year" and taught by the college librarian, which says something about Harvard's priorities at the time. Looking back ten years after graduation, Thoreau wrote a letter poking fun at how little he had learned in college, referring to his "old joke of a diploma," and adding, "Let every sheep keep but his own skin, I say." Yet in that same letter, he rejoiced that Harvard—founded as a seminary—was finally "beginning to wake up and redeem its character and overtake the age" by building a school to teach "the highest branches of science." This was an opportunity he missed, and one he seems to have regretted.[51]

At Harvard, his main focus was on language and literature—Greek, Latin, French, German, and Italian—with a heavy emphasis on the classics. His immersion in languages facilitated his later self-acquisition of scientific languages, especially in the rapidly growing fields of botany and zoology. "One studies books of science merely to learn the language of naturalists," he later wrote in his *Journal:* "to be able to communicate with them." Doing so facilitated "the value of mutual intelligence" because it gave him access to the works of others, and therefore to insights and explanations he would not otherwise encounter. In one revealing case history, he identified a green-red dust within the stones of a wall as "a decaying state of the lichen *Lepraria chlorina,*" writing: "I have long known this dust, but, as I did not know the name for it . . . and therefore could not conveniently speak of it, it has suggested less to me and I have made less use of it." Thus, each new word became a new tool for taxonomic precision, allowing him to identify up to thirty-four types of lichen in Concord alone, all through self-instruction. Thoreau once remarked of Linnaeus that "he appreciated the certain advantage in these hard and precise terms," because they had "preserved anatomy, mathematics, and chemistry from idiots."[52]

Thoreau's emerging psychological profile also prepared him for his self-education in science. His capacity for abstract thought, analysis, and step-wise logic was evident to those who knew him as a youth. He was self-directed in the extreme. He was a visual thinker: when words failed, he sketched. He had intense powers of concentration, able to stay in the groove of close observation long enough to experience "flow," a state of mind so focused on the present that time dissolves. He sometimes had trouble letting go, as shown in his binge of description regarding the spray ice at Clematis Brook in January 1853. Though a passing remark would have sufficed for most naturalists of his era—especially on what must surely have been a bitter cold winter day—he stayed with the task long enough to draw seven technical illustrations and gather enough data to write six pages of highly detailed, largely geometric prose. This gusher of exacting description surpasses those of more widely known cases such as the bubbles in Walden ice and the sand foliage at the Deep Cut.[53]

His fieldwork revealed a physical toughness almost without precedent. Indefatigable, some said. Van Wyck Brooks wrote that when wearing his durable field clothing and well-oiled boots, Thoreau "ranged like a gray moose, winding his way through the shrub-oak patches, bending the twigs aside, guiding himself by the sun, over hills and plains and valleys, resting in the clear grassy spaces." His stamina and apparent insensitivity to discomfort would allow him to witness the height and breadth of outdoor nature in every season under every possible condition. Like field scientists and wildlife photographers who work in remote areas today, his bodily needs were elemental, and thus easily ignored when hot on the trail of nature's secrets.[54]

Thoreau's first post-college foray into nature writing was facilitated by Ralph Waldo Emerson's editorship of *The Dial*. For copy, and to help prod his protégé's writing career, Emerson "begged" Henry to "lay down the oar and fishing line" and "assume the pen" so that he might write a review of the recently published "Scientific Survey of the Commonwealth." The resulting article, published in 1842 as "The Natural History of Massachusetts," was neither a review nor was it scientific. Instead,

Thoreau used it as an opportunity to practice his subjective responses to nature in the Emersonian essay tradition. Though this early engagement with nature "appears prototypical of the best Thoreau work to come," Robert McGregor reminds us that at this stage, Thoreau's tone was still strongly anti-scientific. "Denouncing the Baconian scientific method as false, he concludes that 'we cannot know truth by contrivance and method.' " The following year Thoreau published another essay in the same genre, "A Winter Walk." Again, he focused on the prose-poetry, delving no deeper into science than metaphor.[55]

Geology played a significant part of young Henry's early program in natural history. In 1835, six years before Hitchcock's *Final Report,* Concord historian Lemuel Shattuck included within his *History of Concord* a section on the geology that was typical of the era, being long on classification and short on explanation. He listed the minerals and rocks that could be found about town: "limestone . . . Calcareous Spar . . . Garnet . . . Cinnamon-stone . . . quartz . . . Mica . . . hornblende, actynolite, pargasite, . . . Feldspar . . . argillite, or clay-slate, novaculite, and scapolite. Sahlilte, a variety of augite, or pyroxene . . . Sulphate of iron, or coperas, occurs with a vein of sulphuret of iron, or pyrites in green stone . . . [or] . . . disseminated in clay-slate . . . Lead ore . . . Iron ore." Thoreau may have taken this list—or perhaps the much better one of Edward Hitchcock—as a starting point for building his own mineral collection. Though smaller than his botany or archaeology collections, it was no less distinct or aesthetically important during the Walden Period, being prominently displayed in his attic room of the Thoreau family house on Main Street.[56]

After Thoreau's death in 1862, the bulk of his collections went to the Boston Society of Natural History. His mineral and rock collection, however, went to Frank Sanborn, one of his personal friends. What remains of that collection is contained within a single glass case that was exhibited for at least eight decades in the farmhouse at Fruitlands Museum in Harvard, Massachusetts, before recently being put into storage. This very building had been command central for Bronson Alcott's experiment in communitarianism. That this mineral collection was Thoreau's is verified by a letter dated December 17, 1955, and written by the museum's former director, William Henry Harrison. It came to the museum about 1910 when Sanborn's effects were purchased by the museum's founder, Clara Endicott Sears. Having been part of that Sanborn purchase, Thoreau's collection likely went on exhibit within a decade.[57]

To say that Thoreau's geology collection is poorly known is an understatement. Frank Sanborn, its original owner, didn't mention it in his 1882 biography of Thoreau. It seems to have escaped the scrutiny of Walter Harding, Thoreau's most particular biographer. And also that of Eugene H. Walker, a professional geologist, former president of the Thoreau Society, and author linking Thoreau to the "Minerals of Concord." Using the yellowed and brittle calligraphy labels presently attached to Thoreau's specimens, his collection contains one or more samples of "Pyrite . . . Chalcopyrite . . . Quartz Pyrite . . . Calcareous Tufa Hot Springs Deposit . . . Chalcydony [sic] . . . Granite . . . Mica Schist Concord (?) . . . Feldspar from Pegmatite . . . Magnetite . . . Gypsum . . . Gypsum Var. Salenite [sic] . . . Sulphur."[58]

It's an odd collection. Mundane specimens of local rock are mixed in with gem-like crystals of quartz, gleaming masses of metal, and several curios like sulfur and tufa, all of which apparently can be found within the reach of Thoreau's travels. At one point, at least a few of these specimens probably occupied one of Thoreau's hand-made mahogany "geological cases," two of which he brought to Cohasset when on his way to Cape Cod. These he gave to Joseph Osgood, the local Unitarian minister, who had achieved what Thoreau could not: the hand of Ellen Sewall in marriage. One of Osgood's cases, including its specimens, was later gifted to the Concord Museum.[59]

According Walter Harding, Thoreau's familiarity with minerals "admirably suited" him to join the proposed "government expedition" of Dr. Charles T. Jackson to "survey the mineral lands of Michigan." Though Emerson recommended Thoreau for the job in May 1847, and though he "wanted very much to go along," the opportunity went elsewhere for reasons of political patronage, leaving him stranded in Concord. This little-known historic event became a critical juncture in Thoreau's life, perhaps pre-empting a potential career as an exploration geologist of the western territories.[60]

From his careful reading of Lyell's *Principles* and Hitchcock's *Final Report,* Thoreau learned the basics of geology: for example the volcanic origin of "basalt" and the detrital origin of "yellow sandstone." He applied this knowledge to many details of his local landscape. Some stones, he noted, "split into their laminae," making them easy to "carry away as a specimen." This passage, in Thoreau's own words, proves that he sampled rocks as part of his natural history investigations. Other stones, he noted, were massive, strong, and resistant, being the "temper" against which the stonemason must work. Over time, he learned that "slate stone" suitable for fashioning an Indian hoe might be found in one place, and "white quartz" for arrowheads in another. He saw that stones near Flint's Pond were "comparatively flat," relative to those near Walden, an observation he correctly discerned was related to the contrast in lake shape. Some stones, he knew, were imported from far away, for example the chert with "conchoidal fracture" that was used by natives to fashion their beautiful spear points. "Geologists tell us," he wrote off-handedly, that this type of stone "is not to be found in this vicinity."[61]

During his later scientific sojourning phase (1850–1854), Thoreau became much interested in causes and explanations about geology, reaching a modicum of proficiency. "The most inattentive walker," he remarked with what sounds like a touch of arrogance, "can see how the science of geology took its rise." From "the action of water" on "rounded" hillsides "and marine shells found on the tops of hills . . . the geologist painfully and elaborately follows out these suggestions, and hence his fine-spun theories."[62]

Thoreau documented these "suggestions" in dozens of *Journal* sketches, measurements, and technical descriptions. On April 1852, for example, he noted that "on Conantum Cliffs," the large-scale "seams dip to the northwest at an angle of 50° (?) and run northeast and southwest." His plane geometry records the strike and dip of Nashoba under-thrusting long before it was understood. Though he was silent about the cause of the tectonic grain at the regional scale, he did speculate about rock deformation at the outcrop scale. When standing on "the rocks in the high open pasture" east of Bateman's Pond, he noticed that "the strata are perpendicular, producing a grained and curled appearance . . . like walking over the edges of the leaves of a vast book . . . the strata are remarkably serpentine or waving . . . as if you were upon the axis of elevation, geologically speaking." From this comment it is clear that he was observing tightly folded strata, and correctly understood that they had been created by strongly directed compressional forces.[63]

Cooling and Cracking

As the Acadian orogeny wound down, the New England highlands began to wear away. The bulk of the Nashoba block, however, was still deeply tucked within the middle crust. Though still warm and squishy, it was now permanently attached to the North American craton. In turn, that craton was part of a mega-continent called *Laurasia:* the merger of North America, Greenland, Baltica and Russia. Moving up from the south was an even larger mega-continent called Gondwanaland: a gathering of what are now the separate pieces of Africa, South America, Antarctica, India, and Australia. The docking of these mega-masses against one another led to the third and final collision responsible for creating New England's bedrock. Known as the Alleghenian orogeny, its most superlative result was Pangea, the most super-sized continent ever known to exist.

As with the modern Himalayas and Tibetan Plateau today, the resulting Appalachian highlands became so massive that the over-thickened and over-heated crust could not support the load at depth. As a result, the Nashoba rocks continued to squeeze and shear. At the surface, however, the brittle upper crust was collapsing under its own staggering weight, sagging, stretching, and spreading to create intermontaine basins. These became filled with coarse river sediment being carried by powerful mountain streams and dumped into alluvial fans. When cemented and weakly metamorphosed, these river gravels became the famous Purgatory Puddingstone of Narragansett Bay and similar conglomerates to the north that have long since been eroded away. Between these coarse strata were masses of river sand and seams of coal containing the fossils of primitive plants like seed ferns and giant horsetails, and of giant, carnivorous amphibians and squat, fin-backed reptiles.

Eventually, the Nashoba block cooled to the point where it began to behave as a rigid solid. The coherent mass created earlier by hot compression now broke up along brittle faults aligned by Alleghenian and younger tectonic regimes. Discrete blocks of the upper crust—at the scale of a mile or more—jostled against one another, moving both sideways and up and down. Nashoba had become earthquake country. Large seismic lurches and great landslides became common. Though these younger stresses are now long gone, the legacy of this tearing and stretching remains as a set of well-defined fault zones that crosscut the older grain. Near Walden, they are mostly aligned in a northwest-to-southeast direction.

One of these younger fault zones was later eroded to become the straight, narrow path of the Sudbury River between Fairhaven Bay and its junction with Nut Meadow Brook. There, the contact between the

Andover Granite and the underlying Marlborough Formation has been displaced two miles northward relative to its counterpart on the other side of the valley, through some oblique combination of vertical and horizontal motion. Another younger fault zone aligns the junction of the Sudbury and Assabet Rivers where they meet to form the Concord River at Egg Rock. Many less distinctive faults from this post-orogenic period have also been mapped. In fact, they align much of the nineteenth century transportation network near Walden Pond, each having created its own zone of weakness through the older ridges. The Fitchburg Railroad, Walden Road, Back Road, and Lincoln Road all run roughly parallel to each other along known and inferred faults. In fact, the main basin at Walden Pond lies in a crude triangle between a mapped block fault to the northwest, an inferred block fault to the northeast, and an older thrust fault to the south. The two block faults intersect near the tip of Thoreau's Cove in the vicinity of his house site. Thus, to understand why Thoreau chose to live where he did, we have to wedge our way downward into the deeper reality of the rock.[64]

Finally, Nashoba is the Anglicized spelling of "Nashobah" or "nashope," the native American place name for what is now the town of Littleton, Massachusetts. The source of the name is a prominent hill, whose seismic noises were an important part of native oral tradition and which were well documented by colonial English settlers. Known as the "hill that shakes," Nashoba remains a conspicuous and anomalous seismic zone today. Its occasional noises and rattles involve slight readjustments along the great faults that slivered this beautiful part of rural New England into brittle pieces before the tectonic violence ceased. Ironically, seismologists have confirmed that the "solid bottom" at Walden Pond still shakes, and geologists have confirmed that Thoreau's "rocks in place" came from somewhere else.[65]

Scholars are usually drawn to Thoreau's experience on Mount Katahdin for the psychological drama of his life-changing epiphany. Virtually ignored is his prediction of what will inevitably happen to this titanic scene. The "vast chemistry of nature would anon work up, or work down" the "vast aggregation of loose rock . . . into the smiling and verdant plains and valleys of earth." Despite his earlier allusion that the rock had "rained down" as if "dropped from an unseen quarry," he knew full well that Katahdin granite had emerged slowly from below, "disjointed at the quarry." And that given enough time, it would be transformed into productive lowland soils by means of downward decay and flushing. It is to this story we now turn.[66]

2

LANDSCAPE OF LOSS

"The Musketaquid, or Grass-ground River, though probably as old as the Nile or Euphrates, did not begin to have a place in civilized history, until the fame of its grassy meadows and the fish attracted settlers out of England in 1635." With this nod to the depth of history, Thoreau begins his first book—*A Week on the Concord and Merrimack Rivers.* As it turns

Figure 7. Sudbury River during winter. View is upriver (southward) toward Fair Haven Hill from the bridge at Sudbury Road. Note gentle gradient and land-fast ice. January 2011.

out, the Concord River in its present channel is actually twice as old as the others, having been formed on the bed of an ancient glacial lake that drained about 17,000 years ago, rather than on a pair of estuarine deltas that did not begin to grow until about 8,000 years ago when global sea level stabilized after its post-glacial rise. And upstream from their deltas, the Nile and Euphrates drain highlands much younger than those of the northern Appalachians. The Old World, it seems, is newer than the New.[1]

In Thoreau's day the Concord River was a "broad and beautiful" free-flowing stream "at least ten rods wide and bordered by the most fertile soil in town." Near the village, its path traced a straight line drawn from southwest to northeast. Far to the northeast lay the ancient Puritan settlements along the coast like Salem, and the industrialized lower Merrimack River near Haverhill, sectors already denuded of trees and crowded with houses. Far to the southwest lay a "great tract . . . of unimproved and unfrequented country" in the headwaters of the Assabet River. Also from that direction came peak warmth from the mid-afternoon sun and the crop-growing breezes worshipped by indigenous peoples before European settlement.[2]

In his famous essay "Walking," Thoreau claimed that his sojourns were guided by an inner compass, an "instinct" created by the "subtle magnetism in nature, which, if we yield to it, will direct us aright." More specifically: "My needle is slow to settle . . . but it always settles between west and south-west . . . the earth seems more unexhausted and richer on that side . . . I will walk into the southwest or west." The subtle magnetism he referred to coincided with—and may have even been due to—the tectonic grain of the Nashoba Terrane, which had been etched into the landscape by a quarter-billion years worth of weathering and erosion. On this grain he could walk from Little Goose Pond near Lincoln, past Walden, down to Fairhaven, and then down the Old Marlborough Road and beyond.[3]

In July 1842 Thoreau climbed Mount Wachusett for the first time. Overnighting on its summit, he recognized that "the hill on which we were resting comprised part of an extensive range, running from southwest to northeast, across the country, and separating the waters of the Nashua from those of the Concord." With prescient insight, he discerned that this alignment was present at many scales:

> These lesser mountain ranges, as well as the Alleghanies [*sic*], run from northeast to southwest, and parallel with these mountains streams are more fluent rivers, answering to the general direction of the coast, the bank of the great ocean stream itself. . . . Even the clouds, with their thin bars,

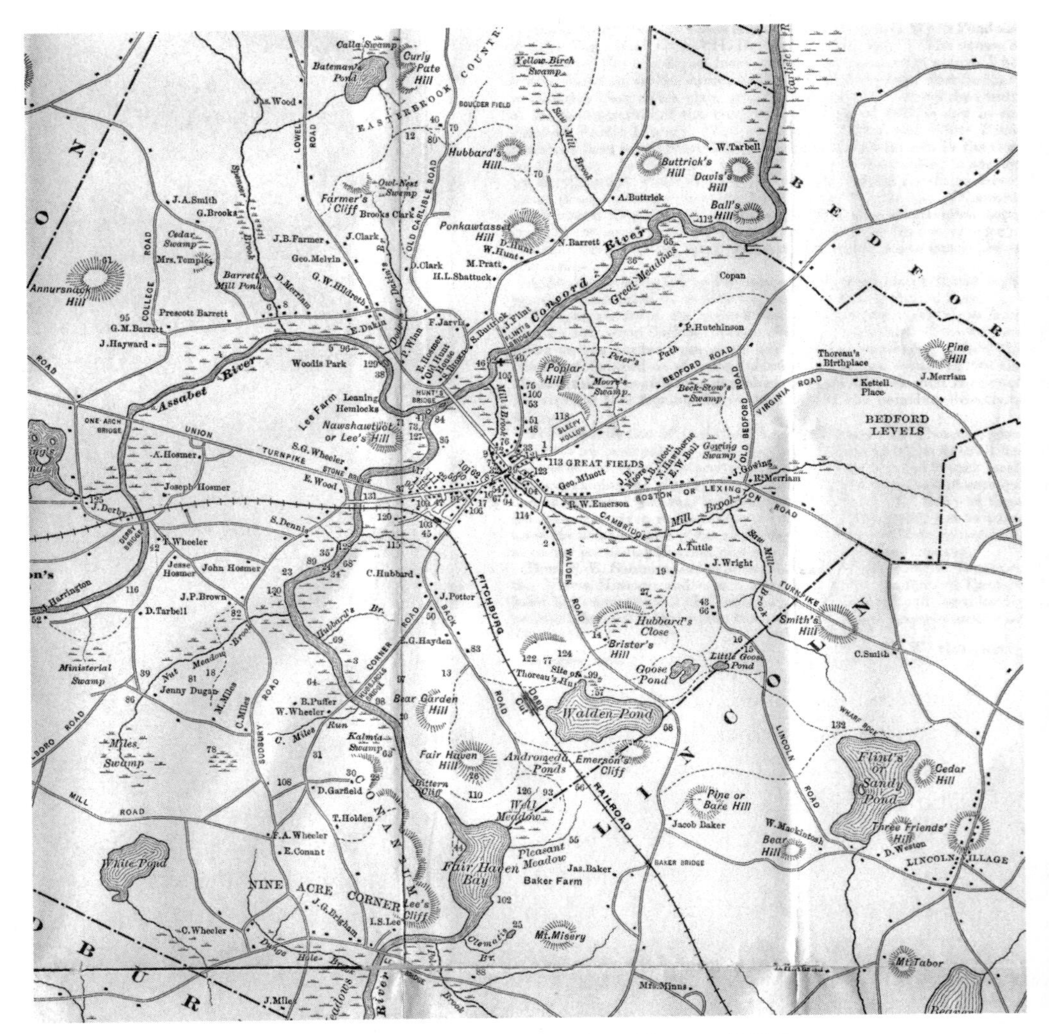

Figure 8. Map of Walden Pond vicinity by Herbert Gleason (1906).
Originally drawn for Torrey and Allen, *The Journal of HDT*, Volume 2 (follows page 1741).
Courtesy Concord Free Public Library.

> fall into the same direction by preference, and such even is the course of the prevailing winds, and the migration of men and birds. A mountain-chain determines many things for the statesman and philosopher.[4]

This passage identifies six phenomena aligned by New England geology: the orientation of the sea coast relative to the main range of the northern

Appalachians; the alignment of subsidiary ranges like the Peterborough Hills; the prevailing direction of streams; the local wind regimes; the formation of clouds; and human transportation. All these features derive from the compressional grain of Acadian tectonism. Crossing this grain from northwest to southeast is the regional slope, which extends from the "spine" of the New England mountains to the down-tilted edge of its continental shelf far beyond Cape Cod. Thoreau country lies midway between this crest and this edge, its local topography governed by countless sub-parallel weak-and-strong layers of rock, the thrust faults between them, and the sheared granite intrusions squeezed up from below.

Rifting and Tilting

"What a piece of wonder a river is," Henry remarked in one of his earliest *Journal* entries, "a huge volume of matter ceaselessly rolling through the fields and meadows of this substantial earth, making haste from the height places . . . to its restless reservoir." His use of the term "matter" reveals his clear grasp of what casual observers usually overlook: that water is not the only thing being shed from land to sea. Also present are dissolved salts, grit suspended by turbulence, sand being dragged over on the bottom, and pebbles bouncing along. From this perspective, rivers operate like self-organized sewers that flush the residues of rock decay to the sea where, after a long pause, they are recycled anew.[5]

This Huttonian worldview renders the word "watershed" a misnomer, because water is merely the agency of change, rather than the object of interest. The most important thing being shed is the land itself, one ion, crystal, or rock fragment at a time. Additionally, most of the water is not being shed. Rather, most of what falls on Walden Woods soaks straight downward, destined either to vaporize back to the sky via the roots and leaves of trees, or to leak from aquifers into the linear "springs" we call streambeds.

"This old, familiar river," Thoreau wrote of the Concord, "is renewed each instant; only the channel is the same. The water . . . may have washed some distant shore, or framed a glacier or iceberg at the north, when I last stood here," before being imported from anywhere and everywhere by the "light-footed air." After falling as rain or snow, liquid water becomes the chief chemical reactant of the geo-biological processes that turn rock back into dust, ensuring that the "stealthy-paced water" of rivers always has something to flush back to the sea. Given "time and perseverance," Thoreau knew that this aqueous half of the rock cycle was fully capable of erasing the mountain chains created by its geothermal first half.[6]

This gradual wearing away of the land is called "denudation": the sum of all mass being lost averaged over the spatial scale of square miles, and over time scales exceeding a century. Local erosion may begin and end with discrete events such as a hill slope being gullied during a thunderstorm, or a riverbank being undercut in flood. But denudation never stops, not even when highlands are actively growing.[7]

New England is thus a landscape of loss. Of denudation. Everything else, including recent glaciation and European colonization, has been a minor detail to this big story. This chapter of the Walden narrative began with a gradual reversal in the direction of plate motion, from collision to relaxation, and then to rifting. Exactly how and why this happened is not known, though geologists have plenty of competing ideas. Regardless of details, early in the process, tectonic rifts like those of East Africa today zippered their way northward from Georgia to Nova Scotia, creating a series of valleys being filled with sediment shed from adjacent highlands. In New England, the softer sediments within the rift zone were later eroded to create the Connecticut River Lowland, the widest part of New England's longest valley, which extends from Long Island Sound to Quebec. After considerable trial and error, a deeper and more permanent rift opened between Nantucket and northwest Africa. There, the supercontinent Pangea was literally torn apart, with the Appalachian side going one way and the African side going the other. Between them was the nascent Atlantic, born when the rift dropped low enough to fill with salt water and create a narrow arm of a new sea resembling the Gulf of Suez. Since then, it has been steadily widening as new oceanic crust was created to fill the gap.

In conventional political geography, New England terminates at the Atlantic coast between Cape Cod and New Brunswick, Canada, passing through Boston, Massachusetts; Portsmouth, New Hampshire; and Eastport, Maine. In geology, however, it terminates at the outer margin of its continental shelf, a permanent structural feature marking the top of the continental slope, which descends steeply for thousands of feet to the floor of the abyss. This true edge of New England traces a line parallel to—but more than a hundred miles south of—its southern, sandy island archipelago between Staten and Nantucket Islands. From there, it continues easterly, and then northeasterly toward offshore Labrador. This line makes the Gulf of Maine and Georges Bank as much a part of the continent as Bangor, Worcester, and Hartford, despite being covered by shallow salt water. This perspective was shared by Thoreau, as his Cape Cod essays reveal. In this context, the present shoreline—where homes are now falling into the sea—is merely the transient edge of the global ocean at this particular geological moment. That edge has risen and fallen nearly a thousand feet against the New England coast, and has moved back and forth hundreds of miles.

Another consequence of sea floor spreading relevant to the story of Walden Pond was the slow cooling of oceanic crust with increasing distance from the volcanic spreading center, the Mid-Atlantic Ridge. As the new oceanic crust cooled and aged, it became denser, sinking further into the mantle. This helped drag the thinned margin of the continent down through the buzz saw of coastal erosion to create an incipient continental shelf. Simultaneously, the residues of denudation being shed from New England's high peaks were being washed southeastward and deposited as massive deltas interfingering with deposits of marine clay, lime, and plankton. This transfer of mass was equivalent to moving a heavy load on a stationary barge from port to starboard. This tipped New England's strong crust down to the southeast and up to the northwest, lifting its terrestrial topography to keep the mass transfer going.

Rivers have flowed down this regional slope and across the tectonic grain of Nashoba ever since the Atlantic was born. Initially, they were rapid and powerful, and drained a rocky, rugged landscape like those of Earth's mountains today. Under such conditions, the turbulent spray of gritty water and the relentless percussions of sand, pebbles and cobbles against a rocky streambed were quite capable of cutting steep canyons. With time, however, the rivers gradually shifted from being agents of active erosion to being agents of flushing, with the task of sluicing the waste products of crustal disintegration down to the sea. Regardless of which mechanism was dominant, mass was always being removed from the top down. As with sun-melted icebergs floating in an icy sea, this loss from the top forced a compensatory rise of the deeper crust due to buoyancy. This led to more erosion, which further diminished the mass, which caused an additional rise, and so forth, *ad infinitum.* Given this powerful feedback loop, rock originally located miles below the surface was slowly brought up to it. Borrowing a term from mortuary practice, geologists say that the mountain root of New England was *exhumed.* In short: disintegration drove denudation, which drove exhumation, which provided fresh rock for more disintegration.

Thus it was that Nashoba—a stable mass of warm, dry rock—was exhumed from a depth of at least ten miles to reach the water-rich and highly corrosive surface of Thoreau's Concord. Though the tectonic grain of this tilted block was a permanent fixture, the specific locations of ridges and valleys at the surface were always migrating to the northwest as denudation bit ever more deeply into the tilted layers. At this time scale, the outer boundaries of watersheds and the lines of flow within them were transient.

Thoreau's Deep Time

Imagine sitting still while your fingernail grows across the Atlantic Ocean from Nantucket, Massachusetts, to Mauritania, West Africa. This is the span of time involved since Concord's exhumation began. Imagine Nashoba being a boulder frozen into a colossal tectonic iceberg somewhere well south of the equator. Since then, that berg has drifted northward about five thousand miles and rotated counterclockwise more than a hundred degrees, bringing the bedrock along for the ride. Imagine being in the deepest diamond mine on Earth and tunneling downward to reach the pressures and temperatures at which the Andover Granite was created. To get there at the average rate of denudation, your speed would have been about one sixty-fourth of an inch per year.[8]

Without such metaphors, it is impossible to grasp the immensity of deep time. This is why Thoreau developed several of his own for the geology section of *A Week on the Concord and Merrimack Rivers*, his first book. My favorite is: "I sit now on a stump whose rings number centuries of growth. If I look around I see that the soil is composed of the remains of just such stumps, ancestors to this. . . . I thrust this stick many aeons deep into its surface, and with my heel make a deeper furrow than the elements have plowed here for a thousand years."[9]

After reading Charles Darwin's *Journal of Researches* in 1851, Thoreau also explored the significance of deep time using emerging ideas about evolution. He wrote in his own *Journal* how individual fitness, adaptation, co-evolution, and competition had, over immense spans of time, combined to shape the animals and plants familiar to him. He came to understand that "life" was a "battle in which you are to show your pluck, and woe be to the coward." The pitch pines fronting Walden pond, for example, were those that had survived being "tried and twisted" on "blowing, stirring, bustling" days. The frogs and toads of his world were those with the best camouflage. He thought it "remarkable that animals are often obviously, manifestly, related to the plants which they feed upon or live among." Indeed, "each creature was fitted to the world not by 'some poor worm's instinct' merely, as we call it, but the mind of the universe rather, which we share, has been intended upon each particular object." Finally, "Nature has produced them. As if a poet were born who had designs in his head." These ideas predate Darwin's *Origin of Species* by at least five years. And though they sound like the "argument from design" from Christian natural theology, Thoreau instead explains them by invoking Hutton's abyss of deep time: "Singular these genera of plants, plants manifestly relate yet distinct. They suggest a history to nature, a natural history in a new sense."[10]

This "new sense" of natural history was paleontology, the most popular part of geology's emerging paradigm. In Thoreau's transcendental language, the modern dead fish he found floating in a Sudbury meadow was a "fabulous or mythological fish . . . not an actual terrene fish, but the fair symbol of a divine idea, the design of an artist . . . as little fishy as a fossil fish." In his later scientific language, however, Thoreau discerned that the dead individual fish was an extant descendant of a body plan so successful that it has been copied by all vertebrates since the dawn of their lineage. More fundamentally, he recognized that the design concept of a streamlined body shape with feeding apparatus at one end arose independently in insects, fish, mammals, lizards, and birds: for example, diving beetles, tuna, dolphins, extinct ichthyosaurs, and loons, respectively. So effective were such designs, that their motifs were "sculpted on ancient monuments," as well as on exposures of rock strata. More generally, Thoreau concluded: "In all her products Nature only develops her simplest germs. One would say that it was no great stretch of invention to create birds. The hawk, which now takes his flight over the top of the wood, was at first perchance only a leaf which fluttered in its aisles."[11]

When Thoreau held *Homo sapiens* up to the fossil record of "antehistoric, geologic, antediluvial rocks," he found our species wanting. Inspired on one occasion by an unusually large fungus, he remarked that "such growths ally our age to former periods, such as geology reveals . . . It suggests a vegetative force which may almost make man tremble for his dominion. It carries me back to the era of the formation of the coal-measures—the age of the saurus and pleisosaurus and when bullfrogs were as big as bulls." During this Paleozoic era, Nashoba lay not near the edge of a continent but in the center of the largest one ever. He enjoyed knowing that a "fossil tortoise has lately been discovered in Asia large enough to support an elephant." I have seen one myself at the Peabody Museum at Yale University where *Archeon ischyros*—a Cretaceous turtle nearly fifteen feet across—adorns the entry wall.[12]

Thoreau knew enough of the Paleozoic to criticize someone for their "trilobite" eyes, and enough of the Pleistocene to know that horses had gone extinct in the New World before being reintroduced by Spanish conquistadors. Once, when walking by a "neighboring stump fence of white roots," he saw within them the bleached "bones of marine monsters and the horns of mastodons or megatheriums." Combining his love of mythology with that of paleontology, he made the obvious connection: "The geologist has discovered that the figures of serpents, griffins, flying dragons, and other fanciful embellishments of heraldry, have their prototypes in the form of fossil species which were extinct before man was created," and which "indicate a faint and shadowy knowledge of a previous state of organic existence."[13]

He envisioned the "order of the world's creation" in that first "spring of the world," when early plants sought out "bare rocks, and scantily clad lands, and land recently bared by water." And before them were "lichens and algae . . . as if belonging to a former epoch." Alder and willow impressed him "as a vegetation which belongs to the earliest and most innocent dawn of nature; as if they must have preceded other trees in the order of creation, as they precede them annually in their blossoming and leafing." He accepted that life originated in the sea. While walking the shore on Cape Cod, he reflected: "Creeping along the endless beach amid the sun-squall and the foam, it occurs to us that we, too, are the products of sea-slime. Stagnant muddy pools were "the seeds of life, the liquor rather, boiled down," the chemical soup from which the first cell was born. And before that was "an inorganic mineral life": the common clay of every living thing.[14]

Thoreau's firm grasp of paleontology laid the groundwork for the climax of *Walden*, which describes the emergence of complexity and beauty from the simple flow of muddy sand at the Deep Cut. It also was the taproot of his lifelong frustration with Christian supernaturalists, who insisted on a fairly brief history of life. Paraphrasing Lyell's *Principles,* he jested: "It took 100 years to prove that fossils are organic, and 150 more, to prove that they are not to be referred to the Noachian deluge." Not everyone believes this, even today. Modern "young Earth" creationists still insist that the Elizabethan-era Mosaic chronology of Archbishop Ussher is the correct one, and that we twenty-first-century scientists are in error.[15]

Breakdown

As Nashoba was exhumed, its hard shell of near-surface rock expanded slightly in response to lower confining pressure, and contracted slightly due to cooling. This created invisible "lines of weakness" at all scales from a single mineral crystal to units large enough to map such as the Marlborough Formation. Deeper in the warmer crust, these stresses and strains had been continuously relieved by plastic flow, molecular diffusion and re-crystallization. But within the outer mile or two of the crust, the cooled, rigid rock strained elastically until rupture occurred. Microcracks lengthened, widened, and integrated into regular sets of planar fractures called joints. The oldest sets were guided by whatever ambient tectonic conditions were present during rock formation, for example the planar fractures of slaty cleavage. The youngest sets were guided by the loss of overburden pressure during erosion, and were thus roughly parallel to the surface: the same process that drives dangerous rock bursts in a quarry. Between these oldest and youngest sets of joints were those expressing the physical memories of intermediate tectonic regimes. Thus, as the shallow crust rose up toward the surface, it became a nested mosaic of blocks bounded by regular fractures, something that can be easily seen in every highway road cut.

Operating below the soil were unrelated physiochemical processes that widened and lengthened existing fractures and created new ones. Seasonal changes in water pressure and thermal changes in volume wiggled blocks apart. Chemical dissolution widened some joints and cemented others. Freezing water heaved blocks aside. Gravity tugged on each mass differently, pulling the nested mass apart.[16]

Within the soil proper, the baton was handed off to biochemistry and biophysics. The "first farmer," Thoreau wrote, was the "lichen which plants itself on the bare rock, and grows and thrives and cracks it and makes a vegetable mould." Next comes "moss on the surface, and starting saxifrage, ferns still green, and huckleberry bushes in the crevices." The beautiful columbine, its "large clusters of splendid scarlet and yellow flowers growing out of seam in the side of this gray cliff," is, at depth, a destroyer because plant rootlets catalyze the chemical disintegration of rock in order to obtain mineral nutrient. Larger roots physically break up the rock as they probe downward for water and structural support.

Given sufficient breakdown and the retention of those breakdown products, any drained surface with sufficient moisture inevitably becomes blanketed with grass, herbs, shrubs, and ultimately trees. Rock has become soil. From the geological perspective, patches of New England forest became thick woody sods, which "from the tops of mountains . . . appear like smooth shaven lawns." Below each patch, granite is being converted: first into a coarse, rusty, granular sand called "grus"; next to an incipient organic layer over stained yellow-brown earth; next to a well-developed soil with distinct horizons; and finally, to a thick, pinkish-yellow or reddish-brown clay- and oxide-rich residue called "saprolite," which translates from Latin as "rotten rock." Such saprolites were the dominant soils above Nashoba for nearly all of its denudation history. In his rock collection, Thoreau had three pieces of pink clay from Martha's Vineyard, all of which were remnants of this chemical breakdown process.[17]

After excavating a muskrat chamber, Henry paused to reflect on what a "wonderful piece of chemistry" the soil actually was. The "very sod is replete with mechanism far finer than that of a watch." He knew that sunlight alone was insufficient for chemical weathering; water was required. He noticed that most stones freshly dug from the subsoil were stained by a faint yellow wash of precipitated iron, but that some could turn the color of woodchuck brown. He knew that granite boulders exposed to the atmosphere turned gray, owing to the condensation of dew, the drenching by rain, and the chemical agents of lichens and microbes. "It seems natural," he concluded, "that rocks which have lain under the heavens so long should be gray . . . Time will make the most discordant materials harmonize." Of all possible earth tones, "a reddish tinge in the earth. . . . An indian hue" was the most "singularly agreeable, even exciting the to the eye. Even the color of the subsoil excites me, as if I were already getting near to life and vegetation." This color foreshadows the *Walden* climax in which life springs from non-life. Thoreau practically worshipped the brownish-black humus of the soil, the "virgin mould, the very dust of nature," the vital key to Concord's landscape of loss.[18]

During each growing season, plants extract carbon dioxide from the air. Using the power of sunlight via photosynthesis, they convert that carbon into organic matter, mostly in the form of leaves and stems. After its return to the forest floor, this organic matter converts back into carbon dioxide, mostly by microbial respiration within the soil humus. When infiltrating water moves downward through this soil gas, it creates dilute carbonic acid, the single most important chemical agent for the weathering of silicate minerals. When this acid encounters a feldspar or mica, it displaces whatever positive ions are present, usually K^+, Na^+, Ca^{++}, and Mg^{++} (potassium, sodium, calcium, and magnesium), causing corrosion. Those not taken up by plants or retained in the soil to make mineral clays and oxides are flushed downward to aquifers where they flow to springs, ponds, lakes, streams, and thence to the sea. Metal ions of iron and manganese (Fe^{++} and Mn^{++}) are liberated to groundwater from darker minerals. If sufficiently concentrated, and if they encounter oxygen, they precipitate to form clots of bog ore, which had been intermittently mined in Concord since 1637. Thoreau saw the precursor to bog ore in the bed of the Assabet River as a "conglomeration and consolidation of sand and pebbles, as it were cemented with oxide of iron (?), quite red with it, iron-colored, to the depth of an inch on the upper side." He seems to have understood the chemistry.[19]

In the more fertile bed of the Sudbury River, calcium ions (Ca^{++}) dissolved from Nashoba's metamorphic minerals were being biochemically precipitated by freshwater mussels (clams). Those same ions in the Thoreau household drinking water became the bones and teeth of the family and their boarders. After death, the hard parts of both clams and humans "turn to dust again" and leach back to aquifers and streams. There, every drop of water flowing toward the ocean is salted to some extent with dissolved ions, prompting Henry to speculate on what "a chemical analysis of the water of the Dead Sea" might reveal. Upon reaching the ocean, they provide the calcium, magnesium, and phosphorus needed for plankton, reef-building organisms, and the bones of whales. All of these eventually get recycled back into rock, as revealed by the marble Thoreau described on Concord's Curly Pate Hill near its border with Carlisle.[20]

Granular residues produced by the disintegration of crystalline rocks inevitably work their way down slopes into stream channels, where they are washed away. The simplest mechanism is direct gravitational fall from the face of a cliff. Thoreau witnessed this only at the Deep Cut: "The sand and stones fall from the overhanging bank and rest on the snow below." More typically, gravity-driven mass movements are invisible, occurring as tiny, ratchet-like expansions and contractions within the soil, whether by roots penetrating and dying back, moisture freezing and thawing, soils wetting and drying, or creatures burrowing and abandoning their myriad pathways. Expansions are perpendicular to the land surface. Contractions are ideally straight down, driven by gravity alone. Thus, each increment of outward expansion followed by downward settling is like one click of a ratchet. Compaction due to animal footfalls also move soil grains downhill because every step pushes more soil downhill than uphill.[21]

During his many sojourns, Thoreau witnessed a variety of slope mechanisms causing soil creep. In plowed fields he saw "asbestos-like ice-crystals, more or less mixed with earth" push tiny allotments of soil upward and outward, which then settled downward..In March 1853 he "opened an ant-hill about two and a half feet wide and eight inches high, in open land . . . which extended to the depth of two feet in the yellow sand, and how much further I don't know." As with frost, ants raise soil grains perpendicular to any sloping surface before they settle vertically, contributing to mass movement. When he described "the paths made by the cows in the sides of the hills, going round the hollows, taking gracefully curving lines in the landscape, ribbing it," he was witnessing both the cause and effect of creep. His own house foundation became "indistinct as an old cellar-hole" within five years, in part because its edges were plowed over, but also because every vegetated slope is being continuously smoothed by creep. Even the weight of sand itself was sufficient to cause settling and downhill motion, as Thoreau witnessed in "a meadow through which the Fitchburg Railroad passes."[22]

Very different mass movement processes operate in the absence of continuous vegetation. In this situation, agitated mixtures of water and sediment create stiff debris flows when moist and slurries when there is sufficient water. If large raindrops strike exposed soil, the "sand and gravel are beaten hard by them." This made early spring "the season to look for Indian relics, the sandy fields being just bared" following a winter's worth of frost heaving and churning. Each large drop of rain falling vertically on an open slope sends more mineral grains downhill than uphill. And if a film of water thickens to the point of flow, it drags grains along with it, a process called wash erosion. At some point, small rills cut down into the earth. And if deep enough, they create miniature badlands resembling "a cave, with all its stalactites turned wrong side outward." Farther downhill on longer slopes, these tiny rills merge progressively into larger "gullies, more or less sandy, where the water has flowed down," such as those he saw cutting the river cliffs at Haverhill.[23]

"We love to see nature clad," Thoreau wrote, "whether in earth or a human body. . . . Nobody likes to set his house under that part of the hill where the sod is broken and the sand is flowing." Indeed, the main difference between a gullied landscape and one with "agreeable" curves is whether a "sod" cover is present or not: that three-dimensional mass of organic fibers that armors the sand from splash erosion and binds it with a root-woven mesh. Within and below that mesh is a zone about a foot thick that is inexorably moving downhill to deliver weathered residues to stream channels. Every slope in a rolling landscape like Walden Woods is thus a three-dimensional conveyor belt designed to ensure that streams never run out of sediment, and that the roots of mountains will always be exhumed.[24]

Thoreau's Day Job

Given Thoreau's legendary skill at pacing distances, finding pathways, pointing directions, and estimating heights, weights, and counts, it is no wonder he "could easily solve the problems of the surveyor," as Emerson remarked. But being able to do something and wanting to do it are not the same, especially for a career. As with many aspiring artists, actors, and dancers working at day jobs while waiting for their big break, Thoreau was an aspiring writer waiting for literary success, and living with a chronic shortage of cash. An abdominal injury during the summer of 1846 made manual labor difficult for a year or so. By 1847 he was dabbling with surveying for pay, listing it third on a list of jobs for a Harvard alumni survey: "Schoolmaster—a Private Tutor, a Surveyor—a Gardener, a Farmer—a Painter, I mean a House Painter, a Carpenter, a Mason, a Day-laborer, a Pencil-Maker, a Glass-paper Maker, a Writer, and sometimes a Poetaster." By July 1848 he had left domestic "service" at the Emerson household and had moved back into his parents' house, where he felt obliged to pay for his keep. "By the fall of 1849," and using borrowed equipment from Cyrus Hubbard, "he was doing enough surveying to justify purchasing a notebook to keep his records straight." By the spring of 1850, he had decided to make it a business. This prompted him to buy a good compass, follow up on several personal referrals, and advertise his services in a printed broadside.[25]

Land surveying would become Thoreau's most remunerative vocation. But it was never his profession, at least if we take him at his word. That he declared once and only once in September 1851: "My profession is to be always on the alert to find God in nature, to know his lurking-places, to attend all the oratorios, the operas, in nature." Yet within a month of declaring his profession, Thoreau was agonizing over his new vocation: "As I go through the fields, endeavoring to recover my tone and sanity and to perceive things truly and simply again, after having been perambulating the bounds of the town all the week, and dealing with the most commonplace and worldly-minded men, and emphatically *trivial* things, I feel as if I had committed suicide in a sense . . . I feel inexpressibly begrimed." Several months later, after "surveying for twenty or thirty days" and "leading a quite trivial life," he tried to balance his day job and night life: "To-night, for the first time, I endeavored to return to myself." This worked for a while. But then things got even worse. Surveying in February 1853 involved "many days of comparatively insignificant drudgery with stupid companions." On April 15 and July 1 came similar complaints that culminated in his broad generalization of August 6: "How trivial and uninteresting and wearisome and unsatisfactory are all employments for which men will pay you money. The ways by which you may get money all lead downward."[26]

But Thoreau had no choice. The debt he had incurred to publish *A Week* went unpaid, and its hundreds of unsold copies would soon come back to fill his garret. Grumbling about his day job intensified during December 1853, and continued steadily until January 1854 when Thoreau confided he felt more sympathy with "rocks or earth" than with his human companions. In his most explicit and extended diatribe against his vocation, he anointed himself as Satan the surveyor, the "Prince of Darkness" standing above his monument in the field.[27]

Thoreau enjoyed many things about land surveying: being outdoors in all seasons, the satisfaction of making successful measurements, and the discovery of new and interesting places. In fact, he found it so enjoyable he did these things for free when he felt like it, the most famous example being his pond survey of winter 1846. What he could barely abide, however, were the social and ethical requirements of the job: the business mandate to placate clients, the close personal contact with strangers for hours at a time, and his personal contribution to the hostile takeover of land for commercial purposes, usually to mark off woodlots for clear-cutting. Despite his social discomfort and the personal hypocrisy involved, Thoreau still preferred outdoor surveying to indoor factory work: the manufacturing of pencils and the grinding of graphite for the family business to produce plumbago or "black lead." Being the lesser of two evils, he begrudgingly stayed with land surveying because it paid the bills.

If scholars insist on calling Thoreau a professional surveyor, then I suggest they fall back on the first definition of this word from the Oxford English Dictionary, which is to "over-see" *(sur- + v(e)ier)* the landscape he loved, "surveyor, if not of highways, then of forest paths and all across-lot routes, keeping them open, and ravines bridged and passable at all seasons." This first definition demotes the second definition: the accurate measurement of land for legal purposes. When Thoreau joyfully proclaimed, "I am monarch of all I survey," he was voicing his declared profession, not his undeclared vocation. In fact, of the twelve times he used the word "survey" in *Walden,* only once did it involve the legal measurement of land and its taking for private gain.[28]

Sediment Transport

In an intellectual feat on par with teaching himself botany, Thoreau taught himself watershed science from scratch, based on direct observations. He investigated every component of the Concord River's water budget, ranging from the diameter of raindrops to the frequency of historic floods at the century scale. Rain varied greatly in its intensity, from mists vaguely realized to "great drops" falling so fast they "flat down the waves and suppress the wind," and feel "like hail on our hands and faces." Dry spells were part of hydrology too, especially the "annual drought" of Concord's deforested agricultural landscape. During the "very dry weather of July," "every traveller, horse, and cow raises a cloud of dust." Also, "the street and fields . . . look more parched than at noon."[29]

Glacially smoothed rock was so impermeable that rain flowed as "a broad thin sheet" of "water gliding over the surface" in what we now call overland flow. Less visible and much more complex were the basics of forest hydrology, which Thoreau seems to have mastered without any coursework or reference books. The interception of water by leaves and stems was straightforward: "If it does not rain more than fifteen minutes," he wrote regarding the wet precipitation that is caught, held, and evaporated, "I can shelter myself effectually in the woods." Dry precipitation, mostly snow, could also be intercepted: "part having been caught in the trees and dissipated in the air, and a part melted by the warmth of the wood and the reflection." Infiltration varied seasonally with the forest cover and the thirstiness of the soil.[30]

Few details of river hydrology escaped his notice. Some streams originated as perennial cold springs from deep "in the bowels of the earth," whereas others of his experience began as ephemeral rills from surface wash. Likewise, some branches of the same river carried a higher proportion of groundwater, and were thus cleaner and steadier, for example the "North" (Assabet) versus the "South" (Sudbury) branches of Concord River. Certain streams responded synchronously to the same storm, whereas others were delayed. Some magnified inputs; others dampened them. Local channels varied in flow rate from rapids and cascades to marshes and miasma. Some streamlines converged to create deep pools, whereas others diverged to create riffles. Turbulence created fantastic forms, including his eyewitness account of an antidune flowing upstream. The edges of channels were shear zones where fast water flowed against pooled slackwater, or where ice floes grated against land-fast ice. Streams like Second Division Brook seemed too powerful for their settings; others seemed too weak.[31]

Winter breakup was a contest between ice floes jamming the river at constrictions, and rising water lifting those jams to break them apart. In Thoreau's words, "the sap of the earth, the river, overflows and bursts its icy fetters." Floods were created both by too little infiltration and by too much input. Annual floods were self-registering, with a flotsam of sticks marking spring high water and of autumn leaves marking the extra-tropical storms (nor'easters) of fall. Some were exceptional, notably the July flood of 1852, the highest in "sixty-three years," based on testimony of "old Mr. Francis Wheeler." Its range between high and low water was "eight feet nine and a half inches," sufficient to transform his "noble stream" into a "chain of lakes." Though dramatic, this range is three feet less than the rise and fall of Walden Pond.[32]

He described a full spectrum of river channel materials and morphology. Though usually dominated by sand, streambeds of gravel or mud were common variations. In "Second Division Brook . . . the ripples cover its surface like a network and are faithfully reflected on the bottom . . . like a golden comb." In the Assabet a "fluvial walk" revealed deep pools alternating with sandy riffles, and the bed materials varied greatly: "your feet expand on a smooth sandy bottom, now contract timidly on pebbles, now slump in genial fatty mud." Elsewhere, the same bottom was "cemented with oxide of iron . . . a hard kind of pan covering or forming the bottom in many places." Nut Meadow Brook was fascinating for its "sudden pitches, or steep shelving places" producing "deep holes in its irregular bottom and the dark gulfs under the banks," a bottom "exceedingly irregular and interesting." Large floodplains were separated from their channels by natural levees composed of fine sand and mud. He saw these above Bedford during flood when the Concord River was "a sheet of sparkling molten silver, with broad lagoons parted from it by curving lines of low bushes. Its meanders were as beautiful and regular as they were common." In the Assabet he noticed that abandoned meanders left "crescent-shaped" lakes now known as oxbows.[33]

Tiny sediment grains like clay, silt, and very fine sand were easily carried in suspension with only minor turbulence. Thoreau enjoyed watching such turbid streams as much as transparent ones, writing, "We love to see streams colored by the earth they have flown over," perhaps because he knew they would eventually clear. "The Merrimack," he knew, which is "yellow and turbid in the spring; will run clear anon." Such streams clarify only when the sediment is flushed through, or when water stands still long enough for the tiniest grit to settle. One such place was Walden Pond, where "only the finest" physical—and metaphorically social—"sediment was deposited around me."[34]

Coarser materials like river sand, gravel, and cobbles move by friction against the streambed. This "bedload" moves intermittently, depending on size: boulders in brief spasms, gravel during peak flows, and sand as long as the supply lasts. In a roadside fieldstone wall near Lexington,

Thoreau noticed "a stone apparently worn by water into the form of a rude bird-like idol . . . [a] very regular pedestal." In Framingham, he was similarly intrigued by "water-worn stones by the gates of three separate houses." Most such odd-shaped stones result from both differential weathering of rock within the soil and the subsequent battering during sediment transport.[35]

Finally, no stream channel ever finishes its flushing job because there is always more sediment arriving from the creeping slopes and gullies above it. Just as there are "rivers of sap" in trees and "rivers of stars" in the heavens, so too were there "rivers of rock on the surface of the earth," fed by "rivers of ore in its bowels." Here, Thoreau demonstrates a profound insight regarding the link between denudation and exhumation. This dynamic view of existence at all scales—for astronomy, fluvial geomorphology, plant physiology, and bedrock geology in the examples above—was refreshingly novel for a historic culture founded on biblical narrative. Surely this is one of the things Thoreau meant when he called his landscape a "living earth."[36]

The shapes and sizes of Thoreau's river channels were also in constant flux as they sought equilibrium: the optimum combination of particle size, channel form, bed slope, and hydraulic roughness. Just as new clay enters and is washed to the sea, so too does new sand enter and leave each bank and bar. "The shifting islands!" wrote Thoreau of the Merrimack River, "The inhabitants of an island can tell what currents formed the land which he cultivates; and his earth is still being created or destroyed . . . the graceful, gentle robber!" As with the water budget controlling the height of the pond, he understood that the shape of each reach of a river was controlled by its sediment budget. Nowhere is this revealed with more clarity than with his scrutiny of meandering rivers.[37]

"What meandering! The Serpentine, our river should be called? What makes the river love to delay here? Here come to study the law of meandering." In this passage, Thoreau sings an ode to joy for river science, in this case to the beautiful curves of the Concord River through its Great Meadows on a beautiful autumn day in 1851. Leeches, he knew, moved in a straight line by oscillating up and down in the vertical plane, producing a series of waves. Snakes, in contrast, move forward by oscillating left and right in the horizontal plane, whether on the ground or in the water. Comparing the two, he induced a more general law of linear forward motion accomplished by tracing sine waves through resistant

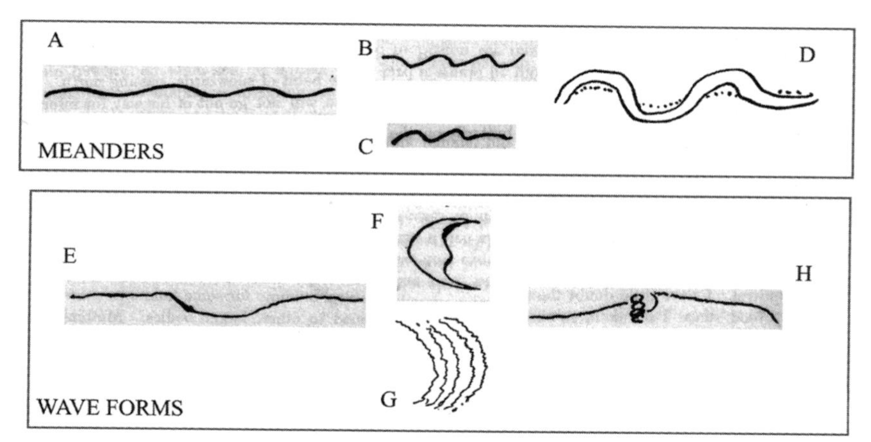

Figure 9. Thoreau's *Journal* sketches of meandering patterns and waveforms. *Meanders:* A, Vertical meanders in Second Division Brook (Mar 28, 1852); B, Vertical undulations of body of great blue heron in flight (Apr 19, 1852); C, Vertical undulations of leech in water; D, Meanders in Sudbury River with point bar lines indicated by blue flowers (pontederia; Jul 18, 1852). *Wave Forms:* E, Scour pool in sand on bed of Second Division Brook (Mar 10, 1853); F, Hog Island migrating near Hull, MA (Jul 25, 1851); G, Lateral friction of river between willow-vegetated banks; H, Snowdrift with vortex (Feb 23, 1854).
Source: Thoreau, *Journal*, edited by Torrey and Allen, 1962.

media in a direction transverse to wave propagation. To these observations he added the flight of the great blue heron, where the sine wave of its trace is a consequence of motion, rather than its cause, the bird's heavy body falling between lifting wing beats. From such examples, Thoreau came to understand that his favorite rivers meandered like snakes in form, but like herons in process.[38]

Thoreau worked on his study of meandering for two full years. By summer 1852, he had concluded that river meandering was a special case of the law of conservation of mass: "on the one hand eating into the bank, on the other depositing their sediment." He also sexualized these curves: "Methinks there is a male and female shore to the river, one abrupt, the other flat and meadowy." Here he refers to the cut bank and point bar, respectively. His choice was arbitrary, and opposite my instinct, which is to fit the inside curve of the point bar into the outside curve of the cut bank. Within a year, Thoreau came to understand that "perfect" meandering requires "a pretty constant difference between the shores" measured relative to a straight line drawn in the average downstream direction. Summing up a summer-long study, he worked his way down to a verbal law of meandering: "Probably all streams are (generally speaking) far more meandering in low and level and soft ground near their mouths, where they flow slowly, than in high and rugged ground which offers more obstacles." This sentence captures four important variables involving a stream's tendency to meander: elevation, bank resistance, stream power, and location relative to base level. His work was far ahead of what was available to him in books, including the four-volume *Final Report* of the state geologist.[39]

Before long, Thoreau realized that his general law of river meandering was, in fact, a special case of an even more fundamental law of meandering in general.

> When I returned to town the other night by the Walden road through the meadows from Brister's Hill to the poorhouse, I fell to musing upon the origin of the meanders in the road . . . [which] . . . pursued a succession of curves like a cow-path. In fact, it was just such a meandering path as an eye of taste requires, and the landscape-gardener consciously aims to make . . . The law that plants the rushes in waving lines along the edge of a pond and that curves the pond shore itself, incessantly beats against the straight fences and highways of men and makes them conform to the line of beauty which is most agreeable to the eye at last.

This passage extends his law of meandering to shoreline edges, walking cows, and creative artists.[40]

In addition to the snake-like horizontal meandering of rivers, Thoreau also recognized their leech-like, vertical motions. "It not only meanders as you look down on it, but the line of its bottom is very serpentine, in this wise, successively deep and shallow." The sketch that accompanies this sentence clearly illustrates the asymmetric pool-riffle sequence of natural rivers, which trout fishermen are well aware of, and which river scientists have since explained. Deep pools created by scour from downward-convergent flows alternate with shallow riffles created by deposition from upward-divergent flows. This sequence emerges spontaneously as a natural harmonic oscillation associated with the great momentum of water working against an adjustable bed.[41]

Finally, in the boldest stroke of his inductive genius during the Walden years, Thoreau linked the side-to-side meandering with up-and-down meandering to recognize an even more fundamental type of three-dimensional meandering known as helicoidal flow. This is a corkscrew motion in which the forward-propagating sine curve of momentum rotates around the line of gravitational flow. In this conception, line, wave, and circle become a single entity. This unification took place in Thoreau's mind on the bank of Nut Meadow Brook on a lovely spring day in 1852 when he noticed the streamlines of flow "meandering as much up and down as from side to side, deepest where narrowest, and ever gullying under this bank or that, its bottom lifted up to one side or the other, the current inclining to one side." At this point, the only thing Thoreau lacked was the explanation for the helical pattern he was seeing. Still searching a year later, he asks, "What is the theory of these sudden pitches, or steep shelving places, in the sandy bottom of the brook?" Thoreau's unwillingness to let go of an observation he does not fully understand brands him as a curiosity-driven scientist, hardly the trope-seeking transcendentalist he had left behind a few years earlier.[42]

Ridges and Valleys

The most valuable thing Thoreau seems to have learned from Hitchcock's *Final Report* was the relationship between the mineral composition of a rock and its resistance to weathering: "that a geologist" can "tell you whether your stones will continue stones and not turn to earth." Simply put, the most resistant rocks will either stand the highest as mountains, or jut out the furthest into the sea as coastal headlands. During his first trip to Cape Cod, Thoreau used Hitchcock's terminology—"hard sienitic rocks"—to characterize the coastal outcrops near Scituate that shipwrecked the "brig St. John, from Galway, Ireland, laden with emigrants." Such rocks, he saw, were resistant to crumbling, even when "laid bare" by the sea. At the other extreme, near Acton he saw "many large stones" quarried for a bridge abutment that, though only a few years old, were "more or less disintegrated and even turned into a soft soil into which I could thrust my finger, threatening the destruction of the bridge." Such rocks, he likely read from Hitchcock, contained a "sulphuret of iron," usually pyrite, which made them particularly prone to disintegration. In his mineral collection, Thoreau's sample of pyrite is a gleaming, beautiful specimen. Put it in moist soil and it will quickly disappear into sulfuric acid, white dust, and rust.[43]

Normally we think of granite—and its close kin, granite gneiss—as a strong rock. And it certainly is, mechanically speaking. But chemically, it can be fairly weak, especially when the crystals are coarse-grained and when it is exposed to a warm humid climate of the sort that characterized Nashoba for nearly all of its geological history. Between resistant grains of quartz and feldspar are books of mica: a light-colored one called muscovite, also known as white mica, and a dark one called biotite, also known as black mica. Both quickly disintegrate into clay, the black one also contributing to the familiar yellow stain of soil. Also present are peg-shaped iron-rich minerals called amphiboles, which dissolve nearly as fast. When these weak mineral links dissolve and heave apart, the resulting mass readily disintegrates into jagged sand of the sort you can find in practically every New England brook.

Mount Monadnock was Thoreau's favorite mountain. He climbed it twice before publication of *Walden* (in 1844 and 1852) and twice afterwards (in 1858 and 1860). It is an Algonquin place name meaning the "mountain that stands alone." This made perfect sense to Henry, who saw it often "fifty miles off in the horizon, rising far and faintly blue above an intermediate range," the Peterboro hills. After Thoreau's era, geologists appropriated the name "monadnock" (in lowercase with a final *k*) for all such erosional residuals, a name now used globally. Hitchcock correctly wrote that monadnocks stand high because the materials from which they are made are somehow more resistant. Indeed, the namesake mountain, though composed mostly of layered rock, is capped by a particularly resistant quartz-rich mica schist similar to that of New England's highest peak, Mount Washington. Topographically, Monadnock is a dome with "spurs or buttresses on every side . . . every way from their center."[44]

In Concord, however, the strong tectonic grain of Nashoba precluded the formation of domes. There, differences in rock resistance created parallel ridges and valleys. For the ridges, Thoreau used the term "iron-bound rocky hills" to connote strength, not composition, because those with the most iron-rich minerals are usually the least resistant. More importantly, ridge rock is usually less fractured, finer-grained, and composed of minerals that were less chemically reactive in the warm humid climates that prevailed during exhumation. Thoreau's understanding that ridge elevation, symmetry, and rock structure were related at a variety of scales is revealed by his *Journal* sketches and descriptions of "Conantum Cliffs" on two separate occasions in the spring of 1852. He also understood differential resistance as a function of lithology, with "limestone . . .

blocks of trap and conglomerate, boulders of sandstone and quartz" all behaving differently.[45]

Growing mountain ranges on active continental margins usually have a positive mass budget. This means that the removal of mass by denudation does not keep pace with the addition of mass by tectonic compression and uplift. However, for passive continental margins such as the New England coast since the breakup of Pangaea, the budget is always negative. They lose more mass than they gain. Early in the process, when the mountains were still high, the rate of loss was fast and the landscape was ruggedly dissected, a pattern similar to what Thoreau saw in the upper Merrimack, where "the River was the only key which could unlock its maze, presenting its hills and valleys, its lakes and streams, in their natural order and position." As the mountains shrink, however, rivers eventually work themselves out of a job. Their watersheds become better adjusted to differences in rock resistance, their channel gradients become lower, the terrain across which they flow gets gentler, and the soils get progressively more clay-rich and thicker. Toward the end of this process, differential rock resistance matters less and less, and regional slope matters more and more. The land approaches what geologists call a "peneplane," a generally flat surface of erosion, usually with a few resistant monadnocks. Taken to the limit, this process culminates in continental "shields," which are vast surfaces of low relief that cut indiscriminately across the jumble of folds, faults, and igneous intrusions of ancient mountain belts. Under such conditions watersheds usually revert to their default dendritic drainage pattern in which tributaries merge hierarchically into larger streams like the veins of a leaf.[46]

Several tens of million years ago, the inland terrain of the Concord River watershed approximated this peneplane condition. Its streams, originating on the worn-down stubs of the formerly grand Appalachian range, flowed southeasterly toward the beveled-flat landscape of the Atlantic Coastal Plain. During this epoch, the landscape of Nashoba may have resembled that of the Appalachian Piedmont of today, an extensive, slightly rolling surface of limited relief on thick, reddish, clay-rich soils lying inland from the coastal plain.

Figure 10. Ridges and valleys looking east toward Concord from flank of Mount Wachusett. Note four parallel ridges as part of the "trellis" drainage pattern. April 2012.

Then came dramatic change. The Antarctic Ice Sheet expanded, the ocean cooled, and global sea level fell. Simultaneously, the bulk of New England passively rose upward due to slight geothermal warming of the mantle and lower crust, which lifted the surface above the prevailing sea. Though the river discharges remained similar, the steeper gradients gave these discharges greater power, especially near the coast. They cut downward and backward into the undulating lowland like a headwardly-migrating waterfall, but more slowly and more complexly. The wave of incision, called "knickpoint recession," worked its way up tributaries, steepening the higher creep slopes. Gradually, the ancient soils were stripped away as the streams settled into valleys located above weak rock, leaving remnants of the former peneplane as broad, flat-topped ridges.[47]

One beautiful day in May 1853, Thoreau walked east from Concord and climbed Smith's Hill in North Lincoln. Facing westward, he overlooked the entire sweep of the Concord River Valley. Inspired, he wrote a "sermon on the mount," one of many literary allusions to the New Testament found in his *Journal*.

> From this more eastern hill with the whole breadth of the river valley on the west, the mountains appear higher still, the width of the blue border is greater,—not mere peaks, or a short and shallow sierra, but a high blue table-land with broad foundations, a deep and solid base or tablet, in proportion to the peaks that rest on it. As you ascend, the near and low hills sink and flatten into the earth; no sky is seen behind them.[48]

The "high blue table land," on which the "near and low hills sink and flatten into the earth" is the ancient peneplane I have just described. Cut below it are the streams of the Assabet Valley, whose interior drainage divides all lie at nearly the same level. From his elevated and panoramic vantage, Thoreau recognized that this remnant plateau is merely the top of a broader and thicker layer, the continental crust itself: the "broad foundation" and "deep and solid base" to the land. "The distant mountains rise," he continues. "The truly great are distinguished . . . You see, not the domes only, but the body, the façade, of these terrene temples. You see that the foundations answers to the superstructure."[49]

Thoreau's lyrical prose captures the essence of central New England topography. His westward horizon was higher because the crust is thicker and tilted up in that direction. Protruding above its ancient eroded surface were the resistant knobs of Monadnock, Ascutney, Cardigan, Kearsarge, and others. Seven years earlier, on his way to a monadnock called Katahdin, he saw Maine's version of the same thing, its peneplane: a "high table land" with "only isolated hills and mountains rising here and there from the plateau." Years later, and while looking outward from the same Smith's Hill, Thoreau saw that the "whole breadth" of the Concord River watershed had been cut down into an ancient surface. To the west it was bounded by the hills of "Berlin, Bolton, [and] Harvard"; to the southwest by the "Assabet Hills,—rising directly from the river," the "highest I know rising thus"; and to the southeast, the "Wayland hills." From Nobscot Hill the entirety resembled "a vast amphitheater rising to its rim in the horizon."[50]

Far to the southwest were the "rounded hills of Stow. A Hill and valley country. Very different from Concord." Here, Thoreau reveals his awareness of the stronger tectonic grain near Concord, where the landscape was more densely corrugated. There, the harder ridges between down-cut stream valleys remained at nearly their former levels, which is why they merge toward the horizon like a series of waves: "The haze makes the western view quite rich, so many edges of woodland ridges where you can see the pine tops against the white mist of the vale beyond. I count five or six such ridges rising partly above the mist, but successively more indistinct, the first only a quarter of a mile off." With each intervening valley, his "western landscape" became "progressively divided." And "of course there are no mountains," he remarked. In their place are "beautiful hills and vales, the whole surface of the earth a succession of these great cups, falling away from dry or rocky edges to gelid green meadows and water in the midst." This pattern was accentuated during winter when "the snowiest view is westward . . . the steep sides of steep hills are likely to be bare," and with "much snow in sproutlands and deciduous woods" below, likely because "the sun has had less chance to thaw the frosts." In such scenes, "there is no perceptible difference as to snow between the north and south prospects" because this view is along the tectonic grain, rather than across it.[51]

Recall that the dominant alignment of streams by the Nashoba tectonic grain is crosscut by shorter segments controlled by younger block faults trending generally parallel to the regional slope. This intersection creates what geologists call a "trellis" drainage pattern, and helps account for the conspicuous difference between the Sudbury and Assabet Rivers where they merge to become the Concord: "We are favored in having two rivers, flowing into one," Thoreau wrote, "whose banks afford different kinds of scenery, the streams being of different characters; one a dark, muddy, dead stream, full of animal and vegetable life, with broad meadows and black dwarf willow and weeds, the other comparatively pebbly and swift, with more abrupt banks and narrower meadows. To the latter I go to see the ripple, and the varied bottom with its stones and sands and shadows; to the former for the influence of its dark water resting on invisible mud, and for its reflections. It is a factory of soil, depositing sediment." Here, he writes of the cross-grain and with-grain segments of the trellis, respectively.[52]

Photographer Herbert W. Gleason's captured this contrast with his famous map of Concord, published in 1906 for the original publication of Thoreau's *Journal* by Houghton Mifflin Company. Beginning in West Concord, and flowing parallel to the bedrock grain, is Nut Meadow Brook. It follows that trend for a mile and a half before joining the mile-long, meadowy lower reach of the Sudbury River. The latter continues on this the trend for another mile and a half to the mouth of the Assabet River at Egg Rock. And from there, it continues on trend for over two miles as the Concord River through its Great Meadows. That is nearly six miles of flow parallel to the bedrock strike, meaning six miles of slow, meadowy, dead streams. The Assabet also flows mostly parallel to this strike within Acton and west Concord, but in its final half mile, links the trellis by cutting southeast across the grain as a "pebbly and swift" stream. Another prominent cross-grain reach is the "notch" of the Sudbury River between Fairhaven Bay and the mouth of Nut Meadow Brook.[53]

Within this trellis landscape was the Walden paleo-valley, a particularly deep and straight tributary of the Sudbury River that never received a proper English name because it lacked a stream. This is because the valley was nearly filled to the brim with glacial sediment and pockmarked with ponds, Walden among them. What had been beautiful brook before glaciation is now a dry "shrub oak plateau" between flanking bedrock ridges. To the southeast, the ridge is surmounted by Mt. Misery, Pine Hill, and Smith's Hill. To the northwest are less prominent ledges: Bear Garden Hill, Brister's Hill, and the ridge near Hubbard's Close. These ancient lines of bedrock, a legacy of exhumation, were the "primeval banks" of Thoreau's "ancient stream."

Dead Stream

"The Concord River is remarkable for the gentleness of its current, which is scarcely perceptible." The boatmen called it a "Dead Stream." In flood, it was a ribbon-shaped lake with a slope so low that Thoreau could easily sail up or down it, and a current so imperceptible that it froze like a sheet of glass, offering good skating for miles in either direction. So gentle was its flow that he noted that "the only bridge ever carried away on the main branch, within the limits of town, was driven upstream by the wind." Having heard that a "descent of an eighth of an inch in a mile is sufficient to produce a flow," Thoreau guessed that the Concord had "probably, very near the smallest allowance." This gradient translates to 0.125 inches per mile, or one part in 506,880, or a dimensionless slope of 0.00000197 or 1.97×10^{-6}. Billiard tables and airport runways are far more steeply inclined than that.[54]

What could possibly explain such a negligible slope? The first answer involves the bedrock grain. The deepest valleys running with the grain are effectively flaring gutters, their sides strengthened by hard rock. When confined like this, rivers with peak historic flood discharges comparable to the Concord's possess enough power to evacuate whatever is in their beds, and thereby keep the overall slope low. Second, the Concord

River flows on the flat, muddy bottom of an ancient lake, which had negligible slope. Third, the former shorelines of that ancient lake—and therefore its bottom—have since been uniformly tilted southward by approximately five to six feet per mile.[55]

Why the tilt? The last great ice sheet over eastern Canada had been a massive dead weight nearly three miles thick. This was enough to flex the rigid membrane of the crust downward and squish the underlying and softer mantle outward in all directions. This created a continent-sized dimple in east-central Canada centered on Hudson Bay that rose outward in all directions. When the extra weight of the ice was removed, this dimple began to heal, quickly at first, and then more slowly. Importantly, a small part of that healing is still taking place. As a result, for the last several millennia, the bedrock knickpoint of the Concord River at Billerica—called the "Falls"—has been rising upward a tiny bit each year relative to land further south. Though immeasurable as an annual rate, during the last several thousand years, this uplift would have been enough to erase whatever gradient the lower Concord River might have managed to produce via sediment accumulation. This back-tilting helped maintain a "dead stream" to the south and a bedrock cascade to the north.[56]

Thoreau recognized this transition between a "living" reach to the north and a "dead" reach to the south: "Just before" the Concord River "reaches the falls in Billerica it is contracted, and becomes swifter and shallower, with a yellow pebbly bottom, hardly passable for a canal boat, leaving the broader and more stagnant portion above like a lake among the hills." It "can hardly be said to flow at all, but rests in the lap of the hills like a quiet lake." During floods, its meadows become "a succession of bays," or a "chain of lakes," or "an endlessly scalloped shore."[57]

Within the first year of English settlement, colonists sought to "abate the Falls in the river upon which their towne standeth," according to a court petition dated September 8, 1636. The problem remained annoying, as Edward Johnson reported in 1650: "The rocky falls causeth their meadowes to be much covered with water, the [*sic*] which these people, together with their neighbor towne [Sudbury] here several times essayed to cut through but cannot." This bedrock knickpoint remained a problem for another two centuries. It was there in 1839 that Henry Thoreau and his brother John Jr. met the mathematically minded canal keeper on their first day of their downriver voyage, and who passed them through the lock on the Sabbath.[58]

The sluggish, back-tilted reach of the Concord River strongly contrasts with the rapid, forward-tilted main stem of the Merrimack River. This became a major dialectal theme in *A Week*. Thoreau described the Merrimack as continuously "falling all the way, and yet not discouraged by the lowest fall. By the law of its birth never to become stagnant" because it is too steep to create lush meadows fertilized each year by overbank "slime." "Unlike the Concord, the Merrimack is not a dead but a living stream, though it has less life within its waters and on its banks." A third stream of importance to Thoreau, the east-flowing lower Assabet River, does not fit in this dialectic. It is aligned nearly perpendicular to the tilt direction, meaning it was neither steepened nor shallowed. Instead, it flows as the archetype of a normal clear-flowing stream, steep enough to avoid becoming "dead" but gentle enough to avoid being a rapid. Nathaniel Hawthorne singled out this reach of the lower Assabet for special praise in *Mosses from an Old Manse*.[59]

Deep loss. Ten vertical miles' worth. Deep time. Two hundred million years' worth. Long enough to bring the deep mountain root of Nashoba up to the surface and etch it into corrugated ridges and valleys. We now turn to the glaciation. It glossed the landscape during the last half-million years, which constitutes the last one quarter of one percent of New England's landscape history.

3

THOREAU'S ARCTIC VISION

Thoreau's inner compass may have been aligned to the southwest, but his imagination consistently pointed north. To the "Esquimaux" he imagined standing silently over their fishing holes on Flint's Pond. To the "Northmen," whose Scandinavian sagas enrich his books, and whose playful god was the namesake for Thor-eau. To the "Land of the Midnight Sun," where "sunrise and sunset are coincident, and that day returns after six months of night." To the aurora borealis, that "burning bush" and "fiery worm" he saw from Concord hilltops during magnetic storms. To the Arctic pack ice and its local surrogate, the drifting floes of Fairhaven Bay. "He seemed a little envious of the Pole," said Ralph W. Emerson as his friend's body lay coffin cold.[1]

Thoreau's most astonishing Arctic vision came on the moonlit winter night of February 3, 1852. After leaving the warmth of his family home, he walked down the railroad tracks and then up toward the cliffs at Fairhaven, "floundering through snow, sometimes up to my middle." With the wasteland of the woodchoppers behind him, and the "hooting of an owl, *Nocturus undulatus*" adding ambience, he looked northwest: "The ground is all pure white powdery snow . . . The moonlight now is very splendid . . . The landscape . . . encased in snowy armor two feet thick, gleaming in the moon and of spotless white. Who can believe that this is the habitable globe? The scenery is wholly arctic. Fair Haven Pond is a Baffin's Bay . . . All is as dreary as the shores of the Frozen Ocean . . . It looks as if the snow and ice of the arctic world, travelling like a glacier, had crept down southward and overwhelmed and buried New England."[2]

Everything about this moonlit vision is technically accurate with respect to what scientists have since learned about Concord's most recent

Figure 11. "Glacier Table on the Mer de Glace, France" by James D. Forbes (1843). Frontispiece of James D. Forbes's *Travels through the Alps*. "Drawn from Nature by Professor Forbes."

glaciation. Everything! The snow falling on the high domes of the Laurentide Ice Sheet were indeed pure and powdery because their summit altitudes were cold and dry, being far removed from oceanic moisture. The snow and ice are listed in their correct order: the former giving rise to the latter after many years of deep burial, consolidation, and vertical pressure, which facilitates the molecular transformation of fluffy snow to crystalline ice. The Arctic world did indeed travel southward as a genuine glacier. Not as pack ice from the "Frozen Ocean," or as icebergs from some catastrophic debacle, but as a solid mass of strain-softened ice that "crept down southward" toward New England. Here, the word "creep" is not analogy, but the correct rheological term glaciologists use for the slow-but-steady, quasi-plastic, viscous flow of ice above a threshold basal shear stress. Also in correct sequence are the words "overwhelmed and buried," which describe the arrival of an unstoppable mass, and its gradual smothering of the terrain. This passage is the "smoking gun" proving that Thoreau understood how ice sheets are born, where they come from, how they move, and that one might have visited Concord in the recent geological past.[3]

Two months later he looked in the same direction from the same spot toward the "Peterboro Hills" to imagine the edge of that sheet retreating northward. By this time, the winter snow had melted north of Concord to reveal a transient regional snowline, "the dividing line between bare ground and the snowclad ground stretching three thousand miles" to the north. By then, Thoreau had learned from Darwin's descriptions that the elevation and latitude of regional snowline was a direct consequence of climate. And by then, he had learned from his own observations that changes in climate could involve changes in the patterns of snowfall; this he documented from the ubiquity of snowshoes gathering dust in the garrets of old men. And by then, he understood that, on a longer time scale, Earth's climate could change—perhaps back to colder, snowier times—as a consequence of the Earth's orbit. True. True. True. Since Thoreau's era, we have learned: that relict ice sheets are indeed continent-sized manifestations of former snowlines; that he lived during a time of changing snowfall patterns near the end of the Little Ice Age; and that orbital changes are indeed the main trigger mechanisms for past epochs of ice. As usual, Thoreau seems to have been presciently, if not mystically, correct.[4]

Thoreau's closest analog for what Concord may have looked like during peak glaciation were the broad expanses of snow he saw across frozen lakes. They provided "new views from their surfaces of the familiar landscape around them." In one specific example, "when I crossed Flint's Pond, after it was covered with snow . . . it was so unexpectedly wide and so strange that I could think of nothing but Baffin's Bay . . . and the fishermen . . . moving slowly about with their wolfish dogs, passed for sealers or 'Esquimaux,' or in misty weather loomed like fabulous creatures." Thoreau knew that fabulous shaggy creatures of the ice-age megafauna had indeed roamed his landscape in earlier times. They had moved southward ahead of the orderly southerly advance of the last ice sheet, and back northward following its much more rapid and ragged retreat. Mammoths, mastodons, saber-toothed cats, dire wolves, wild horses, and, most frightening, short-faced bears had migrated to and fro with the ice margin. And likely following them during the final retreat were fur-clad human hunters.[5]

Laurentide Ice Sheet

Thoreau's moonlit vision from the cliffs approximates the actual reality during the culmination of the last ice age about 22,000–24,000 years earlier. Then, all of New England lay buried beneath the Laurentide Ice Sheet, the largest on Earth, covering much of North America. Created by seemingly endless snowfalls in Canada, it thickened to an elevation more than two miles high. Its perimeter formed an irregular circle drawn clockwise from Nantucket Island, southern Illinois, central Montana, eastern Yukon Territory, the Canadian Archipelago, Baffin Island, Newfoundland, and then back down to Nantucket. Tongue-shaped lobes and interlobes crenulated the perimeter, for example the lobe occupying Cape Cod Bay and another in Lake Michigan.[6]

Within the ice sheet were several major summit domes. From each, ice flowed downward and outward in all directions along broadly curving rays. The ray flowing over Walden began at the summit of the Labrador dome, which lay mostly in northern Quebec east of Hudson Bay. After passing over Concord, it flowed another seventy miles or so to some point near western Martha's Vineyard. What was a slightly curved line in map view was, in the longitudinal view, a vertical slice of ice. Above Concord, it is estimated to have been somewhere between two thousand and four thousand feet thick.[7]

The scene at the top of that slice is easy to imagine because we have exact analogs on Earth today, the existing domes of Greenland and East Antarctica. Like them at comparable distances from their perimeters, the Laurentide Ice Sheet above the Walden paleo-valley would have been a nearly flat plain dominated by dry white snow, extending to all horizons, and descending ever so slightly to the south-southeast. Driven by that slight slope, the ice crept and slid as a single mass in a straight line with perfect laminar flow. Above the ice by day was the blue sky of persistently high atmospheric pressure, except during storms. And above it at night was a magnificently speckled, pollution-free host of stars. Biting winds drifted the snow, perhaps exposing patches of blue ice born in Canada thousands of years earlier. During the short summer, radiant heating on clear days, warm winds, and the occasional Atlantic rain melted the surface slush to ice-cold water, which trickled downward and forward through pores and fractures until it reached the ice-rock interface.

The scene at the bottom is much harder to visualize because nobody has ever seen the base of a thick ice sheet, not even a modern one. Guided by the southeast slope of its surface, however, we know from theory that the ice would have streamed up, over, and around each of Concord's bedrock ridges, sliding and creeping forward with an average velocity of tens of feet or more per year. Flow was faster in summer when there was more water pressure at the bed, and slower in winter when the vertical load was less buoyed up. At time scales ranging from microseconds to hours, debris-laden ice lurched forward in tiny increments over intact bedrock. Each created a tiny seismic event, an ice-quake. And as with geological fault planes in rock, the relentless forward motion snapped, crushed, and abraded rock fragments against one another. Silt-laden icy water gushed through channels at the bottom, some of which were notched down into bedrock by scour, while others were melted up into the ice by the viscous heating of fluid flow. Within these channels, rounded boulders and cobbles tumbled against one another with a muffled clickety-clack, resembling what can be heard on a cobble beach under a strong surf. Southward the stones flowed all year long. Relentlessly toward the edge. Pushed by the hydraulic pressure.

How do we know all this? Engineers have squeezed snow into solid ice and deformed it like putty. Glaciologists have studied the mass and energy budgets of modern glaciers, and have entered their crevasses to install sensitive instruments—cameras, seismometers, thermistors, pressure transducers, tiltmeters, strain gages—at the dynamic ice-bed interface. Software designers have simulated glacier flow. Geologists have studied the landforms and sediments left behind and reconstructed what must have happened.

During the ice sheet culmination, the glacial ray flowing over Concord helped build a boulder-studded moraine on or near western Martha's Vineyard. At the time, this moraine was not on an island, but was one of a series of windswept terrestrial ridges approximately a hundred miles from the sea. They were being buckled up by ice pressure, bulldozed by transient advances, mantled by debris melting down from the ice, and buried by fans of outwash sand.

I will never forget my first view of these materials, which are exposed on the cliffs of Aquinnah, also known as Gay Head. Millions of tourists have seen its colorful sequence of yellow sand, rusty orange gravel, blue-gray mud, drab-colored till, and traces of pink clay. All have been thrust, smeared, or washed into position by a debris-covered ice margin during its many advances. In turn, these colorful earthen strata are bracketed by the green sward of windblown heather above and the white froth of the pounding surf below. And bracketing these are blue sky and blue ocean, respectively. Many of the boulders being scoured by waves today passed somewhere near Concord on its way south before being buried in the moraine far from the edge of the sea, and only recently exposed. Some are composed of Andover Granite, perhaps quarried from Fair Haven Hill.

As with a large blob of soft bread dough rising on the counter, the ice sheet had a dramatic upward slope near its edge. But with increasing distance and thickness, the surface slope diminished exponentially northward, falling to about thirty feet per mile above the Walden paleo-valley. This is equivalent to a yardstick being tipped up on one end by the thickness of a few nickels. Thoreau—with his "surveyor's eyes"—may have been able to see this gentle slope on what would have otherwise been a flat and monotonous plain of snow more than a thousand feet above the highest hills. Sometimes I imagine him there, dressed in a fur parka and walking all the way north to the summit dome in Quebec. There, the surface gradient would have been so flat that not even he could have seen it.[8]

Eizeit and *Etudes*

Our modern theory of ancient glaciation originated in the Swiss Alps. There, Horace-Benedict de Saussure (1740–1799), explorer, scientist, and intrepid mountain climber, described his fascination for glaciers in his *Voyages dans les Alpes*. He gave special attention to colossal boulders transported far from their original locations, which became known as "erratics," from the Latin *errare,* which means to wander. Following in Saussure's footsteps was Wolfgang Goethe, Thoreau's early role model. He imagined an "epoch of great cold at a time when . . . the glaciers of the Savoy mountains went farther down, all the way to the sea." As the nineteenth century dawned, John Playfair described these Alpine glaciers as "the most powerful engines without doubt which nature employs . . . In this manner . . . huge fragments of rock may have been carried to a great distance . . . reached the shores, or even the bottom of the ocean."[9]

A chain of more scientific investigators quickly followed. By 1815, Jean-Pierre Perraudin, a scientifically savvy hunting guide, had stitched together every major idea in the modern glacial theory. These he shared with Ignace Venetz, a highway civil engineer, and Jean de Charpentier, a highly respected Swiss geologist. In 1829, Venetz summarized his fourteen years of field research in a presentation to the Societe Helvetique des Sciences Naturelles at the Hostpice of Great St. Bernard. In 1834, Charpentier presented his more scientific version to the same society in

Lucerne, giving the *first* cogent scientific argument for an epoch of greatly expanded glaciers in the not-so-distant past. In the audience was the young paleontologist Louis Agassiz, who dismissed the idea as absurd. One year later, in 1835, Charpentier published his talk in the French scientific journal *Annales des Mines,* which brought it to the attention of the Geological Society of London through its elected president, Charles Lyell. Another year later, during the summer of 1836, Charpentier invited Agassiz to his research area to show him the field evidence in person. Almost instantly, Agassiz became a zealous new convert, inaugurating his twenty-year campaign to convince the world that massive ice sheets had indeed recently covered much of the globe.[10]

Agassiz clearly had the right approach. Begin with Lyellian "causes now in operation," meaning those processes associated with modern mountain glaciers that could be seen in person: the moraines built up against them, the boulders still rolling off their surfaces, and the polished and scratched bosses of rock emerging from beneath the ice. Next, apply the uniformitarian dictum "the present is the key to past" in order to attach these uncontroversial findings to identical features in the distant foothills of the Alps, far beyond the range of present glaciers. His third step was to apply the same dictum to nearly identical features on both the north European plain and in Scandinavia that had previously been reported in 1832 by the German Reinhard Bernhardi and in the 1820s by the Norwegian Jens Esmark, respectively. Both studies indicated that ice moved down from the north from a place where no glaciers were known to exist.[11]

In 1837, the same year Thoreau graduated from Harvard College, Agassiz gave a surprise lecture on expanded glaciers to a wide circle of European naturalists gathered for the annual meeting of Societe Helvetique, this time in his hometown of Neuchatel. Published later that year as the *Neuchatel Discourse,* this was the *second* cogent scientific argument for ancient ice sheets. Within a year, it had been translated into English by the *Edinburg New Philosophical Journal.* Within three years, and after additional close collaborations with Charpentier and his German companion Karl Schimper, Agassiz pooled everything, wrote furiously, and self-published *Etudes sur les Glaciers* as two volumes in 1840. It was an excellent, well-written summary of the field evidence, and a compelling argument for an epoch of glaciation so recent that its deposits were still unconsolidated and its boulders still strewn about the landscape. Though Agassiz acknowledged his colleagues, Schimper never forgave him for taking sole credit for their joint ideas, and Charpentier never forgave him for scooping his earlier work, which he published, only one year after the *Etudes*. Thus began Agassiz's lifetime of intense personal conflicts with colleagues and contemporaries. One of these would be with a self-taught American field scientist named Henry D. Thoreau.[12]

As with Playfair and Lyell before him, Agassiz's main role was to advance ideas that were already well developed, but not yet accepted by the conservative bastions of institutional science. Given his enormous ego, engaging writing style, boundless energy, personal charisma, and flair for lecturing, Agassiz quickly carried his glacial theory through the scientific courts of mainland Europe, Britain, and finally to America, where he encountered his strongest resistance. His version of the theory involved two related but distinct concepts. First, the already old idea that low-elevation continental ice sheets rather than icebergs or debacles, had been responsible for what we now accept as glacial features. And second, that the creation of these features took place during a global ice age, an *Eizeit,* a geologic epoch associated with an abrupt extinction, ironically based on fossil carcasses from non-glacial sediments in permafrost.[13]

Like all truly important new theories, the acceptance of Agassiz's theory was a function of the national and institutional cultures it encountered, especially the triangle of rivalry between mainland Europe (dominated by France and Germany), the British Isles (dominated by London and Edinburg), and New England (dominated by New Haven, Albany, Amherst, and Cambridge). Agassiz's zealous promotion in Edinburg caught the attention of Scottish physicist James D. Forbes, who became keenly interested in the physics of glaciology, which he described in *Travels Through the Alps of Savoy and Other Parts of the Pennine Chain, with Observations on the Phenomena of Glaciers* (1843). As with Charpentier and Schimper before him, Forbes parted company with Agassiz over the priority of ideas, and the two became bitter academic enemies. In popular culture, Agassiz's ideas are much better known than Forbes's because the cosmic significance of his *Etudes* greatly overshadowed the particular and mechanistic work of the Scottish physicist's *Travels*. Bucking the trend, Thoreau paid more attention to Forbes's work, citing it both in his *Journal,* and in *Walden.*

Agassiz's ideas initially played well with scientists in northeastern America because evidence for continental glaciation was dramatic and ubiquitous. One early advocate was Chester Dewey. His 1839 article "On the Polished Limestone of Rochester, N.Y." was written in response to the 1837 *Neuchatel Discourse*, and was published in a highly visible journal—the *American Journal of Science*—early enough to be cited by Agassiz in his 1840 *Etudes*. Another New Yorker, Timothy Conrad, followed suit with similar support in the same journal. Unmentioned by both was much earlier work by Peter Dobson, a cotton manufacturer from Vernon, Connecticut, who published evidence for North American continental glaciation in 1826. His paper, also in the *American Journal of Science*, predated Agassiz's *Neuchatel Discourse* by more than a decade.[14]

Though quite right about continental glaciation, Agassiz was terribly wrong about something even more important: catastrophism. Astonishingly, he believed that the entire Alpine Range—which stretches over 750 miles and exceeds 15,000 feet in elevation—was violently uplifted *after* the sudden appearance of humans and which had been "immediately preceded" by instantaneous glaciation and global extinction. Quoting from the *Etudes:*

> The development of these huge ice sheets must have led to the destruction of all organic life at the Earth's surface. The ground of Europe . . . became suddenly buried under a vast expanse of ice covering plains, lakes, seas and plateaus alike. The silence of death succeeded to the movement of a powerful creation. Springs dried up, streams ceased to flow, and sunrays rising over that frozen shore . . . were met only by the whistling of the northern winds and the rumbling of the crevasses as they opened across the surface of that huge ocean of ice.[15]

Catastrophism is largely absent from primitive religion and mythology, according to Mircea Eliade's *The Myth of the Eternal Return.* Instead, the ancient world was assumed to cycle through the seasons and generations on a background unchanged since the "Great Time" of creation itself. But with the rise of civilization in general, and writing in particular, one local story of a devastating ancient flood in Babylon became codified in the cuneiform texts about Gilgamesh, and later in the parchment scrolls about Noah. With the spread of the late Roman

Empire and the unholy alliance between its armies and the Latin Church, this provincial idea from the eastern Mediterranean went global. Today, we would say it went "viral." When inserted into the timeless schema of primitive religion, the flood neatly divided all of history into three stages: before the Deluge, during the Deluge, and after the Deluge. In church Latin: *antediluvial, diluvial,* and *postdiluvial* times.[16]

Before geology congealed as a true science, the phenomenon of seashells on mountaintops was open to a variety of explanations. The Renaissance genius Leonardo da Vinci drew the obvious conclusion that they had been lifted up by powerful forces. By the late 1700s, however, this evidence had been perverted to support biblical flood catastrophism. Charles Lyell traced this change back to Scilla, an eminent naturalist and artist from Sicily who, in 1670, published a book on the marine fossils found in the cliffs at Calabria. Because he interpreted them as being *organic* in origin, and because he wanted his work to be accepted, appreciated, and purchased, he argued that marine fossils—exposed in rocks from sea level to the highest summits—were the effects of, and proof for, the Great Deluge, which, according to scripture, had occurred only a few thousand years before. The Church hierarchy greeted Scilla's interpretation as a vast improvement over the competing medieval idea that fossils were *inorganic* "figured stones" introduced by Satan to confuse and test our faith. Thus it was that an artist seeking market approval made, defended, and beautifully illustrated a compelling connection between biblical scripture and the stratified marine record.

Theologians leapt with joy. Within a century, this idea—coming from other quarters as well—had become so deeply rooted on both sides of the Atlantic that Thomas Jefferson felt compelled to comment on it in a letter dealing with Appalachian rocks in Virginia: "Immense bodies of Schist with impressions of shells near the eastern foot of North Mountain recall statements that shells have been found in the Andes 15,000 ft. above sea level which is considered by many writers both of the learned and unlearned as a proof of a universal deluge."[17]

The catastrophist school from which Agassiz graduated came not from the hard-rock basalt and granite of Scotland's Caledonian mountain root—where both Hutton and Lyell had ancestral homes—but from the paleontology of the soft rock, layer-cake strata of the Paris Basin and the nearby Alps, much of which is fossiliferous limestone and shale. The foundation text for this school was Buffon's 1749 *Theory of the Earth,* which linked specific features in the fossil record to a series of catastrophes. Two years later, and at the demand of church clerics, Buffon backtracked to one: "I abandon everything in my book respecting the formation of the earth, and generally all which may be contrary to the narration of Moses."[18]

Following Buffon's footsteps was a French nobleman named Georges Chrétien Léopold Dagobert Cuvier, who was both the father of paleontology and the chief mentor to Louis Agassiz. Cuvier's most famous work, *Discours sur les Revolutions de la surface du Globe,* was published in 1812 and translated into English by Robert Jameson in 1827, Edinburg University's ardent neptunist. It claimed that "The evidences of those great and terrible events are everywhere to be clearly seen by anyone who knows how to read the record of the rocks," times when all organic life "was overtaken by these frightful occurrences . . . engulfed by deluges." Cuvier used the word *"Revolutions"* for what today are called mass extinctions. The last occurred when "glacial conditions came into existence" rapidly, and without mercy. "Life upon the earth in those times," he wrote, "took place five or six thousand years ago, and might be considered as coinciding with the Mosaic Deluge."[19]

Like Hutton, Cuvier was a brilliant thinker who did not accept the literal truth of scripture. Where they differed was in their overall view of Earth history. Hutton's saw steady state recycling and overlooked notions of progress. Cuvier saw the arrow of time as dominant, a narrative of progress marked by chapters in the history of life. Having been successfully tested against the fossil record, Cuvier's catastrophic interpretations were more consistent with the biblical narrative, which is punctuated by disasters.

Enter Reverend William Buckland, the gifted Anglican theologian at Oxford who had inspired Charles Lyell with his lectures. In 1820, he argued "the grand fact of a universal deluge . . . exerting its ravages at a period not more ancient than announced in the Book of Genesis." Cleverly, he invented a new geological unit called "diluvium" that linked the scriptural fact of Noah's deluge with the natural fact of England's unusual glacial soils, most of which showed some evidence of water deposition. Scotland's Sir James Hall took the next step of linking some of the more unusual aspects of Buckland's diluvium—erratics, "bowlder-clay," and scratched rocks—to icebergs moving within catastrophic deluges of unknown causes that he called "debacles."[20]

Meanwhile, back in Thoreau country, Reverend Edward Hitchcock was trying to make sense of it all. For his *Final Report,* which he submitted in 1839, he parsed four different theories of "diluvial action." Of these, he noted that the catastrophic debacle mechanism "certainly quadrates better than any other with the diluvial phenomena of this county." It "supposes that an extensive elevation of the bottom of the Arctic Ocean threw its waters, loaded with ice and detritus, southerly over the American and also the Eastern Continent." This sort of instantaneous change coming from deep within the Earth's crust paralleled the thinking of Louis Agassiz, who also connected massive crustal violence to glaciation.[21]

In that same year of 1839, English geologist Frederick Murchison urged a correction back to the uniformitarian paradigm. He suggested that geologists replace the semantically freighted term "diluvium" with the more detached term "drift" when discussing what we now know are glacial deposits. Icebergs do, in fact, drift over the face of the sea, slowly releasing a mixture of materials from mud to jagged boulders, and occasionally plowing into the shore, scratching rocks underneath. Most English scientists, especially Lyell and his protégé Darwin, embraced Murchinson's "drift" mechanism, a term that remains current today despite its heavy historical baggage. Many scientists on the American side of the Atlantic also followed suit, wrote Canadian geologist Alan Jopling, as "a logical outgrowth—or perhaps an aftermath and anticlimax—of catastrophy [*sic*]."[22]

Louis Agassiz, being a lifelong protégé of Cuvier, was catastrophist to the bone. This set up an inevitable conflict with Charles Lyell, who was gradualist to the bone. When Agassiz's *Etudes* was published in 1840, Lyell's *Principles* was being revised for its sixth edition. In November of that year, and apparently when under the spell of Agassiz's charm, Lyell seems to have temporarily converted to the glacial theory, presenting a paper to the Geological Society of London on the same docket with Agassiz titled: "On the Geological Evidence of the Former Existence of Glaciers in Forfarshire." Shortly thereafter, however, he quietly changed his mind, writing that "the Glacial Theory" of Agassiz is "not applicable to Scotland." On closer examination, his flip-flop may have been a strategic move to preserve the uniformitarian purity of his bestselling *Principles* against Agassiz's frightening global "silence of death" and recent "powerful creation."[23]

Stephen Jay Gould argued in *Time's Arrow, Time's Cycle* that Lyell's treatise was a masterfully argued, well-written, one-sided, and highly biased legal brief for an extremist version of Hutton's theory. I agree. Lyell's case succeeded primarily because it was presented in prose so elegant that it sounds Thoreauvian: "We hear of sudden and violent revolutions of the globe . . . in which we see the ancient spirit of speculation revived, and a desire manifested to cut, rather than patiently to untie, the Gordian knot." It is no wonder that Lyell became the first person to earn his living as a science writer. Charles Darwin was instantly drawn into his orbit in 1831, notwithstanding his pledge never to read another geology book. The same was true for Henry D. Thoreau, who, in 1840, was likewise deeply and permanently smitten by the *Principles*.[24]

Lyell did not work as scientists are supposed to: reasoning downward from observed details to fundamental laws with predictive power. Instead, he worked from the bottom up as a trial lawyer for the plaintiff of scientific deism, accepting *a priori* that he was correct, and staying in that groove until external judgment went either for or against him. Combining the theoretical rigor of William Herschel with the field work of James Hutton, and adding many insights of his own, Lyell began with the *a priori* assumption that earth is a perfect steady state system, in both *agency* of process and *rate* of process. According to his twentieth-century editor Martin Rudwick, this put him "virtually in a minority of one among living geologists." In Lyell's mind, major catastrophes were inadmissible because they were ruled impossible, except locally as volcanic eruptions, earthquakes, tsunamis, and landslides. In text, he was completely committed to the assumption that nothing on Earth can change in a fundamental way, not even organic evolution, writing: "There is no foundation in geological facts, for the popular theory of the successive development of the animal and vegetable world, from the simplest to the most perfect forms." In text, he would not admit into evidence any process or intensity of process not known to be operating on Earth in his lifetime, which disqualified Agassiz's instantaneous deep freeze from the outset. In short, Lyell's lifelong work was "an almost ideological crusade" against catastrophism, with "overtones of cosmic conflict." To fight the good fight against scripturally motivated infidels, he penned: "I must put on all my armour." Sentences like this instantly endeared him to Thoreau.[25]

A scientist is supposed to change his or her mind when new evidence comes along. But a barrister is ethically free to ignore or explain away any new evidence that does not support his case. This explains why Lyell unabashedly cherry-picked the evidence he liked and ignored what he did not. He denied the slow uplift of the Baltic coast from the sea, even though the harbors of early fishing villages had been lifted high and dry in sequence. He insisted that bedrock scratches rising up one side of a bedrock knob and continuing uninterrupted down the other were the marks of icebergs shoaling at the bottom of the sea. He interpreted thick layers of windblown glacial dust in central Europe (called loess) to be abyssal oceanic sediment. And finally, during his celebrated first visit to the United States, he encountered a landscape of large erratic boulders as far south as Long Island but no farther. This manifestly important boundary seems to have had no impact on his thinking.[26]

By the time Lyell's sixth edition went to the press, he had decided to use Murchison's term "drift" for what we now know are glacial deposits. In doing so, he staked out a position he would defend through the next three editions: that erratic boulders, till, and rock striations were the result of icebergs floating and grounding. Though Lyell was ardently anti-catastrophist, old-school diluvialists like Edward Hitchcock sympathized with his interpretations because they required high-amplitude submergences and emergences of the land. Lyell's ninth edition coincided with *Walden*'s publication, making it the last one of concern for this book. In it, he was still rigidly defending a strict version of timeless "revolutions" and remained strongly opposed to the glacial theory.

Ironically, Agassiz was less interested in glaciers as a geological agent than as the killing mechanism for the final global extinction that paved the way for *Homo sapiens* to culminate God's plan. This was in keeping with his lifelong preference for fossils within rock relative to the rock itself, especially those of ancient fish within shale, his most notable area of expertise. It was his knowledge of paleontology—not sedimentary geology or glacial theory—that got him invited to deliver his "Plan of Creation" lectures at Harvard University in 1846, speeches that made Emerson uneasy for their teleology. In fact, Agassiz's zealous promotion of the—Saussure-Goethe-Perraudin-Venetz-Charpentier-Schimper-Agassiz-Forbes—glacial theory during the 1840s was a decade-long detour from an otherwise zoological career that spanned six decades, and which ended sadly on the wrong side of evolutionary theory. To his death, Agassiz believed that God invoked serial catastrophes to wipe out previous stages of life; that his Eizeit was the last of these catastrophes; that Darwinian evolution was in error; and that so-called primitive races had their own creations separate from those of European stock.[27]

Agassiz's 1846 trip to America—funded by Frederick William IV of Prussia—was supposed to be temporary. Instead, this European savant chose to immigrate, owing to political turmoil in Europe, loss of financial backing, family difficulties, and his "rock-star" popularity with geology-craving American audiences and academics. In 1847 he was offered a paid professorship at Harvard and a chance to found its Museum of Comparative Zoology. He accepted and plunged right in, full speed ahead, inaugurating Harvard's permanent commitment to natural science. That same year Agassiz hired a young man named Henry Thoreau to work for him as a paid collector of museum specimens, one who sent the zoologist "many firkins of fishes and turtles" through an intermediary, James Cabot. At some point during his brief employment, Thoreau had a memorable "short interview" with Agassiz in the Marlborough Chapel, one confirmed by a cordial exchange of letters from the summer of 1849. The content of that conversation is unknown.[28]

Astronomical Pacemaker

Earth is often called the "blue planet." This was certainly the case shortly after its origin, when the recently cooled globe was almost entirely submerged by an ocean of distilled water not yet salted by rock weathering. And it remains largely true today, because liquid water still covers most of the globe. On the smaller continents of Earth's previous eons, the colors toggled back and forth between mostly green and mostly brown during wetter and drier paleoclimates, respectively. But on at least one occasion about 0.7 billion years ago, Earth seems to have been entirely white, with thick ice sheets covering all land masses, and pack ice extending into equatorial seas. "Snowball Earth" is what geologists call this condition, which is close to what Agassiz had in mind. Most geologists are convinced that the planetary carbon budget was somehow thrown out of whack due to the arrangement of tectonic plates.[29]

During the grand sweep of Earth's history, five prolonged periods of extensive glaciation have taken place. The most extensive of these so-called "ice-house" periods (as opposed to "greenhouse") was the late Proterozoic event described above, the end of which may have stimulated the dawn of complex multicellular organisms about 550 million years ago. Rocks of this age interpreted to be composed of glacial sediment can be found on the Squantum Peninsula on Boston's South Shore near Hingham. Two later periods of extensive glaciation took place as New England was being created, the first shortly before the Nashoba Terrane accreted, and the second when the northern Appalachians were a rugged and towering range within the middle of Pangaea. New England's fourth and final ice-house period began about thirty million years ago and remains with us today.

This late Cenozoic ice age was ushered in by the randomness of continental drift. The bedrock craton of Antarctica had reached south polar latitudes, allowing it to grow a permanent carapace of ice. This chilled the southern ocean swirling around it, and sent currents of cold water northward where continents deflected the flow. Further north, India's slow collision with Asia had elevated the mountains and the Tibetan Plateau, exposing fresh rock to rapid weathering, thereby sopping extra carbon from the atmosphere and turning down the atmospheric thermostat. Meanwhile, Russia, Scandinavia, Greenland, Canada, and Alaska had converged to bottle up the Arctic Ocean, restricting oceanic heat from the south, and allowing it to eventually freeze. By this time, the Canadian Shield had drifted north to latitudes where winter snows lasted for most of the year. Finally, about two million years ago, North and South America joined hands at the Isthmus of Panama, forcing the flow of sun-warmed Caribbean water northward. Known as the Gulf Stream, this broad current of heat and vapor fueled heavy snowfalls on the vast plateau east of Hudson Bay in Labrador and northern Quebec. Eventually, a summer came when the snow of some previous winter did not completely disappear, a threshold triggered by a slight diminishment of summer solstice sunlight. Being more reflective, and requiring extra heat to melt, this patch of surviving snow begat additional snow the following year, which begat additional snow, thereby setting into motion the glacial narrative of Walden Pond through positive feedback.

The steady rotation of Earth on its axis brings any geographic point toward, and then away from the sun every twenty-four hours. And because this axis of rotation maintains a uniform tilt during Earth's orbit of the sun, we get the seasons of our lives. Almost as simple and steady are three additional periodic orbital cycles that also control the amount of sunlight Earth receives. The top-like wobble of Earth's rotational axis around its celestial pole has a periodicity of about 23,000 years. The axial tilt, now at 23.5 degrees from vertical, steepens and shallows from about 21.5 to nearly 24.5 with a period of about 41,000 years. And the shape of Earth's elliptical orbit ranges from nearly circular to more eccentric with a period of about 100,000 years, bringing Earth closer to and farther from the sun at different times. Each of these periodic cycles operates independently, meaning their solar effects can either combine or cancel each other out. When summed, these cycles trigger regular variations in the severity of glacial stages.

For most of Earth's history, these long-term orbital cycles were operating just as they do today, but the whole-Earth climate system lay too far from the tipping point where ice caps could survive, grow, and shrink. When this threshold was first re-crossed about two million years ago, the climatic fluctuations were governed mostly by the 41,000-year tilt cycle, and Canadian ice sheets were never large enough to expand beyond the St. Lawrence River. Within the last half-million years, however, glacial expansions have been driven primarily by the 100,000-year eccentricity cycle. And they have become much more intense, quite capable of climbing the northern Appalachians and flowing down the southeast regional slope to the northern edge of New England's coastal plain.

There, each successive ice sheet encountered a low, north-facing escarpment composed of firm—but not lithified—strata from the inner continental shelf. Given the backpressure provided by this obstacle, the ice sheet under compression was able to bulldoze, thrust, crinkle, dump, and deposit terminal moraines on what was then dry land, moraines that include slivers of these older strata. Only after the global ice sheets had melted and their waters had been returned to the ocean did the edge of the sea rise to flood the continental shelf and break up the moraine into the sandy archipelago we find today as Nantucket, Martha's Vineyard, Block Island, Long Island, and Staten Island.

The down-glacier distance between Concord and the terminal moraine at Martha's Vineyard is only 2 percent of the total distance to the ice dome divide in Quebec. What this means is that the Walden paleovalley was glaciated only during the most extreme conditions. This also means that after each release from the ice, its local climate would have still been strongly controlled by the continent-sized mass of ice to the north. Based on the best global evidence available (continuous cores from abyssal marine sediments and from the summit domes of the

Antarctic and Greenland Ice Sheets), the Walden paleo-valley was probably glaciated four times. These well-dated global records match the tally of four glaciations present in both the sea-cliffs of Nantucket and and drill cores of Georges Bank. Based on the best astronomical dates on hand, these glaciations culminated about 22, 130, 420, and 620 thousand years ago, plus or minus a few thousand years.[30]

During each of the four ice sheet culminations, the same source of ice moved over the same paleo-valley with the same mechanisms, and reached roughly the same thickness over roughly the same resistant bedrock topography. Even more conservative was the bedrock highland associated with the Bloody Bluff Fault that divides the watersheds of the Sudbury River and Charles River. This suggests that comparable drainage scenarios were present during each ice advance and retreat, meaning that a series of broadly similar glacial lakes existed in the Sudbury Valley as the ice sheets came and went. Using Occam's Razor, the default assumption is that previous ice sheets made previous versions of Walden. Not clones or identical twins but fraternal twins, perhaps the relationship Thoreau had in mind when he linked Walden and White Ponds.

"Mensuration"

Thoreau would have made a fine glaciologist, given his quantitative abilities and his obsession with the physical properties of ice and snow. During the cold winter of *Walden*'s completion, he could often be found walking on the frozen Sudbury River where he observed, described, and classified winter phenomenon. On New Year's Day of 1854 he was enthralled by snowdrifts. His *Journal* account of that experience reveals his insights into the continuum mechanics of fluids at steady state, and with the balance of forces involved: "The drifts mark the standstill or equilibrium between the currents of air . . . The snow is like a mould, showing the form of the eddying currents of air which have been impressed on it, while the drift and all the rest is that which fell between the currents or where they counterbalanced each other." By February he had become intrigued by the crystallography of slab ice, using the mathematical word "tessellation" to characterize the regular spatial packing of large ice crystals on the surface of the Sudbury River, which may have resembled hexagonal floor tiles. In early March, a cold snap caused the frozen ground near his house to contract to the threshold of rupture. He sketched the resulting fractures with detail and accuracy, and with a pattern identical to those mapped by geologists studying fault zones in the Earth's crust. Specifically his sketch includes the fractal scaling of *en-echelon* failures.[31]

These are but three examples—from the final months before *Walden*'s submission—taken from hundreds of such passages written in the preceding few years. Hardwired inside Thoreau's brain at that time was a module for spatial thinking and measurement that sought creative expression. A measuring stick and tape were indispensable parts of his sojourning kit. To say that a dead moose was merely "large" was not enough. It had to be measured, even in units of "umbrella." A flock of geese overhead, to be fully appreciated, had to be counted, mapped, and measured in units of goose: "twelve in the shorter line and twenty four in the longer"; shaped, in "the usual harrow form," [v-shaped] with the junction between lines at the "fourth bird from the front;" and the spacing "between the geese was about double their alar [width] extent." The depth of frozen ground underfoot, to be properly appreciated, required excavation and measurement at five widely spaced sites where the following seven variables were recorded: elevation, aspect and vegetation on each slope and the material layers of canopy, snow, soil, and groundwater. This sort of statistical sampling strategy characterizes much of the best empirical science today.[32]

None of this was lost on R. W. Emerson, who remained perplexed and amazed by Thoreau's habits, even after a quarter-century of friendship. In his eulogy, he wrote of Thoreau's "powerful arithmetic," and his "natural skill for mensuration, growing out of his mathematical knowledge and his habit of ascertaining the measures and distances of objects which interested him." Here, Emerson reveals his fascination for an intellectual trait he obviously lacked, a point perfectly illustrated by the unpublished passages he selected to highlight Thoreau's literary skill. He made no attempt to select passages that highlighted Thoreau's equally remarkable quantitative abilities.[33]

"It was in Thoreau's very nature," wrote Barry Savage, "to begin describing a place he wrote about by giving dimensions first." Merely thinking about the weather precipitated his desire to measure it: "Does a mind in sympathy with nature need a hydrometer?" According to Walter Harding, Thoreau's quantitative compulsiveness even contributed to his death. On December 3, 1860, he became so involved in counting and measuring tree rings on Smith's Hill that he suffered severe exposure, and "acquired a severe cold." The "strain was too much; the cold de-

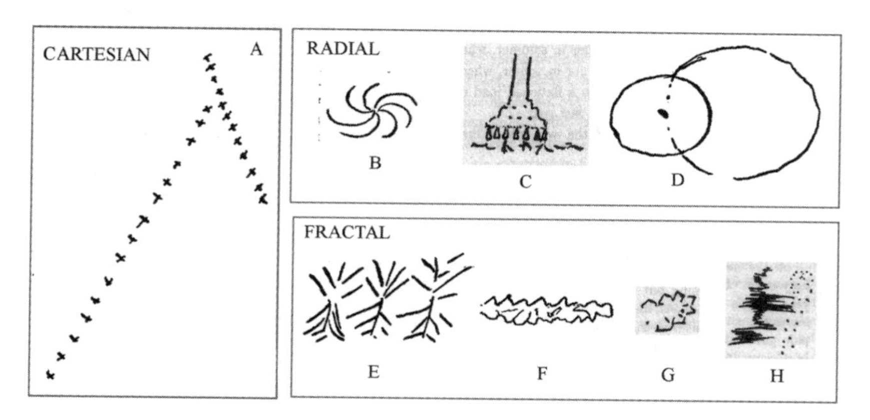

Figure 12. Thoreau's *Journal* sketches featuring geometry. *Cartesian:* A, Structure of goose flock (Nov 23, 1853). *Radial:* B, Spiral opening of pitch pine cone (Mar 27, 1853); C, Splash ice above river (Mar 22, 1854); D, Double halo of half-moon under clear sky (Mar 28, 1852). *Fractal:* E, Reflections of maple sapling twigs (Apr 11, 1852); F, Molars of Mr. Pratt's muskrat showing embedded crenulations (Jul 24, 1853); G, Leaf hoarfrost (Jan 5, 1853); H, Serrated edge of wind-disturbed water (May 14, 1853). *Network:* see Figure 8.
Source: Thoreau, *Journal*, edited by Torrey and Allen, 1962.

veloped into bronchitis and eventually into acute tuberculosis." Henry Canby called such efforts "pathetic—hundreds of measurements of deep snow with a measuring stick, rough dissections, countings of birds, plumbings of the depths of swamps."[34]

I disagree. The act of measurement in and of itself satisfies the need to know something actual and true about a "thing in itself," rather than something subjective and contextual. Any measurement—position, temperature, weight, angle, speed, etc.—is a truth more true than any taxonomic label because the error is never categorical. Only by measuring the depth of Walden Pond could Thoreau "assure" himself and his readers that "Walden has a reasonably tight bottom at a not unreasonable, though at an unusual depth." Merely saying it was deep would have been like saying a moose is large, because it begs the questions: "How deep?" and "How large?" "So with his left hand," as Canby concluded, "he kept firm hold of philosophy and with his right grasped the rule, the measuring cord, and the spy glass."[35]

Again, I disagree. Measurement was part of Thoreau's natural philosophy, his science. Richard Schneider understood that, for Thoreau, "being able to measure depths, distances, and objects, is important as a way of asserting mastery over what has been seen." To experimental physicists, it is the *sin qua non* of what they do.[36]

Thinking physically or geometrically was habitual for Henry even when no measurement was involved. For example, when walking over the irregular topography of Concord, he kept in his mind the curvature of the Earth: "We experience pleasure when an elevated field or even road in which we may be walking holds its level toward the horizon at a tangent to the earth, is not-convex with the earth's surface, but an absolute level." Learning the numerical magnitude of this curvature from Darwin seems to have increased the pleasure: With the "eye being six feet above level," the "horizon is two and four fifths miles distant." Not many hikers I know would keep that sort of thing in mind as they walked along.[37]

Thinking physically extended to every domain of Thoreau's science. In a single year (1853) he used his background knowledge of *optics* to understand the "blue smoky haze" of the March fire season; *heat* to explain something as straightforward as "frost coming out of the ground" or as complex as the relationship between thick dew and grass (the latter is a poor conductor); *mechanics* for the volume strain in lake ice leading to micro-cracks; *hydraulics* for the attenuation of surface waves by emergent aquatic vegetation; and *thermodynamics* for the foundation for everything.[38]

When feasting on the first million words of Thoreau's *Journal* during the two-week read-a-thon in May 2011 that launched this book, I tried to keep track of every entry where Thoreau seemed to be mensurating or thinking spatially beyond what would have been expected for a competent naturalist of his day. Using these admittedly qualitative criteria, and not going back to re-check my identifications, I marked down no cases before November 23, 1850. From that date forward to the close of the Walden period on April 27, 1854, I counted sixty-two cases, yielding an average of about two per month, peaking in the summer, and culminating with rush of eleven in the four months prior to *Walden*'s submission. This last point suggests that mensuration is somehow an important subtext to the book. I arrange them all in order, using one paragraph per year. Skip ahead of you like.[39]

He began on November 23, 1850, with the micrometeorology of air temperature relative to the variables of elevation, shading, and azimuth. On December 17 he summarized the phenology of pond freeze-up relative to their physical attributes and degree of exposure. Later that day, he described the dendritic radial geometry of seepages toward holes punctured in the ice.[40]

In 1851, he begins with a list of six dates on which magnets were used for global exploration. Next were seventeen measurements for the details of a blown-down white oak. The hemispheric difference in latitudes at which icebergs reach the sea. The hydraulic gradient (slope) of the water table in cellars relative to standing water. The depth and differential resistance (shear strength) of swamp deposits. The insignificance of topographic relief on Earth's surface relative to its radius. The geometry of plunging folds within a rock outcrop, whose "strata" were oriented "upon the axis of elevation, geologically speaking." The exact angle at which moonlight reflects, rather than refracts on water of a specified temperature. The 3:1 ratio of down-valley distance to channel length in the naturally meandering floodplain of the "serpentine" Sudbury River. A numerical inventory of trees by species in Conantum Swamp. The "astonishingly sharp and thin overhanging eaves" as details of the "architecture of the snow" drifts.[41]

In 1852: The statistical likelihood of capture of wind-scattered winter leaves by footprint hollows in the snow estimated to be ten inches deep. The snow budget in the forest floor relative to open ground as caused by differences in branch interception (followed by sublimation and melting) and the reflected heat absorbed by the darker color of the vegetation. The discharge of stream flow relative to its channel size and watershed area. The date of "ice-out" for Walden relative to other years. The apparent increase of cloud velocity caused by curvature of the Earth near the horizon. The geological strike and dip of foliation in the Conantum cliffs. The relative temperatures and variations in groundwater temperature as reflected in springs and wells. A calculated minimum estimate of "16,335 clams to twenty rods of shore (on one side of the river)" at Fairhaven Bay. Lampooning the cranial measurements of Dr. Samuel Morton as a means of measuring knowledge. Using the apparent density of a haze as a surrogate for the distance to a distant mountain. The precise mathematical curve of nearby Round Hill. The sympathy between the flood stage of Walden Pond and that of Lake Superior, indicating a regional climate cause. The duration to maximum canopy storage during a summer rainstorm of 15 minutes. The visual perspective of distance near and far. The timing of high water at Walden at the decade scale based on re-growth of trees.[42]

In 1853: The acoustic, optical, and thermal properties of ice with respect to the pattern of its bubbles. The spiral rays of the basal section of a pitch pine cone. Mapping the shafts and adits of a woodchuck burrow. The contrast in the sine wave made by a leech in water, relative to a snake on land. The fractal serrations on the edge of a moonlight reflection. The exact dimensions of an "Indian hoe" described as semicuculus in shape. The angle for an inclined plane marking the base of clouds as a likely marker for weather front. The distance in miles that Bangor, Maine, would lie above Concord, were they both on the same line of longitude: the answer, converted to three significant figures, was 162 English miles. The pyramidal form of crystals in black ice on the river.[43]

In 1854. The snow depth after a New Year's Day storm, based on sixty-five systematic measurements and subtracting the tare of snow held by trees. The complex geometry of cliff icings, dubbed "ice organ pipes," described with words like "triangular projections . . . perpendicularity . . . various angles . . . solid pipe-like icicles . . . rows of pillars or irregular colonnades." The perfect geometric figures made by snowflakes, whether "six-rayed stars or wheels with a centre disk." The mechanical sensitivity of lake ice to imperceptible changes in weather, making it an acoustic barometer of sorts. The relationship between ground elevation and depth of freezing and the dimensions of cut blocks of frozen ground. Estimating the water pressure beneath the winter ice of Walden versus Fairhaven using the measured ratios of ice thickness to its buoyancy.[44]

The onset of habitual mensuration at the close of 1850 immediately follows the culmination of Thoreau's breakup with Emerson, and immediately precedes what Thoreau biographers have noted as his transition toward scientific thinking, what Bradley Dean called his "dramatic reorientation." This habit of mensuration immediately follows the rapid emergence of Thoreau's vocation as a surveyor. Based on the particulars of Walter Harding's biography, it began in early 1850 and culminated near the end of that same year when Thoreau printed an advertising broadside that claimed—unabashedly—that he was "accurate within almost any degree of exactness." By year's end, precise outdoor measurement and mapping had become an almost daily habit, one that I suggest he subconsciously carried forward into his sojourning science when not engaged as a land surveyor. This, I further suggest, was the first of two transitions that turned Thoreau toward science. The second, coming only seven months later, was his close encounter with the journal of geologist Charles Darwin.[45]

The "Great Year"

In *The Days of Henry Thoreau,* Walter Harding concluded "the most frequently recurring symbol in *Walden* from the beginning of the book to the very end is the symbol of birth and renewal." If one of his *Days* can be an epitome of his year, then so, too, can one of his years be an epitome of the planetary eccentricity cycle, which, for the purposes of this book, I will call the "Great Year." It has four "Great Seasons," capitalized to distinguish them from the conventional solar year.[46]

The *Summer* of each Great Year is the stable interglacial world like the one at Walden today, with at least several millennia of relative warmth, and with ecosystems adapted to dry sandy soils, probably the same pitch pine and white oak species present today. During this shortest Season, lakes and ponds fill with the organic sediment before and after its time of peak warmth. The *Autumn* of each Great Year is its longest Season, lasting tens of thousands of years, which is longer than the entire history of human habitation in North America. It begins with the early growth of a newborn ice sheet far up in Labrador. As it gathers volume over the next several tens of thousands of years, tiny glaciers develop on summits of the New England highlands, especially in the White Mountains and on Katahdin, carving hollows called *corries, cirques,* or *cwms* by the Scots, French, and Welsh, respectively. By the time the ice sheet nears Concord, these tiny glaciers have long since been engulfed by Canadian ice, and the interglacial ecosystem near the Concord replaced by more cold-tolerant and wind-resistant species living south of the ice sheet.[47]

Winter begins when the tapered edge of the south-flowing ice sheet reaches the present site of Walden. It culminates when the edge is hovering near the terminal moraine. The turning point toward withdrawal is forced by a slight, but sustained increase in annual summer heat, which causes the enormous glacier to lose mass, and its edge to migrate northward. Winter ends when the receding edge of the reaches the Walden paleo-valley. By that time, any vestiges of a previous Waldens have long since been scraped, plowed, and washed away.

The Pond we know today and all of its potential predecessors were born in the *Spring* of their respective Great Years. Each began with the withdrawal of active ice from the paleo-valley, which was at least several millennia after the recession began. Each ended when the ice sheet to the north became too small to significantly chill the local climate, allowing *Summer* to return to start the cycle again.

Theoretically, Walden I, Walden II, Walden III, and today's Walden IV may have followed their respective culminations of the 100,000-year solar cycle, with several cycles missing. In the culminating section of Walden, Thoreau wrote: "This is frost coming out of the ground. This is Spring. It precedes the green and flowery spring, as mythology precedes regular poetry." This same passage can be used at the multi-millennial scale: the withdrawals of glacier ice during Great Springs precedes the lushness of interglacial woodlands during Great Summers.[48]

Glacial Action

Broadly speaking, the timing of the Laurentide maximum coincided perfectly with astronomical theory. At the millennial scale, however, earthly causes were involved: its outer edge oscillated back and forth in response to processes like the buildup and release of meltwater at the base of the sheet. Variations at the century scale involved the sluggish response times of something so massive, and variations in local physical conditions associated with subglacial topography and glacier-bed materials, for example the presence of permafrost. Through some combination of these oscillations at all time scales, two major moraines were built. The outermost connects Nantucket, Martha's Vineyard, Block Island, the South Fork of Long Island, and Staten Island like irregular beads on a regular string. Ten miles north of that margin is a similar, younger, and more massive moraine, the remnants of which are Cape Cod, the Queen Elizabeth Islands, the hills of Charlestown, Rhode Island, Fisher's Island, New York, and the North Fork of Long Island.

During all this back-and-forth near the ice-sheet terminus, it was business as usual at the Walden paleo-valley. Every year, an excess of ice from the north streamed above it to replace the deficit to the south. At the surface, glacier flow was uniform, and in an unwavering line to the south-southeast. Deep below the surface, at the elevation of Concord hilltops, it also flowed directly on line, smearing till on the up-glacial flanks of bedrock prominences and creating cavities on their down-glacier sides. But within the vertical topographic interval between the highest hills and the lowest holes, the direction and intensity of ice flow was progressively controlled by local bedrock topography.

Recall that at Walden, the flow of ice was mostly across and somewhat oblique to the bedrock grain. Hence, wherever a blocking ridge was encountered (for example, Pine Hill in Lincoln) the ice was forced sideways toward the nearest gap in the direction of prevailing flow, which was preferentially to the southwest. The Sudbury River north of Fairhaven Bay was one such prominent gap. So much flow was forced between the cliffs of Conantum and Fairhaven that the valley's cross section resembles a scaled-down version of one of the famous, U-shaped notches of the White Mountains: Pinkham, Grafton, Dixville, and Franconia. Other trough-like outlets for subglacial flow in the vicinity of Concord include the lowermost mile of the Assabet River and the northwesterly flowing reach of the Concord River in Bedford.

Embedded near the base of the moving ice were billions of jagged rock fragments ranging in size from garages to grit. These had been quarried from bedrock to the north by mechanisms involving ultra-local differential stress, non-inertial collisions, and subglacial freezing. Depending on size and concentration, these fragments became the cutting, grinding, and polishing tools that streamlined Concord's ledges and smoothed its rock eminences with grooves, scratches, and general scour. Such features were spectacularly exposed on the broad summit of Mount Monadnock, which was low enough to have been deeply buried by the ice, and yet high enough to maintain an artificial tree line after its forest was burned off to eliminate the wolves, and its thin soil washed away as a consequence. On that summit Thoreau saw that "the rocks were remarkably smoothed, almost polished and rounded; and also scratched. The scratches run from about north-northwest to south-southeast."[49]

He noted similar, but less spectacular, erosional features in Concord as smooth bosses of rock in upland pastures. When steeper and jagged to the south, these are known as *roches moutonnees* for their resemblance to a rock sheep standing in the grass, a term Thoreau may have encountered in Hitchcock's *Final Report.* He noted one such outcrop at James Baker's farm: "hillocks, which are stone-capped," with "rocky summits," . . . "brows of earth, round which the trees and bushes trail like the hair of eyebrows, outside bald places, templa, primitive places, where lichens grow." On Wachusett, he saw the bald summit as a "clear space, which is gently rounded" and "is bounded a few feet lower by a thick shrubbery." Throughout Concord, he puzzled over "broad and flat surfaces" that sharply truncated the visible internal layering, wondering "how these rocks were ever worn even thus smooth by the elements." The modern answer is they represent the grand sum of scouring erosion

Figure 13. Glacial smoothing and striations on Mount Kearsarge, New Hampshire. Thoreau described similar features on nearby Mount Monadnock. Note chiseled graffiti dating back to at least 1915.
July 2010.

at the base of ice flowing relentlessly forward with a toolbox full of rock chisels, files, scribers, routers, planers, and sanders.[50]

When concentrated beneath moving ice, the pulverized residues of crushing and grinding were often smeared—or more technically "lodged"—to the bed. This process deposited what Concord townsmen called "hard-pan," a sediment they found firm enough to thresh grain upon. Technically, the material is "till," a term introduced by Charles Lyell from his native Scotland for an "obdurate" soil. When it is thick, lodgment till landscapes exhibit "broadly undulating curves" or "swells" because the till caulks depressions like stiff putty but is too weak to maintain a sharp edge against moving ice. Instead, it becomes molded, smoothed, and often streamlined. At Round Hill the profile traced what Thoreau called a "mathematical curve." Today, these "very graceful" slopes often qualify as drumlins, whose slopes often approximate a curve called a lemniscate loop. On one of these hills "north of the [Cambridge] Turnpike near the Lincoln line," he was perplexed by the "greenness of the sward there on the highest hill." This situation usually indicates finely crushed lodgment till, a material famous for its luxuriant, "ever-green" pastures.[51]

Basal meltwater also flowed to the south-southeast, but on a much faster schedule. Technically, much of the water did not come from ice melting, but from annual rains and snowmelts. Regardless, all of it ended up at the bottom, having worked its way down through rills and canyons melted into the ice, many of which drained into entry portals called "moulins." At some depth this liquid water pooled up to create what is effectively a water table within the ice sheet. It rose and fell with the seasons, helping to relax and reload some of the dead weight through partial buoyancy, thereby changing the forward rate of glacial flow. The grand sum of all this basal water was pushed relentlessly south-southeast by some combination of water pressure and elevation.[52]

At Concord, practically all of the subglacial drainage took place in discrete channels. Most were melted upward into the ice as tunnels, especially when bedrock topography sloped in the same direction as the ice surface. Less commonly, especially when flow was being driven uphill, the channels were cut downward into rock, creating notches. Regardless of morphology, every drop of water flowing within these tunnels and notches contained the rinsed effluent of the glacial crushing mill, a turbid mixture of clay, silt, sand, pebbles, cobbles, and small boulders being washed toward the moraines and the sand plains beyond them. Clay, silt, and sand were easily and continuously washed forward at a scale of mass transfer that far exceeded that of moving ice. Boulders and coarse cobble gravel moved during surges of flow.

Today, most of the length of the Merrimack River flows down the regional slope in south-southeast direction before taking a sharp left turn at Lowell to follow the northern edge of the Nashoba bedrock grain. During the Laurentide culmination, however, this lower reach was blocked by thick ice, forcing the subglacial Merrimack River to keep flowing down the regional slope directly over Concord. And as with the movements of the basal ice, the flow paths of basal meltwater also became progressively more complex at progressively lower bedrock elevations. The end result was a colossal system of anastomosing channels carrying sand and gravel over Concord from as far away as Hudson's Bay. As with the ice, this system linked slower-flowing segments parallel to the bedrock grain with faster-flowing segments across it. In what is now Concord, the flow would have been generally southeast into the paleo-valley, southwest down it to the Sudbury River at Fairhaven Bay, southeast and uphill toward Lincoln, up and over the bedrock divide with the Charles River at Weston, and then onward toward the Vineyard.

During winters, the rate of surface meltwater production dropped to nearly zero. As a consequence, tunnels near the margin were squeezed partially shut by the creep of the ice. When surface melting returned in the spring, the volume of liquid water initially had trouble escaping, forcing its retention. At some critical threshold, however, the rise in pressure would have strengthened the outflow. Heat released by the associated turbulence would have enlarged the tunnels through an unstoppable positive feedback until the stored water was released, not as a giant flood, but as a strong annual pulse of flow. From the ice margin each year gushed a torrent of previously retained water mixed with a winter's worth of rock residues. Feeding that torrent were the subglacial tunnels flowing over the bedrock beneath what is now Walden Woods. With the flow concentrated in narrow channels, and with the water rendered dense by high turbidity and low temperature, boulders the size of watermelons were easily bounced along before coming to rest. Many of the stones that lined Walden's shore were delivered through such tunnels, rather than more directly from the ice.

Sometimes, stones being carried by subglacial channels became trapped in stationary vortices to drill bedrock potholes, which are legendary for their curiously smooth, cylindrical, overhanging, and even corkscrew shapes. Thoreau knew how they formed, and was aware of their great antiquity. "A stone which the current has washed down, meeting with obstacles, revolves as on a pivot where it lies, gradually sinking in the course of centuries deeper and deeper into the rock. Drilling took place perhaps before the thoughts began to revolve in man," far beyond the era of "Hindoo and Chinese history." From Hitchcock's *Final Report,* he learned of one "in the town of Canaan in this State, with the stones still in them on the height of land between the Merrimack and Connecticut, and nearly a thousand feet above these rivers." There, as with the much lower divide between the Charles and Sudbury Rivers in Weston, the subglacial meltwater had been forced up and over the watershed divide by the southeasterly gradient of water pressure being driven by the southeasterly slope of the ice surface, which was flowing down the southeasterly trend of the regional slope, which was tectonic in origin.[53]

At this point of my narrative, Earth's largest ice sheet has reached its maximum extent and thickness. Its debris-laden ice was streaming beneath what is now Walden Pond, smoothing and polishing the underlying granite. Within the next few millennia, however, that ice flow would slowly decelerate before stopping completely, leaving an enormous stagnant block behind. This would be the mold from which Lake Walden would be cast.

4

AFTER THE DELUGE

Ralph Waldo Emerson's Phi Beta Kappa address to the graduating class of Harvard on August 31, 1837, is considered by many to be America's intellectual declaration of independence. Published as the essay "The American Scholar," it proclaims: "We have listened too long to the courtly muses of Europe . . . please God" our age "shall not be so. We will walk on our own feet; we will work with our own hands; we will speak our own minds." This is exactly what patriotic members of the Association of American Geologists did three years later during their inaugural national meeting in Philadelphia.[1]

Led by James Hall, and in unified opposition to the British system of naming and correlating rocks, they gathered in 1840 to create their own system, believing that "an American stratigraphic classification was best for America," and that "European names like Cambrian and Silurian—now the global standard—were considered undesirable." Hall, the most ardent spokesman for the American view, was a force to be reckoned with, as much feared for his violent temper tantrums as he was respected for his stamina and brilliance. Ten years earlier, at age nineteen, he had walked over 200 miles from Hingham, Massachusetts, to Troy, New York, in order to study geology at Rensselaer under the famous neptunist Amos Eaton. From that year forward, Hall was a man with a mission. For American geologists, he was the equivalent of Thomas Paine and Samuel Adams rolled into one.[2]

Conservative Blunder

At their second national meeting in Philadelphia in 1841, Hall's Association of American Geologists spoke their own minds on very different issue:

Figure 14. Ice floes on Walden Pond. View across the eastern, central, and western basins from the swimming beach. Floes are much thinner than those Thoreau described as deforming the shoreline.
January 2012.

the radical ice age theory of Swiss paleontologist Louis Agassiz, whose book *Etudes sur les Glaciers* had just been published. After hearing a fairly neutral review by Massachusetts State Geologist Edward Hitchcock, they appointed an expert committee to investigate the matter and report back the following year. Chair of that committee was Dr. Charles T. Jackson, brother-in-law to Ralph Waldo Emerson and the former state geologist of Maine. His position on the theory in question was eminently clear to those who had chosen him, based on his earlier reports of Maine geology: "Diluvial blocks, pebbles, sand, gravel, and clay, mingled in confusion, attest their origin and deposition amid the waters of the deluge."[3]

Thus, it came as no surprise to those attending the third national meeting of the Association when Dr. Jackson rose as chairman to put a beneficent spin on his "deluge" deposits: "The geologist sees in the diluvium not proofs of Divine vengeance, but evidences of the highest wisdom and goodness of the Creator in thus preparing and commingling the soils of the earth." This sentence is vintage "natural theology," then the ruling scientific paradigm in America. Its goal was to link God's revelation through nature to his revelation through Christian scripture to create a whole stronger than the sum of its parts.[4]

Ostensibly, the study committee found Agassiz's theory "utterly inadequate" because a uniformitarian "cause now in action" had recently been "discovered, which was deemed sufficient to have produced the various effects which are termed diluvial, or drift phenomena." The effects to which he referred were the familiar soils and landforms of Walden Woods: its stony mud, erratic boulders, isolated ponds, abundant gravel, clay pits, streamlined hills, and smooth bosses of rock. And the new "cause now in action" Jackson referred to was actually a new compilation of observations—"an examination of more than eighty persons, principally masters of vessels"—applied to an old cause: "fixed icebergs or glaciers of the Arctic and Antarctic shores" commonly drifting to latitudes of 40° north and 36° south with embedded stone debris. Presenting these eyewitness accounts of icebergs as positive evidence for the "drift" theory was a classic case of cherry-picking data to support a preexisting bias. And the negative evidence presented against the competing theory of ice-sheet glaciation wasn't evidence at all, but theory: "the grand difficulty" of conceiving a glacier "wide enough to reach from Newfoundland to the Rocky Mountains." This was a classic case of putting the cart of theory ahead of the horse of observation, in this case arguing that something cannot happen because it is too big to happen. After "an animated and pleasing debate," the co-leader of that committee, Reverend Hitchcock, rose to the podium and "remarked that so disastrous had been his experience in respect to the glacial theory of Agassiz, that he was almost afraid to say any thing more on the subject."[5]

When Thoreau sauntered to Cape Cod a few years later, he became fascinated by the "rounded hills and hollows" of Truro, Wellfleet, and Barnstable that ranged up to two hundred feet high. Ditto for the enormous erratic boulders he saw sprinkled around a landscape otherwise dominated by loamy sand, one of which was "forty-two paces in circumference and fifteen feet high." To state geologist Edward Hitchcock, however, the "hills and hollows" were "diluvial elevations and depressions" that mirrored the "chopped sea" creating them, and the boulders were related to the same cause. "That remarkable and very powerful currents of water have swept over this continent from the north and northwest," he wrote, "I cannot doubt." But with respect to the details, however, Hitchcock remained troubled: "No theory of diluvial action hitherto proposed, is so free from objections that I feel satisfied with it." And "I confess I am not prepared to explain the *modus operandi.*"[6]

Whereas Hitchcock was confused, Thoreau was amused. This is why he played the jester during his summary of the *Cape Cod* geology: "There is a delineation of this very landscape in Hitchcock's Report on the Geology of Massachusetts, a work which, by its size at least, reminds one of a diluvial elevation itself." By "elevation," Thoreau probably meant flood depth. He was also perplexed by Hitchcock's suggestion that Cape Cod's huge "bowlders" could have been carried by water. Without naming names, he suggested that someone with a "smattering of mineralogy" look north-northwest to "Cohasset, or Marblehead" for "some interesting geological nuts for him to crack, if he should ever visit the mainland." Based on modern glacial mapping, these locations lie directly up-glacier from the erratic boulders being discussed, suggesting that Thoreau may have known where they came from and how they got there, and was teasing Hitchcock once again.[7]

Hitchcock was a prolific promulgator of geological natural theology. In practically every map, report, lecture, letter, committee transcription, and speech of his career, he insisted that what was true for "diluvium" of Cape Cod was also true for the whole region: "In New England, the greater part of it certainly appears as if the result of powerful currents of water, rushing over the surface in the manner of a deluge." This interpretation mistakes the prevailing southerly flow of countless glacial meltwater channels draining the ice for a single, south-directed, tsunami-like flood. Hitchcock's error was consistent with both his scriptural bias and the training he had received in the 1820s, when neptunists ruled the geological classrooms of America. Though he specifically ruled out Noah's flood as the cause of his deluge, he did so not because the idea was far-fetched, but because the diluvium did not contain the fossil bones of the human sinners who would have been drowned. Reading this twisted logic was a real jaw-dropper for me.[8]

Hitchcock, like Lyell, had initially been receptive to Agassiz's ideas but later recanted. He openly acknowledged that Agassiz's "theory of glacial action has imported a fresh and lively interest into the diluvial phenomenon of this country" and gave him full credit for creating a "new *geological sense* . . . for the '*tout ensemble,*' of diluvial phenomena." Nevertheless, when all was said and done, Hitchcock refused to endorse it. He insisted on "retaining the term *Diluvium*" as an official map unit, despite its scriptural connotation. He did this knowing that he would "probably be thought by some, either ignorant of the present state of geology, or unreasonably tenacious of former opinions." Instead of accepting the term "glacial action" he obfuscated the debate by inventing his own term, "glacio-aqueous action," because he believed it "more accurately expresses the agency." Finally, and without naming names, Hitchcock claimed "he was happy therefore to find his views in accordance with the whole geological world." This denial, obfuscation, and overstatement strikes me as face-saving wordsmithing by a man whose ego was nearly as large as his rival's. Embedded within this specific conflict between Hitchcock and Agassiz was the larger national one between a patriotic American and a privileged European.[9]

Agassiz's theism put the supernatural ahead of science. Hitchcock's theism put it ahead of both science and government. To his state-funded *Final Report* (submitted to the printer on December 1, 1839) he added an eleven-page postscript (dated April 1, 1841 and printed in the location of a preface) containing information that had come to light during the previous twenty-eight months. In a note explaining this postscript, Hitchcock states that he had "not hesitated to incorporate in the work" the many new facts that had appeared since its initial submission, "as the reader will see, without consulting the government." Having declared himself immune from state oversight, and having acknowledged that glaciers were important agencies in mountains, he refused to use the term "glacier" in the textbook section of his government report (Part IV: Elementary Geology). Instead, he treated them as slump-landslides consisting of ice, retaining his term "landslips" for what everyone else was calling a glacier. He also refused to alter his chronology for the diluvial epoch, which he dated to the last of "six demiurgic days of scripture." In a second jaw-dropping piece of twisted logic, he suggested that some antediluvial species rendered extinct by the flood were "recreated" from whole cloth exactly as they had been before the deluge. So much for the separation between church and state, and between scripture and science, during Thoreau's era.[10]

Thus it was that an expert study committee convened by a national organization, and led by two government-appointed geologists, officially and erroneously discounted Agassiz's glacial theory in 1842. According to the eminent historian of geology George White, this decision "set back glacial geology in America by almost a generation." Historically, Dr. Jackson faded from the New England scene. Hitchcock carried on, fighting for his term "glacio-aqueous," by which he meant any permutation of diluvium, drift, deluge, and debacle, provided there were no real glaciers involved. Even Emerson sounded like a member of this diluvial crowd when he wrote of Concord: "Here once the Deluge ploughed,/Laid the terraces, one by one."[11]

Individually, some American geologists had been promulgating the glacial theory since 1826. This was ten years before Agassiz's conversion and twenty before he first set foot in America. Many more American naturalists became converts after 1846 when Agassiz began lecturing to audiences of thousands. By the mid-1850s, the glacial theory had become almost mainstream among American geologists, and Agassiz had long since returned to his zoological-paleontological career. But within the conservative bastion of organized American science, the 1842 rejection of the glacial theory remained in force until 1862, the year of Thoreau's death.

Thus it was that Thoreau's social experiment at Walden Pond was conceived, planned, and implemented, and the resulting book was written, rewritten, and carefully edited, during a time of great confusion between wildly different ideas about how the New England landscape came to be. In one corner of this scientific brawl was the cosmic catastrophism of Agassiz's instantaneous deep freeze and ice-sheet glaciation. In a second corner was the debacle favored by Hitchcock, possibly caused by instantaneous uplift of the Arctic. In a third corner was drift deposited by the armadas of icebergs favored by Charles Lyell. And in the fourth was the slow-but-steady advance of subarctic ice sheets described by James Forbes. Had Walden Pond been a volcano or a coral reef, this boxing ring of competing ideas would have made little difference. But because Walden Pond is diagnostically glacial in origin, it was a great enigma to New England natural science during the Walden years from 1845–1854. This wide-open intellectual context allowed Thoreau to be more self-reliant with his own field geological interpretations than he might otherwise have been, and forced him to be more cautious about what he wrote. *Walden* became a very different book for these two reasons.[12]

Thinning to Stagnation

After creating its outermost moraines, the Laurentide Ice Sheet began to die by mass starvation. Though it continued to gain mass during most of the year, heavy losses from warmer and longer summers more than offset those gains. For the first few thousand years, this deficit had little effect on the edge because the great white blob continued to ooze down from the north as the dome lost elevation. But eventually the ice sheet thinned to the point where the surplus inflow from the north became too small to prevent accelerating retreat toward Concord. By the time the margin crossed through Middlesex County, the average annual rate of down-melting likely approximated ten feet per year, translating into a northward rate of back-melting of about one hundred feet per year.[13]

One popular mistaken image for New England glacier recession—courtesy of media confusion between tidewater retreat and terrestrial retreat—is that of an icy cliff pulling back like the blade of a bulldozer in reverse. The actual picture for Concord was far more prosaic: a gently tapered wedge of dirty ice melting mostly downward. In fact, the moment of deglaciation on such irregular terrain would have been hard to pin down. A mile or two to the north, the tapered wedge would have been mostly ice, with perhaps only the highest hills poking through it, and the mass still thick enough to deform under its own weight. Such ice is said to be *active* because, through brittle at the surface, it remains ductile at depth, and therefore able to shear forward like very cold putty. To the south, the tip of the wedge would have been mostly melted, with detached remnants present only in optimal settings. Such ice is said to be *stagnant* because it is too thin to deform and is immobile from top to bottom. This tapered pattern of retreat yielded two quasi-parallel boundaries that migrated north together, though not in lockstep: a northern boundary between active and stagnant ice, and a southern boundary between stagnant ice and bare land. Between them was a band called the "stagnant fringe," ranging from half a mile to a several miles in width depending on local conditions.[14]

This doublet of quasi-parallel margins passed over the Walden paleovalley about 16,500 years ago, plus or minus a few centuries. Early in the passage, the hills flanking Walden such as Pine Hill and Fair Haven Hill would have been rubbly islands of land surrounded by debris-covered ice. Very late in the passage, the islands would have been composed of stagnant ice surrounded by land. The last to go were either remnants in shaded valleys, or—more commonly—buried beneath the milky-gray waters of icy glacial lakes. Near the middle of the passage, large masses of debris-covered stagnant ice lay scattered on the landscape above the level of lakes. All were laced with layers of crushed debris, especially in the lower hundred feet or so. When this debris became concentrated as a surface layer, it insulated the ice, slowing its rate of disappearance for a century or more. This layer thickened as the ice melted downward until it was finally let down on the land like a stony shroud.[15]

Where this "meltout till" was thick, coarse, and uniform, the result was dry stony ground like "Channing's 'moors,'" being "extensive and rather flat rocky pastures without houses or cultivated fields." Where it slid sideways into cavities, it created irregular hills and ridges. Where it was thick enough to preferentially insulate the ice, it created small kettles, now dry hollows. Where large boulders or rock slabs were present, they were released as erratics protruding above the rest of the glacial litter. In passages reminiscent of Katahdin, Thoreau described these "great gray boulders . . . fifteen or twenty feet square" as "slumbering, silent, like the exuviae of giants . . . brute life, they seem to have." In Natick, he sketched "a boulder some thirty-two feet square by sixteen high, with a large rock leaning against it." Such perched and leaning stones are the hallmark of down-melting from stagnant ice. Thoreau's word choice of "exuviae" was perfect. These massive boulders were indeed "sloughed off" the dying glacier during its last gasp.[16]

Thoreau and Darwin

Mount Tabor, in Lincoln, is hardly a mountain. Rather, it is a resistant knob near the drainage divide between the Concord and Charles Rivers. Standing there on May 1, 1851, Thoreau looked southeastward toward "Waltham Hill" to the harder, more ancient rocks of the Avalon Terrane on the African side of the Bloody Bluff Fault. "I had in my mind's eye a silent gray tarn which I had seen the summer before high up on the side of a mountain." The word "tarn," "from the Old Norse *tjorn*," refers to a small bedrock lake carved by a cirque glacier at the valley head. This word choice likely reflects his recent reading of William Wordsworth's *The Prelude*. The landform, however, was one he was quite familiar with, having seen them on Bald Mountain in the high Catskills and in Tuckerman's Ravine during his climb of Mount Washington. He probably also caught glimpses of the "remarkable semicircular precipice or basin" on Mount Katahdin's "eastern side." Tarns, he knew, were special features associated with ice and rock.[17]

One month later, Thoreau encountered the rocky glacial landscapes of the southern hemisphere through the writing of another Englishman, Charles Robert Darwin. The book was *Journal of Researches Into the Natural History and Geology of the Countries Visited During the Voyage of H.M.S. Beagle Round the World, Under the Command of Capt. Fitz Roy, R.N.,* revised in 1845 and printed in New York as two volumes in 1846. More widely known titles for the same work include *Voyage of a Naturalist Round the World* and *Voyage of the Beagle.* This would become Thoreau's favorite book within his favorite genre of travel literature. Its author was not the iconic, bushy-bearded evolutionist in his fifties, but the less-well-known, clean-shaven geologist in his twenties. According to his award-winning biographer Sandra Herbert, Darwin was, for the first thirty years of his career, "known both popularly and professionally as a geologist." In February 1859, and to recognize him for his many contributions to the field, the Geological Society of London awarded him its highest honor, the Wollaston Medal. Publication of his *Origin of Species* only eight months later in August 1859 instantly eclipsed Darwin's prior major achievements. Because *Walden* was published five years before the *Origin,* Thoreau was able to do something we cannot, which is to read Darwin as the geologist he was at the time.[18]

When Darwin left England for his round-the-world voyage in 1831, he carried with him a departure gift: Volume I of Lyell's *Principles,* published in its first edition the previous year. Before reaching the Cape Verde Islands, he had already been swept into Lyell's orbit. Thrilled, he preordered copies of Volumes II and III for pickup in ports of call as they were published. So influential was Lyell's thinking during the voyage that Darwin dedicated his *Journal of Researches* to him with this comment: "The chief part of whatever scientific merit this journal and the other works of the author may possess, have been derived from studying the well-known and admirable *Principles of Geology.*" This dedication may have jumped out at Thoreau when he read it in 1851, because he, himself, had been smitten by Lyell's great book in 1840, eleven years earlier.[19]

Lyell never voyaged to latitudes high enough to see the actual processes by which glaciers and icebergs create drift. Darwin, in contrast, witnessed them in the southeastern valleys of the Andes, Tierra del

Fuego, and the Patagonian Ice Cap. "Glaciers were for Darwin one of nature's grand spectacles," wrote Sandra Herbert, after reading Darwin's effusive descriptions: "It is scarcely possible to imagine anything more beautiful than the beryl-like blue of these glaciers, and especially as contrasted with the dead white of the upper expanse of snow," and the icebergs "floating away . . . for the space of a mile, a miniature likeness of the Polar Sea." In contrast to the European Alps, where glaciers lay nested in high, rugged valleys, in Argentina and Chile Darwin saw "tremendous and astonishing glaciers" extending "along the Pacific coast for 650 miles northwards." These were fed by an ice cap along the spine of the southern Andes: "vast piles of snow, which never melt, and seem destined to last as long as the world holds together." Thoreau took careful notice, quoting several of these passages and adopting the color "beryl-blue."[20]

Darwin also recognized evidence for an older, "ice-transporting boulder period," now known as the glacial period, or late Pleistocene. He distinguished two mechanisms for "Erratic boulders." Those located near "lofty mountains have been pushed forward by the glaciers themselves," whereas "those distant from mountains, and embedded in subaqueous deposits, have been conveyed thither either on ice-bergs or frozen into coast-ice." Because Darwin was able to witness sediment deposition from icebergs as a "cause now in operation," he was, by geological convention, allowed to lump them under the general term "alluvium," thereby avoiding the drift-diluvium debate. With respect to striated and polished rocks distant from glaciers, however, Darwin remained a Lyellian loyalist, judging all to be the product of glacially generated icebergs being grounded against land. "Only later, gradually, and through a tortuous route," wrote Herbert, "did glaciers become for him a geological agency of major significance."[21]

For someone looking to replace Emerson as a role model, Darwin would do very well. Biographer Robert Richardson concluded that Thoreau's "copious notes attest to his extraordinary prescient sympathy with many of Darwin's interests, including his minutely detailed observational techniques, his fascination with change in nature, even his writing style," which "is often very close to that found in Thoreau's journal." I completely agree. Like Thoreau, Darwin was a restless young naturalist, a college-educated sojourner who shunned the established professions others had in mind for him, and who seemed uninterested in personal wealth. Here was someone who cared deeply about nature, paid attention to her finest details, and was openly willing to express his emotional response to nature, whether from the grand sublimity of landscapes, or from the intellectual delight of working out nature's mysteries. Here was a secular soul who used "providentialist language" such as "Author of Nature," "God of Nature," or the "Creation" to describe what lay beyond the physical realm, but who never invoked "God" as an active or meddling agent. Here was a gifted writer with the technique of being constantly present as an observer: the Darwinian "I saw, I beheld." Loren Eiseley, who has long been one of my own favorite writers—and whom I suspect Thoreau would have greatly enjoyed—recognized Darwin and Thoreau as genuine *sympaticos*.[22]

Although Darwin's circumnavigation on the HMS *Beagle* "was, in a sense, a natural history walk around the world," his greatest scientific contribution—natural selection—emerged from the details of local place: the changing shape of finch beaks across the Galápagos archipelago. Ditto for Thoreau, who made a deliberate and conscious choice to become a geographically restricted generalist, rather than a global specialist in some narrow topic. He feared that an excess of distant travel "might completely dissipate the mind." Hence, Thoreau's gratitude: "I cannot but regard it as a kindness," he wrote in November 1853, "in those who have the steering of me that, by the want of pecuniary wealth, I have been nailed down to this my native region so long and steadily, and made to study and love this spot of earth more and more." Had Darwin's father not bankrolled his *Beagle* adventure (complete with a paid manservant), Darwin might have followed the path of Gilbert White—one of Thoreau's earliest role models—to become a country vicar with a principal interest in domesticized nature.[23]

Darwin's *Journal* struck Thoreau's genius like a gong. Landscape, he could now see, was something to be explained rather than something to live on as background. Its most important clock was the extremely slow, metronomic tick of uniformitarianism superimposed on what Darwin called the "endless cycle of change, to which the earth has been, is, and will be subjected." Thoreau quickly adopted Darwin's glacial lexicon. He developed a new enthusiasm for scientific rigor, using words like "hypothesis" and "theory" as Darwin had adapted them from John Hershell's *Preliminary Discourse on the Study of Natural Philosophy*.

Like Darwin, Thoreau had been inspired by the works of Alexander von Humboldt—especially *Cosmos* and *Anisichen*—both of which contained thrilling and detailed descriptions of landscapes. Unfortunately, Humboldt was a self-described "man of the equator" whose descriptions of coastal, tropical, and volcanic scenery had little diagnostic relevance for the terrain informing Thoreau's *Journal* and *Walden*. In contrast, Darwin's descriptions from the glacial latitudes of the southern hemisphere dealt with features Thoreau encountered every day in his northern world. In fact, Thoreau never set foot beyond the glacial border prior to the publication of *Walden*.[24]

Thoreau's *Journal* abruptly changed on June 11, 1851, the day he began to enter Darwin's ideas. For the very first time, he mentions an unnamed bird, perhaps emulating Darwin's quest for unknown species. Also for the first time, we get a clear statement of Thoreau's commitment to develop a rigorous scientific phenology, rather than merely romantic descriptions of seasonal change: "No one to my knowledge," he wrote on June 11, "has observed the minute difference in the seasons—Hardly two nights are alike—. . . A Book of the seasons—each page of which should be written in its own season & out of doors or in its own locality wherever it may be." After an unexplained hiatus of four days, Henry resumed his Darwinian splurge: "June 15. Sunday. Darwin still:—Finds runaway sailors." Then, without dropping a beat, he picked up the *Beagle* narrative where he left off and carried it forward to the end of the voyage. One final point. *Journal* expert Sharon Cameron observed that "after June 1851 entries occur almost every day," though "they have no consistent length." This was Darwin's pattern.[25]

Glacial Lake Sudbury

During glacial recession from eastern Massachusetts, the higher, active part of the ice sheet to the north maintained a smooth southerly slope as required by viscous theory. In contrast, the recently deglaciated terrain to the south was an irregular landscape of bedrock hills, ridges, watersheds, and valleys, and having about five hundred feet of topographic relief. At higher elevations, the recently exposed stream networks were able to flow downward and outward within the original trellis pattern toward all points of the compass. At lower elevations, however, these ancestral valleys encountered one of two situations. Those draining away from the ice sheet could carry on as before. In contrast, those with a northward component of flow found their outlets blocked by the massive dam of ice pulling slowly back to the north.[26]

This latter situation was true for the whole of the Concord River watershed. Upon reaching their lower valleys, the free-flowing tributaries of the Sudbury River entered a glacial lake that was expanding northward as the tapered wedge of the stagnant fringe melted downward. Draining into that enlarging lake was the grand sum of all water from the recently uncovered watershed to the south, and from the vast ice sheet to the north. Given this impregnable dam, every drop was forced to flow southward over the drainage divide with the Charles River through an overflow channel or spillway located in what is now the highest reach of Cherry Brook in Weston.[27]

This spillway was prepared for this role, having already been notched by powerful subglacial flows during the glacial culmination. Immediately after being exposed, any loose gravel and sand within this channel would have been swept away by torrential overflows, leaving a concentrated lag of boulders similar to what Thoreau saw in Saw Mill Brook: "a brawling mountain stream, as much obstructed by rock—rocks out of all proportion to its tiny stream . . . as if a torrent had anciently swept through here." After its initial flushing, however, the rate of bedrock channel erosion would have quickly slowed to nil because the overflow had been decanted of anything coarser than "glacial flour," the clay-sized residues of abrasion and crushing. Thus, the Cherry Brook spillway held the glacial lakeshore to a fixed elevation during deglaciation. Any remnants of stagnant ice above the level of the spillway were candidates for exposure to warm sunlight, rapid melting, the release of stony debris, and wash erosion. Conversely, remnants below it were candidates for rapid burial by copious meltwater sediment pouring into the lake from subglacial tunnels, and from the bare, rocky hills surrounding the lake.[28]

Thoreau saw very clearly that the heart of his sojourning country had been inundated by a vast lake in the not-too-distant past: "The inland hills and promontories betray the action of water on their rounded sides as plainly as if the work were completed yesterday." In Concord, the dense fogs of late summer created "ghost" lakes that filled the Sudbury Valley to the level of its former strandlines: "The fog is a perfect sea over the great Sudbury meadows in the southwest, commencing at the base of this Cliff [Fairhaven] and reaching to the hills south of Wayland, and further still to Framingham, through which only the tops of the higher hills are seen as islands, great bays of the sea, many miles across, where the largest fleets would find ample room and in which countless farms and farm houses are immersed." Looking downstream to "where the village of Concord lay buried in fog," he "thought of nothing but the surface of a lake, a summer sea over which to sail." These passages echo Darwin's use of fog to reconstruct comparable inundations of bays and coves of coastal Chile.[29]

In these lyrical passages, Thoreau word-mapped a glacial lake whose "primeval banks make thus a channel which only the fogs of late summer and autumn fill." Half a century later, this same ancient lake was geologically mapped and carefully reconstructed by a young Harvard graduate student named Richard P. Goldthwait. His thesis, "The Sand Plains of Glacial Lake Sudbury," published in 1905 as Volume 42 of *Bulletin of the Museum of Comparative Zoology at Harvard College*, remains the definitive reference for the origin of Walden Pond. He seems to have been unaware of Thoreau, whose works were then little known beyond a small coterie of admirers. For example, Goldthwait used " Lake Walden" as the proper place name, in contrast to Thoreau's "Walden Pond." And for the shape of the lake, he used a poorer-quality map than the one Thoreau had produced. One year later, however, in 1906, Houghton Mifflin's publication program raised Thoreau's literary profile to the level where Goldthwait probably would have noticed. Had this been done just a year or two earlier, the geology of Walden Pond might be far better known than it is today.[30]

Thoreau's pioneering techniques for reconstructing his unnamed version of Glacial Lake Sudbury were similar to what Goldthwait used a half-century later. At the highest level Henry recognized both the conspicuous lake shorelines and the deltas flowing into them during "some other geological period." At the middle level were prominent terraces, especially the one above Concord River being "the ghost of the ample stream that once flowed to the ocean between these now distant uplands in another geological period, filling the broad meadows." This terrace, we now know, constitutes the reorganization of the drainage immediately "after the deluge." At the bottom level were the evenly laminated layers of lake mud with a "delicate stratification . . . like the leaves of the choicest volume just shut on a lady's table." These were almost certainly varves, a diagnostic signature of deep glacial lakes, in which

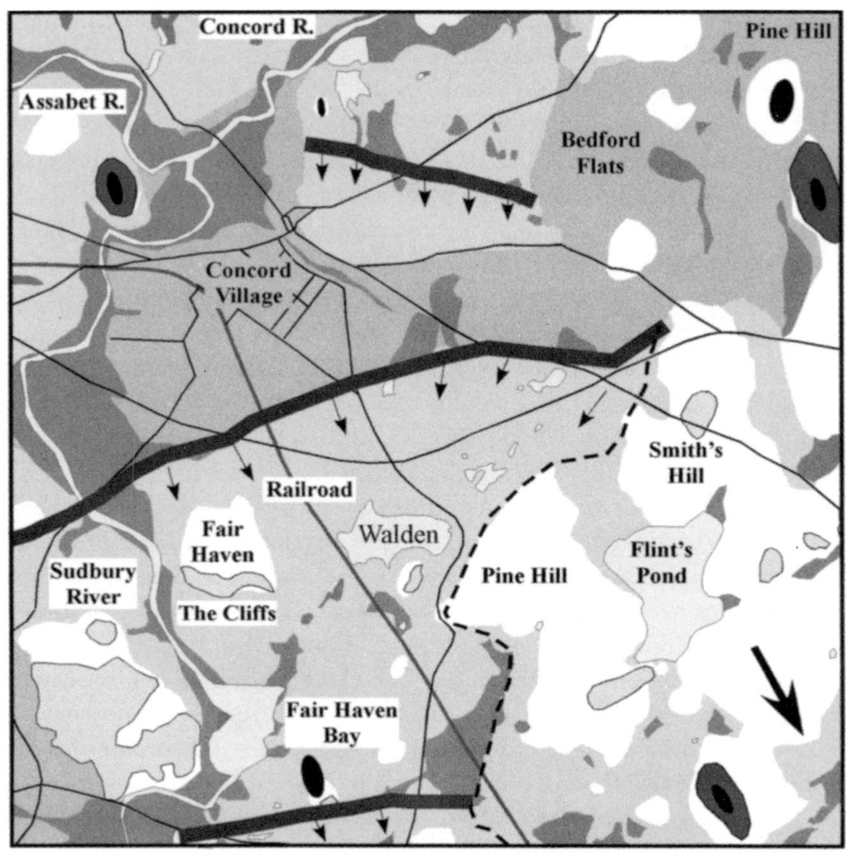

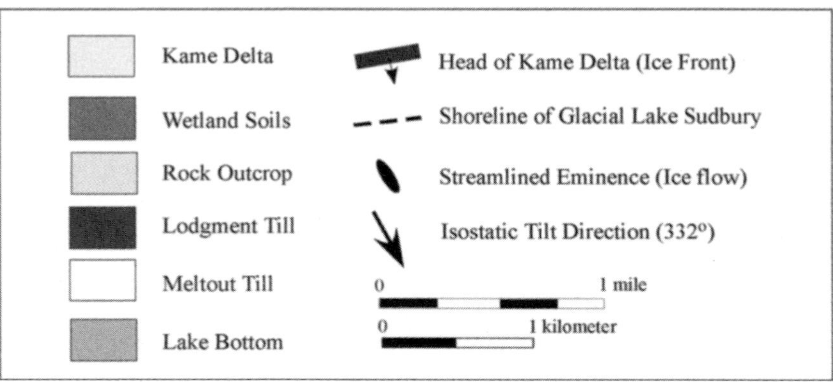

Figure 15. Glacial geology of the vicinity of Walden Pond.

Sources: Map units from Stone and Stone, *Surficial* and Koteff, *Concord Quadrangle*. Ice front positions and shoreline of Glacial Lake Sudbury from Koteff, *Glacial Lakes*. Isostatic tilt from Koteff, *Delayed Postglacial*.

bands of summer silt alternate with thinner, darker layers of winter clay. Varves are the perfect analog for tree rings, which Thoreau loved counting as well. Something about their cyclic rhythm appealed to him.[31]

Thoreau's Transition to Science

When Thoreau set up housekeeping at the pond on July 4, 1845, he gave no indication that he would write a book about his experiment in deliberate living. Rather, his main agenda was to memorialize his deceased brother John Jr. by writing *A Week on the Concord and Merrimack Rivers,* described by Walter Harding as "one of the most complete failures in literary history." Seven months later on February 4, 1846, Thoreau experienced another disappointment while giving a lecture on the English writer Thomas Carlyle in the town of Carlisle. What his audience really wanted to hear was why their speaker had moved into the woods to live alone near shore of a pond. How did he live? What did he eat? Was he lonely? Was there some secret involved? Thoreau responded by writing a separate lecture titled "A History of Myself."[32]

The text of that second lecture, read to the Concord Lyceum on February 10, 1847, became the opening salvo of *Walden,* as published seven years and many drafts later. Lyndon Shanley, in his pioneering literary dissection *The Making of "Walden,"* concluded that the basic structure and theme of Thoreau's book was established at the outset, albeit with a great expansion of its nature writing in years to come. This certainly applies to the consistent literary voice and the book's main philosophical message: a wake-up call for personal spiritual rebirth set within the context of nature. But with respect to practically everything else, recent scholarship concludes that *Walden*–Part I (Versions I–III, 1846–1849) and *Walden*–Part II (Versions IV–VII, 1852–1854) are distinct in content, chronology, purpose, and technique. "Not by any standards," conclude Stephen Adams and Donald Ross "is the first version of *Walden* a unified text." Sharon Cameron agreed that the final book is a diptych hinging on "The Ponds," with Part I dealing with man *and society* and Part II dealing with man *and nature.*[33]

The chronological hiatus (1849–1852) between these two distinct periods of work on the *Walden* manuscript, and the dramatic dichotomy in content before and after it, have long puzzled Thoreauvians. Leo Marx concluded that this "change in mid-career is the most controversial episode of Thoreau's intellectual life. To many who knew him well,

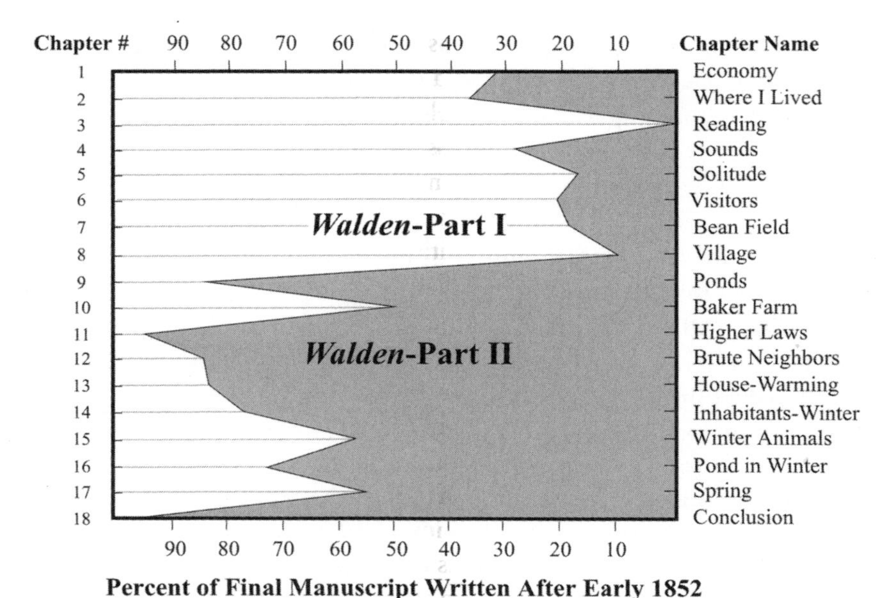

Figure 16. Proportional content of *Walden* before and after early 1852.
Simplified from Figure 2 of Adams and Ross, *Mythologies,* 58, using percent of total manuscript at "D" Stage (Version IV) or later.

including Emerson, it seemed puzzling, misguided." Midway through the transition Thoreau began to intuit a lasting psychological change within himself, confiding in his *Journal:* "I seem to be more constantly merged in nature; my intellectual life is more obedient to nature than formerly, but perchance less obedient to spirit." Explicitly, "The character of my knowledge is from year to year becoming more distinct & scientific," which is another way of saying it was becoming less transcendental. During this transition, his *Journal* became so overstuffed with technical observations that, by the spring of 1853, he considered renaming it "Field Notes."[34]

Alan Hodder called this transition a "watershed in Thoreau's life," a "time of new beginnings," separating the "cosmopolitan and more open-endedly subjective reflections of earlier years to the detailed, empirical, and at times dry formulations of fact after 1852 or so." Bradley Dean called it a "dramatic reorientation." Rochelle Johnson saw a "general shift away from metaphor and toward conveying the literalness of natural phenomena through description." Adams and Ross concluded that

Thoreau shifted his intellectual focus from being a neoclassical scholar and transcendental novice to being a naturalist with an unusual mix of romance and science. Robert Sattelmeyer highlighted Thoreau's "recently acquired habits of careful observation and description." Taking the long view, Donald Worster summed it up nicely: "after several decades of happy innocent wandering . . . Thoreau now began at the age of thirty-five to approach nature with more method and precision. He became a self-educated naturalist."[35]

That the change took place is undeniable. But scholars remain at loggerheads over why. H. Daniel Peck argues for one particular breakthrough insight, the realization that "time was a circle." Robert Richardson argues that it was broad exposure to the works of Linnaeus, Darwin, and Gilpin, among others. Laura Dassow Walls argues it was Alexander Humboldt's influence following his readings of *Cosmos* and *Anisichen* in 1849. Bradley Dean came closest to my interpretation when he wrote that Thoreau's shift followed a conscious realization that "he would need to master a scientific study of nature so that he could follow and articulate the steps by which fact flowered into truth."[36]

To illustrate Thoreau's shift from "science light" to "science heavy," I offer before and after *Journal* entries on the same topic, likely from the same place. In August 1845, having lived at Walden Pond for about a month, Thoreau heard "the distant rumbling of wagons over bridges,—a sound farthest heard of any human at night." Later in that *Journal* entry, he converts the sound into a transcendental symbol and literary trope, a "divine dictate" to the ear analogous to the ethical dictate in his mind. Six years later, on June 11, 1851—the very day he began to enter material from Charles Darwin into his journal—Thoreau's response to the same sound was a sophisticated acoustic analysis: "I hear from this upland . . . a wagon crossing one of the bridges . . . this still and resonant atmosphere tells the tale to my ear. Circumstances are very favorable to the transmission of such a sound. In the first place, planks so placed and struck like a bell swung near the earth emit a very resonant and penetrating sound; add that the bell is, in this instance, hung over water, and that the night air, not only on account of its stillness, but perhaps on account of its density, is more favorable to the transmission of sound."[37]

Thoreau, writing like a physicist, was correct on every point. On that calm summer evening, air directly above the bridged stream had been chilled by conduction because the stream was cold because it was draining aquifers recently recharged by spring snowmelt. This cold air formed a stable mass between the banks because it was a denser fluid than ambient air and occupied a topographic notch below prevailing breezes, technically a zone of aerodynamic flow separation. Sound travels faster and more efficiently in such a chilled corridor, creating a linear "wave guide" cutting through the humid and muffled ambience of the meadow. Introduced into this wave guide directly above the acoustic reflector of the cold-water surface was the sharp percussion of a hardwood plank being struck against the bridge beams by the weight of a large draft animal concentrated on one hoof and levered along the length of the plank. This hammer-like blow thus became the "sound farthest heard of any human at night."[38]

After his permanent transition, Thoreau walked like a field scientist, carried the equipment of a field scientist, read scientific articles and texts on field science, and made important new discoveries based on field evidence. Here, I use the adjective "field" as Thoreau did when he wrote: "a field of water betrays the spirits that are in the air." It is an outdoor setting where things happen and where observations can be made. I avoid the adjective "physical" because, in Thoreau's day, physical science connoted bench-top laboratory research on fundamental phenomena such as mechanics, thermodynamics, electromagnetism, and chemistry. I avoid the labels "naturalist" and "natural historian" because—in modern usage—they discount Thoreau's lifelong interest in physics and engineering, and are today heavily skewed toward *biophilia,* our instinctive love for living

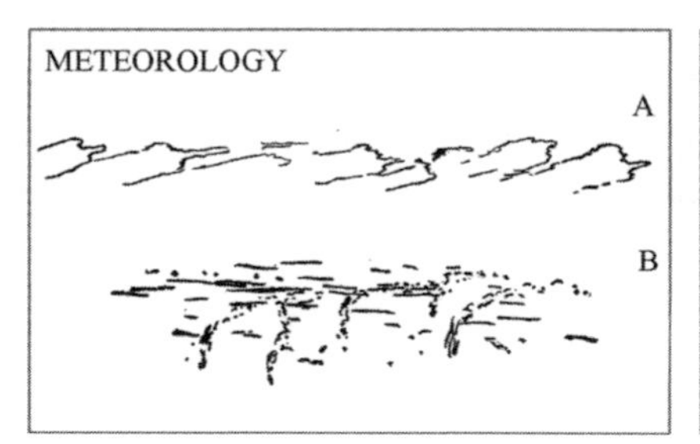

Figure 17. Thoreau's *Journal* sketches of meteorological and geologic phenomena. *Meteorology:* A, Dark clouds aligned under red sky (Jul 10, 1851); B, Chimney smoke rising through haze to reach temperature inversion (Oct 31, 1853). *Geology:* C, Cuesta structure at Fairhaven Hill from Hubbard's Bridge (May 3, 1852); D, Terrace tread, scarp, and angle of repose on "glorious" sandy banks at Cochituate (Nov 7, 1851); E, Erratic boulders at Nonesuch Pond shown in frontispiece of this book (Nov 7, 1851).

Source: Thoreau, *Journal*, edited by Torrey and Allen, 1962.

things. This evolved trait biases our attention toward Earth's biosphere, which, by thickness, is smaller in proportion to the whole of nature than a lichen is to a cliff. The biosphere is also trivial with respect to power, extracting and exchanging less than one percent of the total solar and orbital energy driving other equally beautiful natural processes that interested Thoreau at least as much as did botany and zoology: notably the flow of rivers, the roar of the sea, the color of a sunset, and the dance of the clouds. Finally, the word "naturalist," as used today, connotes a kinder, gentler, more amateur scientist than your average professional geneticist, physiologist, sociobiologist, seismologist, or metallurgist.[39]

Consider Thoreau's field observations at the Deep Cut during the spring thaw of 1852: "When the flowing drop of sand and water in front meets with new resistance, or the impetus of the water is diminished, perhaps by being adsorbed, the drop of sand suddenly swells out laterally and dries, while the water, accumulating, pushes out a new sandy drop on one side and forms a new leafy lobe, and by other streams one is piled upon another." Ignoring tense, there are ten physical action verbs—flow, meet, diminish, absorb, swell, dry, accumulate, push, form, and pile—in this single sentence that are every bit as natural as the blossoming of a flower or the basking of a tortoise. Thoreau wrote about such things—soil mechanics, rock mechanics, fluid mechanics—with a *joi de vivre* equal to those he used to describe plants and animals. Seldom, however, do his sentences about these phenomena find their way into ecocriticism, because authors of this genre are biased by their biophilia: emphasizing the prefix "eco," rather than the prefix "geo." They overlook the fact that writing like a civil engineer was a routine part of Thoreau's sojourning practice. Writing with such "precision," concluded Bradley Dean, "enabled him to generate an impressive amount of viable scientific knowledge because it employed precisely the same rigorous methods as standard science."[40]

Landscape Revealed

June 11, 1851—Darwin Day—inaugurates 363 days of stepwise revelation in Thoreau's *Journal* that culminate with his conceptual model for the origin of Concord's landscape. On that first day Thoreau used the word "erratic" for the first time in its proper glacial context. The entry for June 12 contains only three sentences, the longest being, "There would be this advantage in travelling in our own country, even in our own neighborhood, that you would be so thoroughly prepared to understand what you saw you would make fewer traveller's mistakes." Here, I believe Thoreau is resolving—or perhaps rationalizing—his decision to concentrate on his local landscape in order to take advantage of his great familiarity with it. Depth, not breadth, would be his goal.[41]

The entry for June 13 begins with a moonlight walk south from the village along the railroad tracks:

> As I entered the Deep Cut, I was affected by beholding the first faint reflection of genuine and unmixed moonlight on the eastern sand-bank while the horizon, yet red with day, was tinging the western side. What an interval between those two lights! . . . There the old and new dynasties opposed, contrasted, and an interval between, which time could not span. Then is night, when the daylight yields to the nightlight. It suggested an interval, a distance not recognized in history. Nations have flourished in that light.

This, I suggest, was a reflection on the depth of geological time, which is so protracted that "old and new dynasties"—each of greater duration than all human history—were constantly being exchanged as natural cycles.[42]

June 15 brought into Thoreau's *Journal* another flurry of excerpts from Darwin's *Journal*, including this revealing quote: "Daily, it is forced home on the mind of the geologist, that nothing, not even the wind that blows, is so unstable as the level of the crust of this earth . . . the great continents are, for the most part, rising areas; and . . . the central parts of the great oceans are sinking areas." This is Thoreau parroting Darwin, parroting Lyell. When he applied this concept to his local landscape, he realized that it, too, had experienced great changes in water level, most notably the deep inundation and subsequent drainage needed to produce visible shorelines left high and dry.[43]

On August 6, Thoreau returns to the contrast between Darwin's distant travels and his own: "How often it happens that the traveller's principal distinction is that he is one who knows less about a country than a native! . . . It takes a man of genius to travel in his own country." Revisiting this point on August 19, he wrote: "As travellers go round the world and report natural objects and phenomena, so faithfully let another stay at home and report the phenomena of his own life,—catalogue stars, those thoughts whose orbits are as rarely calculated as comets."

Here, the contrast between Darwin and himself could hardly be more striking. On August 21 Henry concluded: "The discoveries which we make abroad are special and particular; those which we make at home are general and significant. The further off, the nearer the surface. The nearer home, the deeper." With this global-local contrast in mind, Thoreau finally declared his true profession on September 7, from which he never wavered: "to attend all the oratorios, the operas, in nature." Darwin had declared much the same when he stepped on to the deck of the HMS *Beagle* for his five-year voyage around the world.[44]

That autumn continued with a flurry of connections made between Concord and Darwin's landscape. September 8 brought the realization that the Concord River Valley had higher ancient shorelines like those Darwin described for the broad fjords of Patagonia. Of his local landscape, Thoreau wrote that "an inquisitive eye may detect the shores of a primitive lake in the low horizon hills, and no subsequent elevation of the plain has been necessary to conceal their history." Continuing on October 2 he saw, but did not quite understand the "hard-pan" of glacial lodgment till. On October 5, pasture outcrops "worn smooth by the elements." On November 4, the torrential channel of Saw Mill Brook. On November 7 he applied the diagnostic term "moraine," as used by Darwin, to a ridge near Long Pond in Wayland. In early November, and following Darwin's example from Patagonia, he went back to Concord's boulder field to measure the size of glacial erratics and to greatly elaborate on their description. On December 5 he used one of Darwin's phrases—"chain of ponds"—for the first time in his *Journal* to describe those in the Walden paleo-valley. This flurry culminated with Thoreau's vivid descriptions of the clay-rich lake-bottom sediments and the resulting "sand foliage" at the Deep Cut on December 29–31.[45]

Five days later, on Jan 5, 1852, Thoreau's imagination jumped the voltage gap across a Huttonian "revolution." The thawing, sliding, and flowing then taking place at the Deep Cut may have helped him see the much larger, and much older, thawing, sliding, and flowing responsible for creating Fairhaven Bay. He was very specific:

> Sitting on the Cliffs, I see plainly for the first time that the island in Fair Haven is the triangular point of a hill cut off, and forty or fifty rods west, on the mainland I see the still almost raw and shelving edge of the bank, the raw sand-scar as if sodded over the past summer . . . as if the intermediate portion of the hill had sunk and left a cranberry meadow. It is with

> singular emotions that I stand on this Cliff and reflect in what age of the world this revolution, the evidence of which is of to-day, was evidenced by a raw and shelving sand-bank.

The "raw sand scar" that caught Thoreau's attention is still there, easily seen when the leaves are off the trees. It is the relict headwall of the collapse slope that "sank" downward to create the kettle of Fair Haven Bay, leaving the island as a less collapsed portion. This mass-movement may have been inspired by Darwin's account of a similar "great slip" on a steep slope in Tierra del Fuego. Even more important than the "slip" mechanism for Thoreau was the antiquity of the event, dating to a previous geological "revolution" that was older than the known history of all human nations.[46]

Nine days after this insight at Fairhaven Bay, Thoreau extended the same idea to the origin of Walden Pond. Overlooking its "level shrub oak plain" from the moraine of Heywood Peak, he saw that "Walden and other smaller ponds, and perhaps Fairhaven, had anciently sunk down in it" because the "level is continued in many cases even over extensive hollows." That level, he knew, had been built by an ancient stream that "flowed thither" toward the high "primeval banks" of the Sudbury Valley, during "some other geological period," and whose ancient shorelines were mirrored by fogs of late summer. Thoreau's "sinkhole-in-ancient-delta" origin for Walden Pond was the same one Goldthwait improved on half a century later, thanks to a guiding theory, better nomenclature, and much better evidence.[47]

Thoreau's sinkhole origin for his chain of ponds was almost certainly aided by his previous experience of seeing—but not quite understanding—similar features on Cape Cod three years before. The plateau below which Walden was sunk resembled what sailors called the "Table Lands of Eastham," which is also covered by shrub oak and riddled with kettles. Both terraces, he knew, were underlain by horizontal seams of finer-textured sediments dating to an earlier part of the geological epoch. Both contained "circular" hollows. In Eastham he compared one of them to the conical depressions the ant lion digs in loose sand to create its predatory death trap, a description that can be found in Darwin's *Journal* as well. Walking toward one of these "hollows," Thoreau feared that he "might tumble . . . and be drawn into the sand irrecoverably." Though he knew the immediate cause of these sinkholes—loss of support at depth—he did not speculate on the cause of that loss of support.

Certainly, he did not accept Hitchcock's official interpretation that they were diluvial swirl pits from great eddies churned by a catastrophic "chopped sea."[48]

Within a month, and aided by deep snow, Thoreau finally arrived at an explanation that had the potential to account for all of these odd features: "It looks as if the snow and ice of the arctic world, travelling like a glacier, had crept down southward and overwhelmed and buried New England." To this Arctic vision, and from the same spot, he soon added "towering icebergs . . . with cavernous recesses." Prompting both passages—iceberg and glacier—may have been the recent return to New York of Arctic explorer Elisha Kent Kane on September 30, 1851, whose experiences were then making a media sensation. Unfortunately, nowhere does Thoreau explicitly invoke the loss of buried glacial ice to explain the sinking of Walden Pond, or the related land-sliding at Fairhaven Bay.[49]

Robert Richardson argues that it was Thoreau's experience of watching the sand flowage at the Deep Cut in late December 1851 that "provided the impetus" to "return and this time not just to revise but reshape" the *Walden* manuscript, beginning a continuous process that would carry, almost without interruption through Versions IV, V, VI and VII to the spring of 1854. Could the impetus have been not only the culmination of the sand foliage, but the entire two-month chain of insights leading up to it?[50]

Four Deltas

During most of its history, the elevation of the Glacial Lake Sudbury shoreline was firmly fixed by the Cherry Brook spillway. Not until the tapered ice edge receded to a position near Brister's Hill (north of Walden) did it uncover a slightly lower east-draining spillway to the Charles River through Kiln Brook (Tophet Swamp). At that point Glacial Lake Sudbury ceased to exist, and a short-lived phase of Glacial Lake Concord came into existence at nearly the same elevation. Shortly thereafter, additional northward recession opened up a much lower route to the sea through the Shawsheen River in Bedford, which allowed the high-level glacial lakes to drain and a lower one to replace it. The historic floods of Thoreau's era re-created this lower glacial lake to a "level water-line" against the trees, "as if the water were just poured into its basin and simply stood so high." Each episode of "water on the meadows," became an inland "mimic sea," complete with cranberries on the bottom.[51]

Throughout the existence of the glacial lake sequence, an astonishing flux of sediment—grit, sand, granules, pebbles, cobbles, and boulders—was being washed through tunnels from active ice to the north. On flat topography, this material gushed southward and outward from the stagnant fringe to create outwash fans like those fronting modern Icelandic ice caps, and those fronting the Laurentide culmination on New England's southern archipelago. When the copious sediment entered a confined valley draining away from the ice, there was not enough room for it to spread out as a fan. Instead, it filled the valley wall to wall with a flat floodplain of braided channels dominated by sandy gravel, like the flat-bottomed glacial rivers of Alaska today. When the sediment flux entered standing water, however, the power to move sediment abruptly dropped. Instead of flowing away from the ice, coarse sediment accumulated against it and near it, creating a situation where stagnant ice, strong currents, standing water, and an almost unlimited supply of sediment all met in the same place.

Boulders and cobbles being transported by surges of meltwater flow were immediately dropped where the subglacial tunnels entered standing water, creating complex and overlapping piles full of washouts and collapses. Sand, which constituted the great bulk of incoming sediment, kept going, being washed outward and away by steadier currents of ice-cold water. This sand created broad delta lobes that overlapped with one another like a broad stack of pancakes leaning gently away from the ice. As these lobes built upward and outward, the space below the water surface was filled from the bottom up, especially in narrow bays of "arms." The Walden paleo-valley was one such "arm." Hence, it quickly filled with sediment gushing in from the northeast near Goose Pond, probably over the time span of several centuries. This "fill" was the underwater part of a delta, the front of which was advancing southwesterly toward the main lake in the main valley. Behind the advancing front was a delta plain just above the level of the lake. This well-studied, and very predictable sequence of delta deposition happened all over New England in hundreds of places, one of which just happened to be at Walden Pond. There is absolutely nothing special about its geology.[52]

More broadly, the valley of the Sudbury River between Wayland to the south and Fairhaven Hill to the north contains a succession of three massive glacial deltas. Each formed where the stagnant edge of the receding ice became temporarily stabilized by the underlying bedrock topography. With the ice still strongly active to the north, the relentless flux of subglacial sediment was sufficient to plug the entire width of the valley with delta deposits. And because these were banked up against stagnant ice to the north, they are properly called "kame deltas." In each case, the delta plain—composed of glacial outwash—rose slightly northward until it terminated, either abruptly against a local moraine ridge, or tapering above the sloping edge of stagnant ice. In the latter case, subsequent melting of that ice and the subsidence of the delta plain above it, created the jumble of gravel knobs and small kettles called dead ice terrain.

The most extensive of Glacial Lake Sudbury's three kame deltas is what Goldthwait called the "Lake Walden sand plain," or the "Walden delta" for short. This distinctive landform is the eastern portion of a much larger mass extending all the way across the Sudbury Valley to beyond White Pond. At Walden Pond State Reservation headquarters, the "sand plain" is so large, flat, and well drained that a visitor parking lot for three hundred and fifty cars was built with minimal grading. To the west there was plenty of room for the former Lake Walden amusement park. Beginning at its northeastern limit at the former ice front, and guided by the gradient of the paleo-valley, the delta plain declines steadily in elevation from about 220 feet near Goose Pond to an elevation of about 210–200 feet just east of the shore of Walden Pond, to between 200–190 feet on the pond's west shore, and then to an elevation between 190–180 feet south of Fairhaven Bay. None of the ponds or hollows sunk below it could possibly have existed during its active growth, because they would have been quickly filled by the unstoppable flux of sediment. Projecting through the delta plain were the bedrock-cored hills of Conantum, Fairhaven Hill, Emerson's Cliff, and Mount Misery.[53]

The high ground just east of Concord's ancestral village constitutes the fourth kame delta of Thoreau country. This one was built into the short-lived high stand of Glacial Lake Concord. Though this delta is the smallest of the four, it is the most obvious one, being an isolated plateau of gravelly sand standing above lower, wetter ground on all sides. Its eastern part is a collapsed mass of sand and gravel containing Sleepy Hollow, which Thoreau surveyed before his townsmen converted it to a cemetery park. Further west, "Squire" Emerson "hired an Irishman to dig" material from it to "fill low ground on his property." Near the delta's western tip, Concord's first settlers built their semi-subterranean houses, as described by Puritan writer Edward Johnson in *Wonder Working Providence* (ca. 1650). "After they have thus found out a place of aboad, they burrow themselves in the earth for their first shelter under some hill-side, casting the earth aloft upon timber, they make a smoaky fire against the earth at the highest side. And thus these poore servants of Christ provide shelter for themselves, their wives and little ones."[54]

Standing above all four delta plains are higher ridges of coarser material that accumulated directly from the ice, largely by processes associated with gravitational sliding and flowage. Revolutionary Ridge lies above Sleepy Hollow. Heywood's Peak lies above Walden. These may properly be considered moraine ridges, even though no glacial bulldozing or thrusting was involved. Between these two named ridges lies the bulk of Walden Woods, a chaotic terrain full of unnamed ridges, lumps, bumps and twenty-eight random kettles that bottom out at various depths in various materials. Thoreau loved this patch of pervasively collapsed terrain—centered west of Brister's Hill—for the ecological variety set up by the topographic chaos. One special delight was "frosty hollow," a deep and elongated "pond-hole behind my dwelling" where geese occasionally came to feed, and where Thoreau soaked his axhandle. Another was the residual hill excavated by the railroad to make the Deep Cut. A third was the Boiling Spring, whose artesian pressure was likely created by blocks of sediment juxtaposed by meltdown collapse.[55]

Located at low elevations beyond the delta plains and the dead ice topography are the lake-bottom sediments. These were the flat drapes of clay-rich mud that settled from suspension out of standing water in the open lake. Today, these materials underlie all of Concord's major wetlands including the Great Meadows, Bedford Flats, and Beck Stow's Swamp, one of Thoreau's favorites. When properly drained, they became profitable agricultural soils for tillage, what Brian Donahue called the "heart of husbandry in Concord." When dug and baked, they made wonderful bricks, hundreds of thousands of which were made during some years. Sandwiched between the clay and overlying peat was abundant "bog iron ore" that, "as early as 1660," was "smelted and wrought in bars for the customary purposes of life," wrote Lemuel Shattuck, an early historian.[56]

Concord's first Puritan settlers apparently experienced the poor drainage of Concord's lake-bottom sediments. Though their semi-subterranean burrows into the delta sand kept "off the short showers from their lodgings . . . the long rains penetrate[d] through to their great disturbance in the night season." This likely refers to the rise in the water table above the glacial lakebed clay after sustained rain and snowmelt in late winter and early spring.[57]

Because glacial debris becomes progressively broken down during its transport, the smallest particles tend to come from farthest away. The mineral composition of lake bottom mud integrates every stone crushed from Labrador to Concord. The average sand grain at Walden probably traveled a hundred miles or more from its source rock; the average pebble ten miles or more; and the average small boulder a few miles. Thus, it would not surprise me if more than half of the bulk sediment surrounding Walden Pond came from New Hampshire, brought there via the maze of anastomosing tunnels beneath the ice.

Theory Vindicated

The story I have just told of the four kame deltas in the Sudbury Valley would have been inconceivable to both Edward Hitchcock and Charles Lyell. Neither of their theoretical schemas allowed the possibility of an ice dam to the north. Lyell would have explained the inundation by gradual subsidence beneath the sea and the arrival of debris-rich glacial icebergs. Hitchcock would have invoked a violent, tsunami-like flood of icebergs and water coming from the north. Charles Darwin, the third geologist to whom Thoreau turned for ideas, would have been right by the time *Walden* was published, but had been wrong earlier in his career when he studied a broadly comparable situation at Glen Roy, Scotland. There Darwin initially mistook its high-level shorelines as a consequence of Lyellian submergence followed by emergence. He soon corrected this error, realizing that an ice dammed lake was involved. "This paper was a great failure," Darwin wrote of his mistake "and I am ashamed of it." Shortly thereafter, on June 1, 1842, he experienced a "Paulinian" conversion to the glacial theory within a tarn named Cwm Idwal in north Wales. He documented this psychologically intense experience in a footnote to the 1846 edition of *Journal of Researches*, the one Thoreau read so carefully in June 1851.[58]

Beginning in 1855, most Americans scientists would convert to the glacial theory, thanks to American naval surgeon and Arctic explorer Elisha Kent Kane. During his second Arctic voyage in search of Sir John Franklin's lost expedition, he and his crew became stranded in the frozen seas northwest of Greenland. Some survived to tell a tale of great endurance and heroism. While searching for a way out, Kane encountered the main mass of the Greenland Ice Sheet "moving onwards like a great glacial river seeking outlets at every fjord and valley . . . Here was a plastic, moving, semi-solid mass, obliterating life, swallowing rocks and islands, and ploughing its way with irresistible march through the crust of an investing sea." After his celebrated rescue, Kane returned triumphantly to New York on October 11, 1855, and began granting interviews, lecturing, and writing up his account. This, of course, was one year too late for *Walden.*[59]

Publication of Kane's *Arctic Explorations* in 1857 moved the idea of ice-sheet glaciation from theory to fact. Twenty years after the *Neuchatel Discourse,* Kane gave Louis Agassiz a modern analog for his theory, but at the cost of his cosmic catastrophism. He gave Charles Lyell admissible evidence for a "cause now in operation," but at the expense of proving him wrong about iceberg drift. He gave Charles Darwin a better explanation of the landscape features he had seen in Patagonia, and reminded him, once again, never to put theory ahead of observations. He gave his old geology teacher, William Barton Rodgers, founder of Massachusetts Institute of Technology, a student to be proud of, in spite of his mentor's catastrophist views. He gave Reverend Hitchcock a permanent heartburn that lasted until he died.[60]

Three years after publication of Kane's *Explorations,* Hitchcock wrote: "No geologist, . . . it seems to me, can read Dr. Kane's description of the glacial phenomena along the coasts of Greenland and Grinnell Land, and not feel that if a similar state of things once existed in Canada, New England, and other northern parts of our country, it would satisfactorily explain the phenomena of drift. Hence it is fair to presume that such a state of things did once exist here, and that drift has resulted from ice floes, icebergs, ice belts and glaciers." Note the conditional "if." Also note that "glaciers" were last on his list of potential agents. And in his final summation, Hitchcock wrote in 1860: "On such icebergs and ice floes, for the present I take my stand."[61]

Lyell did eventually convert. In the year *Arctic Explorations* was published, he privately admitted his error in a letter to his father. But publically, he held on until 1863. This was one year after the American Association for the Advancement of Science embraced the glacial theory, thereby correcting the error made twenty years earlier by its antecedent organization. Led by John Strong Newberry of Ohio, glacial geologists from the Midwest finally convinced their more conservative eastern colleagues that solid ice sheets, rather than debris-laden ice floes, had indeed covered much of North America, and not just once, but several times. Not even the strongest of debacles or the largest armada of drift icebergs could explain the depth of the Great Lakes, the uniform sheets of glacial till pasted over the central plains like coats of paint, the straight-line scratches on curved bedrock humps, and colossal erratics left as "rocking stones" on narrow ridges. By this time, Charles Whittlesey—Newberry's mentor and America's first glacial geologist—had drawn a map of the United States showing the former ice sheet border east of the Dakotas. On it is a large arrow drawn directly over Thoreau country showing the direction of glacier flow. Ironically, his arrow was based on evidence described in Hitchcock's 1841 *Final Report.*[62]

One year after Walden's publication, a young caricaturist named Frank Bellew looked out over Walden Pond and saw what "appeared to be an extinct gravel pit." This snap judgment was actually pretty good, except for one detail: the direction of gravel removal. It is upward for an excavation and downward for kettle subsidence. As with abandoned and revegetated gravel pits, the water table in kettles often stands midway between their flat bottoms and flat tops, and their surrounding banks often lie at the angle of repose. The details of how this happened at Walden is the subject of the next chapter. But for now, the answer is simple. The bottom of Walden dropped like an icy piston pulled downward below its delta plain to create a void deep beneath the water table.[63]

5

MELTDOWN TO BEAUTY

"A vast blue fort or Valhalla . . . thirty-five feet high on one side and six or seven rods square . . . and estimated to contain ten thousand tons." That was Thoreau's description of the colossal stack of lake ice cut from Walden by Irish laborers "in the winter of '46–7." After stacking it up, they had to insulate it against the coming heat of summer to minimize melting before shipment. So "they began to tuck the coarse meadow hay into the crevices," and then covered it all with hay and boards. Though the stack was uncovered and exposed to the sun in July 1847, it still took more than two full summers and intervening calendar year for it to disappear. Not until September 1848 did the last bit melt and trickle away.[1]

With this account, Thoreau gives modern readers a *faux* block of stagnant ice large enough to help us comprehend the meltdown of the actual block in Walden's western basin. Both masses—of stacked lake ice and crevassed glacial ice—were enormous and immobile. Both were extracted from something far more dynamic and interesting. Both were greatly delayed in their melting by thick insulation.

Picture the ragged tapered edge of the Laurentide Ice Sheet melting downward and separating into residual blocks. Those above the level of Glacial Lake Sudbury and exposed to the bright summer sun on flat ground disappeared within decades, or a few centuries at most. In the submerged Walden paleo-valley, however, everything was ideally situated to detach, isolate, bury, and delay the melting of stagnant ice blocks beneath a rapidly advancing meltwater delta. Eventually, these blocks would create the "chain of ponds" sunk into its delta plain between Little Goose Pond and Fairhaven Bay. Walden, the deepest of these, formed above the thickest ice block in a place shaded by Emerson's Cliff.[2]

Figure 18. Tranquil beauty at Walden Pond. Early morning conditions in fall often show morning mist, no waves, smooth ripples, and a deeply shaded beach (foreground). Note distant boat, barely visible to right. View is west along the length of the lake.
September 2009.

After its burial, neither the sun nor the lake had access to the Walden block. Downward melting through the delta plain was limited because the summer heat had to be conducted through dry gravel, a moderately good insulator. Inward melting from the sides was even slower because the saturated pores were contiguous with the ice-cold water of Glacial Lake Sudbury, whose shoreline was partly rimmed by glacial ice, and whose surface was frozen solid in the winter and covered by floes of pack ice during brief summers. We cannot know how long it took the

Walden block to melt, but a millennium or two is quite realistic based on modern analogs and heat flux calculations.

Thoreau knew perfectly well that his "chain of ponds" had been sunk below an ancient river bottom, having said so in his *Journal*. What he lacked was the mechanism for creating the void into which they sunk. Had he been reading the shipping logs of English and Russian vessels exploring the Alaskan Pacific coast, however, he would have likely discovered the mechanism in time for *Walden*'s publication. There, Captain Sir Edward Belcher in 1837 and Captain Michael Tebenkof in 1848 provided eyewitness accounts of debris collapsing downward as stagnant ice of the Malaspina Glacier melted. And also by that time in Ohio, geologist Charles Whittlesey had correctly induced the origin of kettles based on eroded exposures showing them in cross section. Thoreau almost certainly saw such a cross section near the "Highland Light," on Cape Cod, but did not recognize it for what it was. "In one place," he wrote, "the bank is curiously eaten out in the form of a large semicircular crater." Based on my own beach walking experience in that vicinity, his "crater" was probably a partially filled kettle sliced open by coastal erosion. If so, it would have shown signs of downward collapse on its flanks. Contemplating this feature may have been Thoreau's closest brush with understanding how Walden Pond was formed.[3]

Four Kettles

Depending on the height of the water table, what is officially known as "Walden Pond" is actually a coalesced lake created by the filling of four separate kettle basins with groundwater. Each basin was created when a discrete block of ice became isolated from the others during the final stage of glacial meltdown. If we raised the water table high enough, Lake Walden would coalesce with dozens of other nearby depressions before losing its identity within a vast, inland sea shaped like its antecedent, Glacial Lake Sudbury. If we dropped the water table low enough, Walden's main basin to the west would become a solitary oasis in a landscape so dry that the beds of all local rivers would be drained and dusty, like desert arroyos. Between these two extremes, the size and shape of Lake Walden is governed by the height of the water table against the irregular and porous topography of a collapsed glacial delta.[4]

In one of *Walden*'s blandest descriptions Thoreau wrote that Walden's "shore is irregular enough not to be monotonous." He is much more

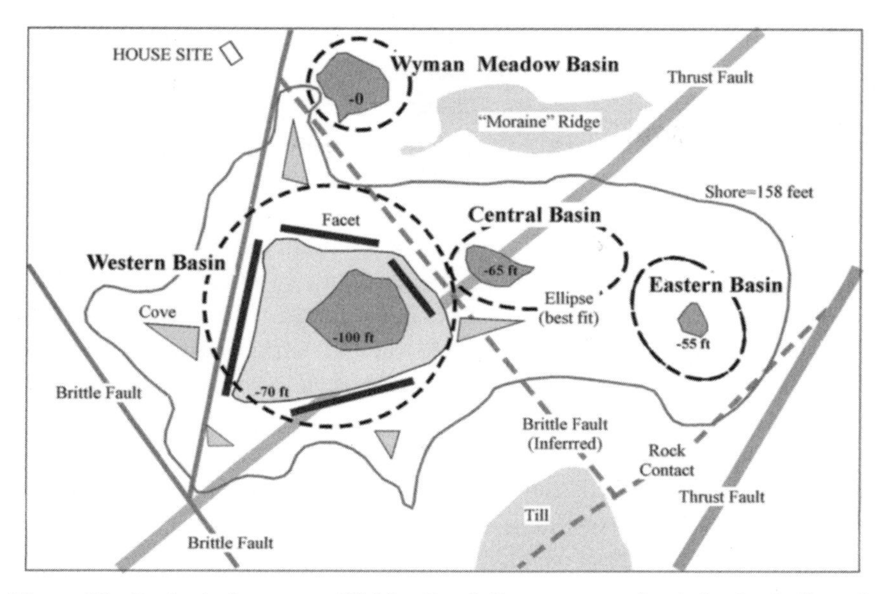

Figure 19. Geologic features at Walden Pond. Four separate kettle basins indicated by dashed ellipses, with their lowest closed contour shown in gray and depths indicated in feet. Triangles show the stylized locations of the floors of coves above their truncated terminations, as defined by adjacent contours. Heavy black lines show straight edges (facets) of inner basin interpreted as sediment-mantled submerged talus slopes, which generally parallel rock structure. "Moraine" ridge and Till "Island" rise above level of otherwise flat kame delta plain (white). Thick solid gray lines shows mapped thrust faults. Hachured dashed line shows a mapped rock contact within the Andover Granite. Solid and dashed thin gray lines show mapped and inferred block faults, respectively. Bathymetric contour of –70 feet outlines the polygonal inner basin midway through its depth.
Base and topography from Colman and Friesz, *Geohydrology*. Geology from Barosh, *Bedrock*.

specific later in the text, and even more so in his *Journal*. There, he differentiates the western shore "indented with deep bays," the "bolder northern shore," where collapse slopes are highest, and "the beautifully scalloped southern shore, where successive capes overlap each other." In this source passage he specifically differentiates the "graceful sweeping curve of the eastern" shore which, when seen from a boat to the west, struck him as an "amphitheatre for some kind of sylvan spectacle."[5]

This "amphitheatre" is the heavily wooded eastern rim of an elliptical kettle whose stabilized collapse slope is high, steep, and uniformly curved. Of Walden's four kettles, this one most closely resembles the "circular" kettles he saw on outer Cape Cod. The western margin of this eastern kettle, though submerged, is plainly revealed by the bathymetry. Below a depth of twenty-five feet, the *amphi*-theater becomes a *full*-theater in the round bordered on all sides by collapsed sand and gravel. At its bottom at minus 50 feet is a flat, muddy surface that, if drained, would make a fine central stage for the three act play told by its sedimentary archive at depth: collapse, stability, human impact. As with Walden's western basin, its bottom is flattest and deepest where the basin's length and width intersect. In other words, it, too, exhibits Thoreau's "Law of two Diameters."[6]

Adjacent to the eastern basin is the central basin. Though completely submerged today, it is plainly visible on the bathymetric map and is completely surrounded by collapsed sands and gravels, confirming its separate origin.

The largest, deepest, and most important of Walden's four kettle basins combines Thoreau's descriptions of his "western shore" and the "beautifully scalloped southern shore." At about forty acres, this western basin is the true twin of White Pond, also "about forty acres." Significantly, this was the only basin visible from Thoreau's house. Logistically, it was the one closest to Concord Village via the railroad tracks, the cart path behind his house, and the trail to the Bean field. My point? The vast majority of observations and written reflections of Walden the pond in *Walden* the book came from this basin alone. Unlike the three other kettles at Walden, the western basin has four roughly triangular coves radiating outward from its center. The apex of each resembles the point of a stylized star. Running clockwise are Deep Cove at 5:00, Railroad Cove at 7:00, Ice Fort Cove at 9:00, and Thoreau's Cove at 12:00 noon. A fifth cove, though completely submerged, nicks the central basin, as shown by the bathymetry at about 2:00.[7]

Below a depth of about sixty feet is a deep, central, inner basin. Above its flat bottom, this basin approximates the shape of an imperfect, low-angled, faceted cone with four straight segments joined by rounded corners. In map view, every contour line between minus 60 and minus 90 feet encloses a crude, four-sided polygon. Thoreau saw this from Heywood's Peak on a hazy, late-summer afternoon, describing it as: "a yet smoother and darker water, separated from the rest as if by an invisible cobweb, boom of the water nymphs, resting on it." Here, the words "cobweb" and "boom" prove that he saw either this submerged polygon, or some surrogate of it. The optical conditions allowing this memorable view were the exceptional clarity of the water at this time of year—due to minimum phytoplankton concentrations—and to an "autumnal

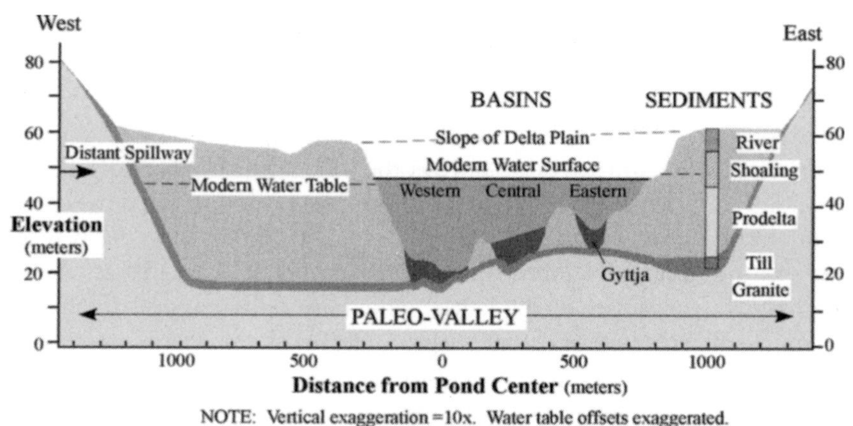

Figure 20. Geologic cross-section of Walden Pond.

Adapted from Figure 2 of Colman and Friesz, *Geohydrology,* 5. Line of section from their Figure 4, 8–9. Lake spillway elevation shown +2 meters above Koteff's reported elevation of 47 meters for the Cherry Brook spillway to account for channel stage, but with no correction for isostatic tilt. Delta sediments from borehole CTW214. "Shoaling" indicated by coarsening upward transition between foreset beds of the prodelta and topset beds (river) of the delta plain.

haze" that rendered the shoreline "more indistinct" and the "heavens more reflective near the center."[8]

More than five years earlier, Thoreau had found at the flat bottom of this inner basin an area "several acres more level than almost any field," where the depth varied less than "one foot in thirty rods," and where the "greatest depth was exactly one hundred and two feet." Using reflection seismology to "see" through the soft sediments beneath this surface, scientists have since added an additional twelve feet of depth to the basin. Below that, there is only a thin veneer of coarse glacial sediment above the rock surface, which appears quite flat. This surface is the glacier-scraped bedrock floor of the Walden paleo-valley, through which the ice was streaming before it gradually decelerated to the threshold of stagnation. Glacial polish, grooves, and striations are almost certainly present. Though the veneer of coarse sediment has not yet been sampled below the lake muck, it was characterized as "gravelly sand" and labeled as "till" in a borehole below park headquarters. Above this continuous layer of basal debris, but present only between the separate basins, are thick deposits of collapsed sand and gravel. This distribution requires that the individual kettles were created from separate residual ice blocks before meltdown was complete.[9]

Walden's western basin has a concentric radial symmetry. To illustrate this, I drew three circles, distorting them slightly into ellipses to improve the fit. The outer circle encloses the tips of the coves, which resemble five points of a stylized star. The middle circle follows the outer edge of the inner basin, which mirrors the shore at the base of the bold headlands and the mouths of the coves between them. On the headland portions of this perimeter, Thoreau found the bank to be "so steep that in many places a single leap will carry you into water over your head." These banks are the original collapse slopes of the kettle, which are only scarcely notched by wave action. The innermost of three circles in the western basin is the boundary between the steep facets of the inner basin and its flat floor. Importantly, all three circles nest inside one another symmetrically, and all surround the central nadir. This bull's-eye symmetry proves that a single, thick block melted inward and downward from all sides. Aside from their mouths, the morphology of the coves appears unrelated to the morphology of the deep inner basin, suggesting a separate origin during an earlier, higher, and more random stage of the meltdown process.[10]

The last of Walden's four kettle basins holds Wyman Meadow, named after "an old man, a potter, who lived by the pond before the Revolution." I suspect that a basal lens of silty clay near its edge gave him access to the raw material he needed. Depending on water level, this basin toggles back and forth between being a part of the confluent lake system, and being a separate pond covered by standing water. It is linked to Thoreau's Cove by a narrow passage during times of high water and by a small bar during times of low water, over which drains a trickle. During the summer of 1852, this kettle was a "secluded cove" of the larger lake where Thoreau caught "fish from a boat" far inland from the shore. By late 1853 it was "stagnant and drying up," with "dense fields of light-colored rattlesnake grass." During the thirty years I have been intermittently watching the treeless "Wyman Meadow," there has always been some standing water in it," qualifying it as a marsh. Thoreau concluded that this basin did "not properly belong" to the others. This is true for the criterion of continuous confluence, but false for the criterion of geological origin.[11]

"About the year 1824," the water level of the coalesced system was extremely low. During an outing with his family, Thoreau as a seven-year-old boy "helped boil a kettle of chowder, some six rods from the main shore," on a "narrow sand-bar running into" Thoreau's Cove. Imagining him standing there as a child reminded me of times I spent on sand bars with my own young sons, suggesting a thought experiment. I am there with young Henry, standing on that loose sand with a shovel and two blocks of ice, each the size and shape of a shoe box. We dig two holes in the loose dry sand: one about three feet deep and another barely deep enough to hold the block. We place one block in each hole, smooth the surface, and take a look before walking away. Both are identical. We return the following day to check them out. The outline of the deeply buried block is shaped like a broad elliptical bowl. The outline of the shallowly buried block looks like a box in profile, its slopes being straight and steep and its bottom flat near the middle. The deeply buried case parallels the rounded kettles of Eastham and Wellfleet, where we know that ice-block burial was deep and in uniform materials. The shallow case parallels the irregular kettles of nearby Brewster, which mirror the shape of larger, jagged stagnant blocks where the burial was thin, if present at all. The kettles at Walden were intermediate between these two cases.[12]

To sum up, Walden Pond is not a landform. Rather, it is a lake occupying four porous landforms at high water. Drop the water level a few feet, and Wyman Meadow is shunted aside as a marsh, bringing the count down to three. Drop the level an additional forty feet, and the count becomes two: a small lake above the eastern basin and a much larger one coalesced over the western and central basins. Wyman Meadow has become a seasonal vernal pool and dry hollow. Drop the level to minus 60 feet, and these two lakes become three, each in its own kettle. At this point, Lake Walden has become a chain of three ponds draining westward via groundwater, one into the other. This internal chain is a miniature version of the larger, external "Chain of Ponds" that links Walden to Fair Haven Bay in one direction and Goose Pond in the other. Controlling both chains is the Nashoba tectonic grain.[13]

Thoreau's Observatory

Walden opens with this line: "When I wrote the following pages, or rather the bulk of them, I lived alone in the woods, a mile from any neighbor, in a house which I had built myself, on the shore of Walden Pond." The "when," is unambiguous. Thoreau lived there from July 4, 1845, to September 6, 1847, a span of two years, two months, and two days. The "lived alone" is also unambiguous, based on multiple historical documents, though he clearly had plenty of visitors and went to town almost daily. The delightful ambiguity of "where" is my main interest here. In this single sentence, one of Walden's four basins becomes three places all at once: "In the woods, . . . in a house, and . . . on the shore."

The phrase "in the woods" is usually interpreted as a literary flag being waved—once in the title and forty additional times in the text—to emphasize Thoreau's societal remove, perhaps disingenuously, as Leon Edel suggests. Indeed, Prudence Ward did remark that his house was easily visible from the public road, and lay within a mile of the village, hardly the "grip and wild" of the Maine wilderness with its "stern or savage" forests. In fact, Thoreau's own commodity farm—an agrarian commune of one—lay just a few hundred feet away, and his "arrowy pines" were sprouting after a recent clear-cut for firewood within a "young forest of pitch pines and hickories." His "clearing" was large enough and sufficiently sunlit for open-meadow weeds, "johnswort and goldenrod," the former being a European invasive. The underbrush was thin, owing to the sterile dry soil, a far cry from the moss-covered dank of northern New England.[14]

The social ambiguities of the phrase "in the woods" are of no concern to this book because I have exempted this issue from my table of contents. With respect to forest hydrology and lake chemistry, however, Thoreau most certainly did live in the woods. Even with many of the trees cut down, the fibrous forest soil still effectively armored the surface against rainsplash and gullying, held loose particles in its root mesh, and facilitated infiltration through the acidic humus. And Walden's shore remained protected by its selvage of roots, more so than during the bungled twentieth-century engineering "improvements." A lake bordered by anything else—asphalt, tillage, pasture, condominiums—would have been a very different place, perhaps with an inflowing stream, a muddy shore, and murky water, respectively. Geologically, Walden was undoubtedly a woodland lake, even when shorn of its trees.

The phrase "on the shore" greatly exaggerates the proximity of Thoreau's house to the lake. It also contradicts his alternative claim in "Sounds" that his house was "half a dozen rods from the pond," or ninety-nine linear feet. And even this contradicts the distance based on Bronson Alcott's memory of the structure he helped raise, and on Roland Robbins's archaeological excavation a century after Thoreau's house-raising party. Both of their reconstructions put the house at more than a dozen rods away, or 204 feet. This is significantly farther inland from the shore than is typical for the zoning setbacks for lakeside cabins today, suggest-

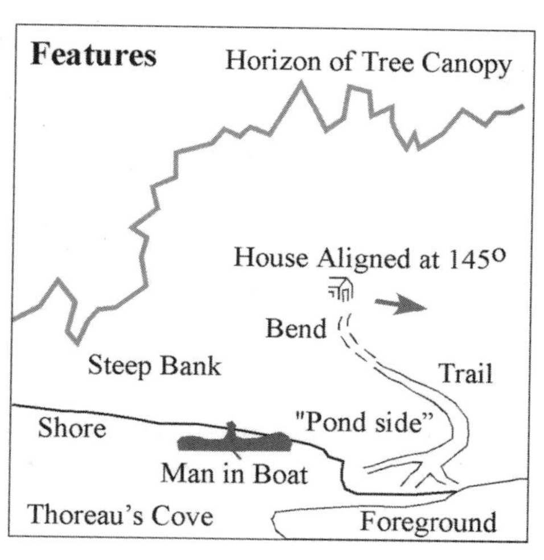

Figure 21. Drawing of Thoreau's "pond-side" by May Alcott. Diverging trails near the lake indicate Thoreau's intensive use of this outward shoreline convexity as his "pond side." Bend in the trail near house is consistent with Sophia Thoreau's sketch for the title page of *Walden*.
Courtesy Concord Free Public Library.

ing that building on the shore was not part of his plan, perhaps because it would attract too much attention.[15]

So, to what does "on the shore" refer? I assert that it was his default point of contact on the western side of Thoreau's Cove. By trail, this is about four hundred feet to the south-southeast and, by elevation, about forty feet lower than his doorstep. There, a small gravel bench curving outward from the otherwise straight edge of the cove served as a natural dock for Thoreau: a place to dip drinking water, launch a boat, and greet the morning. He pinpoints this "pond-side" in "Solitude" with a panoramic description of his "little world all to myself." Specifically: "I have my horizon bounded by woods all to myself; a distant view of the railroad where it touches the pond on the one hand, and of the fence which skirts the woodland road on the other." Only at as this precise location does the cone of vision he describes terminate at the tracks to the west and at the Lincoln Road to the east.[16]

This precise locus is strengthened by Thoreau's comment that "I had not lived there a week before my feet wore a path from my door to the pond-side." The beginning of this path is shown on his sister Sophia's sketch for *Walden*'s title page. It angles to the right of the doorway (southwest) before curving down the contour to the shore (south-southeast). The end of this path is shown on a pencil sketch by May Alcott. Emerging from the hillside woods, it branches into distributary trails on the convex-outward projection of the shore. This "dock" location, rather than one at the head of the cove, is supported by Thoreau's description of swimming "across the cove" during his ritual baths, rather than outward from its tip, where the shore is mushier, the air more stagnant, the brush closer in, and the view more constricted. Thus, Thoreau's word images of the lake in *Walden* would have been hyper-concentrated from this piece of shore, because this is where he kept his boat, and baptized him self twice per day in the summer. These activities require a double-crossing of the shore and a pause in between. Assuming this was indeed Thoreau's "pond-side," then his routine access to the lake took place on land owned by Cyrus Hubbard. This made him the "squatter" he claimed to be for his lakeside activities.[17]

Only by locating his house higher on the "side of a hill" and back from the shore was Thoreau able to build on land owned by Emerson, while also maintaining a southerly prospect across the main part of the lake. Though moving back contracted his cone of vision behind the prominent bluff to the east (where the sand bar attaches), it raised its height to an elevation where he could almost see the gently sloping tread of the ancient delta terrace across the lake.[18]

Finally, we consider the phrase "in a house." Economically his domicile was a *hut* or *shack,* built inexpensively for a temporary abode. Historically, it was a *shanty* or *shantee,* something built for a temporary purpose, in this case, for an experiment lasting two years. Spiritually, it was a *lodge* for inviting woodland visions. Artistically, it was a *studio* or *home office,* where the solitude for reading and writing could be sustained: Ellery Channing called it a "wooden inkstand." Socially, it was a *cottage,* given its pastoral remove from the bustle of town. Metaphorically, it was a *cabin,* not a rude log structure, but a ship's cabin for an extended voyage to navigate the "Atlantic and Pacific" oceans of his own consciousness. Architecturally, it was a *house,* complete with a cellar dug seven feet deep into sand, foundation, brick fireplace, casement windows, a tight-fitting door, wooden siding, shingling, and a lath-plastered interior. Thoreau clearly wanted his readers to know he was living in a conventional house like theirs, but one shrunk down to an appropriate size. He deliberately misleads us with his pioneering word imagery of felling trees with an axe. The timber he cut was used only for framing the structure: posts, beams, rafters, and studs. Logs were never part of the deal.

The frame building Thoreau constructed was indeed all of these—hut, shack, shanty, lodge, studio, office, cottage, cabin, house—plus one more. For the purposes of this book, it was a scientific observatory. *Meteorologically*, it lay just above a watery cove indented into steep woodland. This would have amplified slight changes in air pressure over the water, converting them to fragrant breezes nearly as sensitive as those on the water. *Acoustically*, it was backed up by a wall of trees and located part way down the south-facing slope, which muffled sounds coming from the northerly direction of town. It fronted a stable layer of cool air above the water, which enhanced the transmission of sound from that direction, especially from the railroad. And living near the tip of a triangular cove was analogous to listening through an "ear trumpet" because it magnified far-away sounds like church bells. *Optically*, the longest line across the western basin coincides with its deepest point, and nearly bisects his cone of vision. *Astronomically*, the zenith of the sun at high noon lay just west of this bisecting view over the pond.[19]

The doorway of Thoreau's observatory contracted and focused his images like the frame of a camera: "Though the view from my door was still more contracted, I did not feel crowded or confined in the least. There was pasture enough for my imagination." By my count, fourteen snapshots within *Walden* take place through this visual frame, one of which, ironically, begins his chapter "Sounds."[20]

> Sometimes, in a summer morning, having taken my accustomed bath, I sat in my sunny doorway from sunrise till noon, rapt in a revelry, amidst the pines and hickories and sumachs, in undisturbed solitude and stillness, while the birds sing around or flitted noiseless through the house, until by the sun falling in at my west window, or the noise of some traveller's wagon on the distant highway, I was reminded of the lapse of time.

Had Thoreau been at the threshold of his house, he could not have seen any part of the house interior, including the first flash of light through its "west window." Indeed, the vertical, south-pointing corner of his house acted like the style of a sundial, preventing light from falling on his writing desk until exactly noon.[21]

Based on the compass bearings of Thoreau's chimney foundation taken by Roland Robbins, the house doorway faced southeast with a bearing of 145°. Within measurement error, this azimuth aligns with the eastern edge of Thoreau's cone of vision over the lake and with the eastern facet of its deep basin defined by its submerged talus slope. In turn, both of these are likely consequences of the rock structure at depth, probably a post-Acadian block fault running parallel to the railroad alignment. When the rising sun reached this line of sight—as seen through the constriction of the doorway—the astronomical, glacial, and bedrock features were triple-aligned. This may have been coincidence. But it is also plausible that Thoreau, being so gifted at spatial thinking, worked this out ahead of time. Regardless, *Walden*'s final line, "The sun is but a morning star," has special meaning for this azimuth of 145°.[22]

Ice Block Meltdown

It's no accident that Walden's western basin is the deepest part of the deepest lake in the state. Geometrically, the northwest-facing slope of the bedrock hill of Emerson's Cliff would have helped maintain a stable southeastern edge of the block during deglaciation. Stratigraphically, the eastern edge of the paleo-valley below Smith's Hill and Pine Hill would have shunted meltwater sediment above the block, speeding up the rate at which the kame delta would have advanced. Thermally, slow melting was assured. Temperatures within the aquifer during the time of Glacial Lake Sudbury would have been close to freezing because the water table would have been graded to an icy shore only about ten feet below the delta plain, and greatly diminished the circulation of groundwater adjacent to the block.[23]

Early in the process, any sag on the surface of the delta plain above the melting blocks would have been filled by the high flux of sediment pouring toward Lake Sudbury as glacial outwash. This scenario is supported by the morphology and sediment texture of Henry's Bean Field, which occupies the base of a high-level, west-flowing, and flat-bottomed meltwater channel that Thoreau called a "level upland near Walden Road" . . ."strewn with little stones." In sequence, this broad channel extends straight across Thoreau's "frosty hollow," part of Wyman Meadow, and the northern tip of Thoreau's Cove before continuing northwesterly. These depressions must post-date channel flow because they are sunk below the base of the channel.[24]

As the ice margin retreated northward, the outwash sediment feeding the Walden delta was progressively shunted elsewhere. One year, the sag above melting ice would have gone unfilled. Before long, it would have been deep enough to pool water above seasonally frozen ground during the spring melt season. That water would have absorbed more solar heat than an equivalent presence of rock or soil. When the seasonally frozen ground thawed, this sun-warmed water would have been free to trickle downward, transferring its heat to the block at depth, causing melting, and creating the birth of a shallow subsidence crater. As it deepened, slumps and debris flows would have been directed radially inward toward the deepening and widening hole, especially in the western basin. At times, seepage pressures from shallow groundwater may have been high enough to cause headward sapping, perhaps inaugurating cove formation. Seepage pressures would have been especially high when Glacial Lake Sudbury abruptly drained. This dropped the water table by at least thirty feet throughout the aquifer, steepening its westerly gradient. This new regime would have strengthened the circulation of groundwater infiltrating through a greater thickness of sun-warmed gravel, thereby increasing the melting rate of buried ice. Deep collapse was inevitable.[25]

Bare earth. Flowing mud. Boulders toppling. Pebbles clattering. Cracks forming and healing. Gullies serrating the banks into badlands. The crescendo continued as the sediment-covered ice in the deep basin dropped like a piston being retracted downward. The coves, created earlier by some combination of collapse, seepage, and erosion, were chopped off from below, left to dangle like hanging valleys, now submerged. Between them were the higher bluffs, which kept lengthening beneath the water as collapse continued. Surges of runoff sent plumes of sediment into the water-filled part of the void. Turbidity currents hugged the bottom. Gray swirls of suspended silt and clay migrated away from the shore into open water. Eventually, the meltdown crescendo gave way to a decrescendo, as thickening sediment helped insulate the last remaining ice and as the circulation became more stagnant. Inevitably, all four kettles ran out of ice: Wyman Meadow first, and then the central, eastern, and western basins in sequence. Had the block in the western basin been thicker, the encroaching talus slopes on its four sides might have converged to finish their faceted cone.[26]

"Ere long," Thoreau mused, "not only on these banks, but on every hill and plain and in every hollow, the frost comes out of the ground . . . Thaw with his gentle persuasion is more powerful than Thor with his hammer. The one melts. The other but breaks to pieces." This description fits the "Spring" of Walden's creation. On the far side of the geochronology, varve dates indicate that this took place roughly 17,000 years ago. On the near side are a variety of so-called "bog-bottom" dates of kettle swamps and ponds in eastern Massachusetts, which reach a maximum age about 15,000 years. The final stages of collapse likely occurred near the end of this two thousand-year interval, making it a good round estimate to use. This later date is also supported by the morphology of nearby Fairhaven Bay, a similar sized kettle. Because it remained unfilled with sediment, meltdown there must have occurred after pioneering tundra vegetation stabilized the watershed. This suggests that meltdown took a millennium or more, rather than centuries.[27]

A Lake Is Born

During the first few centuries after deglaciation, almost every patch of land surrounding Walden Pond would have been producing loose sediment. With hardly a sprig of grass and no soil mulch, the ground froze deeply each winter to a condition bordering on permafrost. Drifted snow and frozen slush lay thickly in protected hollows. The wet snows and drenching rainstorms of March–April would have caused intense runoff over a surface that was simultaneously being heaved by frost and blasted by wind. Sheetwash, rill erosion, and gullying would have been chronic. Fans of sediment would have spread out below every growing ravine.

Though Concord had been liberated, the Laurentide Ice Sheet remained a looming presence to the north. Its great dome remained over eastern Canada while its margin lingered in northern New England. During summers, a thin layer of cold, dry air was constantly being chilled by surface conduction. Being denser than the ambient atmosphere, it flowed downward and southward over the snow-covered ice sheet while simultaneously being steered to the right by the Coriolis force. This created a potent summer anticyclone with wind speeds commonly in excess of twenty-five miles per hour. The air dried out as it descended, blasting loose sediment beyond its edge with gusts like that of March days in New England when the dust swirls over the daffodils and the fire danger is surprisingly high. At such times, wrote Thoreau, "there is the least possible moisture in the atmosphere, all being dried up, or congealed" and "the earth has to some extent frozen dry."[28]

The anticyclone diminished in power as the edge pulled northward and the dome descended. This allowed greater penetration by the prevailing westerly winds blowing eastward over the braided outwash plains of the Mississippi, Ohio, and Susquehanna River systems where dunes and drapes of dust up to a hundred feet thick were being deposited. Dunes formed throughout New England at this time as well, usually near the leeward margins of large glacial lakes, whose flat deltaic shorelines provided an abundant source of sand. Much of the loamy coarse dust mantling the soils of Walden Woods today came from such local sources. Thoreau spotted actual sand dunes east of Concord Village, "along Peter's path," hills that reminded him "a little of the downs of Cape Cod, of the Plains of Nauset."[29]

Regardless of direction, these stronger deglacial winds would have brought stronger waves to early Lake Walden. Sweeping across the treeless delta plain with greater force, they would have gripped the water surface more tightly, creating higher waves than those of today. These would have swashed the shore more vigorously, undercut the banks more deeply, initiated slumping more frequently, and kept the sediment coming. Additionally, ice on the pond was probably twice as thick than that of Thoreau's era, meaning it would have gouged the unprotected shore with greater force. After a few centuries of exposure to this wave lashing and ice-thrusting, a lag of stones would have developed within the shoreline's vertical range, thereafter limiting the sediment input. The matrix sand would have been pulled outward, mantling wave-cut shallows. So little sand remains that Thoreau had to import it from across the pond in order to plaster his house, likely from the western shore where it was imported during railroad construction.[30]

Gradually, the odds tipped in favor of hardy pioneering vegetation. Lichens began to dot a moonscape of stone. Tufts of grass poked from protected sites. Prostrate tundra species like those of the heath family—cranberry, blueberry, bearberry, Labrador tea, wintergreen—and shrubs of willow, alder, and birch moved in on open dry surfaces. Horsetails, mosses, and sedges pioneered the shore. Diatoms—one-celled phytoplankton—pioneered the cold and often muddy water. Colonizing vegetation on its bordering slopes began to buffer the extreme exposure to windstorms, rain, shoreline waves, and frost. Gullies healed. Slumps were anchored. The surface was armored. Organic residues of dead leaves and twigs decomposed through the action of microbes and fungi, thickening the humus and increasing the soil's ability to hold moisture. The roots of plants, now well nourished, wove a mesh through the subsoil, creating a protective sod. The burrowing of organisms, especially ants, loosened and concentrated granular materials below the root layer, making it more pervious to infiltrating water, and began the process of burying the larger stones. Patch by patch, the highly visible surface wash processes of the earlier stage gave way to less visible creep processes operating within the soil, rather than above it. Slope angularity gave way to hill curvature. The sharp brinks surrounding each kettle collapse became convex, the base of their talus slopes concave. Walden Woods assumed its present topographic form.

During this transition, any glacial grit that reached Lake Walden, regardless of source—distant wind, bank collapse, wave erosion, surface wash—was suspended in the water before settling to the bottom. Gradually, all four of Walden's kettles became sealed with a layer of mineral sediment shaped like a contact lens, being thickest and sandiest at the base, silty near the middle, and clay-rich near the tapering edges. The vernacular term "clay pit" often refers to excavations into such layers, which are almost always present beneath hollows, making them responsible for many vernal pools. Thoreau noted several on Cape Cod that perched rainwater well above local aquifers. As this clay stratum thickened in Walden's basins, it also rose upward on the margins, sealing away potential groundwater exchanges at depth and forcing flow to a higher level. This concentrated the springs and zones of seepage to near the zone of wave action, helping to tighten the link between aquatic and groundwater regimes.[31]

Physically, Lake Walden was cleared of mud. Metaphorically, "God's Drop" was "clarified in his thought." By about 13,000 years ago, the clarity of Walden water would have reached that of a high mountain lake because the grit was gone and the biological nutrient in the soil remained low, thereby limiting the concentration of phytoplankton at the base of the food chain.[32]

Secrets in the Muck

Pools, ponds, and lakes occupy closed topographic depressions on the earth's landscape. If one is not in the process of being created, it is in the process of being filled, usually with organic matter: residues from the land, plankton from open water, and the remains of animals and microbes that live on the bottom. In lakes, everything eventually settles downward, "leveling the inequalities," as Thoreau put it, to create a bottom "as flat as any level field." Unless decomposed, this material steadily thickens upward from the soupy muck of deep water to the soft shallows of protected bays where emergent aquatic vegetation takes root, for example lily pads, wild rice, and rushes. Gradually, even these shallows are shoaled by organic detritus, giving rise to the fibrous peat of marshes, bogs, and fens. These, in turn, yield to the woody peat of swamps before "terrestrialization" is complete.[33]

The downside to losing a lake to deposition is gaining an archive of aquatic memory via sedimentation. Present in all four basins of Walden today is the pollen that drifted in from the cultivated meadows of colonial times, the soot from the wood-burning locomotive blown easterly over the pond, and the silt washed from the shoreline engineering of the early twentieth century. Though such deposits are continuous in deep water, what they archive is highly selective. No smells are preserved. No laughter. No sunsets. What remains are the physical, chemical, and anatomical residues of whatever ended up at the bottom and escaped being consumed by benthic life.

Thoreau explored this organic archive "as if it were water," using a straight sapling, peeled and sharpened in order to push it through the woody peat of swamps and bogs. He also examined the fossils it contained. On one occasion, an "irishman digging" in Moore's Swamp showed him "small stumps,—larch methinks,—which he dug and cut out from the bottom of the ditch,—very old ones." Larch was very prevalent during the interglacial transition, when kettle hollows were less fertile and bog-bottoms were colder. Ditching at Dodges Brook produced even more diagnostic fossils from greater depth, in this case, "spruce logs as big as his leg, which the beavers had gnawed, with the marks of their teeth very distinct upon them." These came from "nine feet deep," and "soon crumbled away on coming to the air," indicating their great antiquity. Both cases provided clear evidence for one vanished ecosystem giving rise to another. This was a familiar refrain for Thoreau who noted in the second *Journal* entry of his life, that, "every part of nature, teaches us that the passing away of one life is the making room for another."[34]

The flat patch of muck at the bottom of Walden's deep hole was first sampled in 1939 by the limnologist Edward Deevey, who called it "typical ooze," meaning the residue of dead plankton raining down to the bottom. When he washed and inspected the sample, he found a high proportion of fragments from the leaves of deciduous forest trees, which makes sense to anyone who has visited in the pond in mid-autumn, when leaves float away from the shore. Such leaves gradually decompose with depth, leaving softer material with varying amounts of identifiable tissue called muck by swimmers, detritus by ecologists, and *gyttja* by palynologists (pollen scientists). This has been seen and photographed by scuba divers who have descended there with floodlights to reveal its dull colors, the gray, black, and dark olive green of chemical reduction.[35]

Such lake bottoms integrate everything reaching it. Falling in from the shore are the leaves, wood, and seeds of terrestrial vegetation that become waterlogged and sink. Blowing in from the land are pollen, spores, dust, and the residues of burning. Washing in from the shores are mineral soil, humus, and charcoal. Generated within the aquatic ecosystem are plankton and macroscopic organisms that sink when they die. Arriving fastest were heavy objects like Thoreau's allegorical "fifty-six," a cannonball weighing that much in pounds. Passing most slowly are the tiniest flecks of organic tissue with near-neutral buoyancy, for example, a red mite, the down of a goose feather, a pollen grain, or a single diatom. These might take weeks to settle, depending on the time of year. Descending at intermediate speeds are flat stones skipped on the surface by children, mineral grains of silt from distant dust, chunks of organic matter like a dead minnow, and the charcoal from firebrands tossed by the Thoreau brothers who made "a fire close to the water's edge, . . . and . . . far in the night, threw the burning brands high into the air like skyrockets, which, coming down into the pond, were quenched with a loud hissing."[36]

Given sufficient oxygen, anything organic arriving at the sediment interface is decomposed back to the elements and recycled within the lake. Thoreau witnessed this exchange of state: "I remember that when I first looked into these depths there were many large trunks to be seen indistinctly lying on the bottom." This was likely in May 1841. "But now they have mostly disappeared," he wrote thirteen years later in the 1854 *Walden*. Though oxygen is usually available on the bottom, it does not reach more than an inch or two into the muck. The lowest penetration of this gas is where ecology passes the baton to geology. From there on down, there is nothing but anaerobic sediment becoming firmer and firmer with depth. Under the right conditions, it could become coal or black shale.[37]

Postglacial *gyttja* has been a gold mine for natural scientists wishing to reconstruct the past from its remains. The most widely studied evidence comes from pollen. Learning to identify the different pollen grains is hard enough. Much harder is interpreting the results, because the abundance of pollen in lake sediment is a function of many variables. But the basic idea is pretty simple. Take a sample from the core. Date it with radiocarbon or some other means. Concentrate its resistant pollen by chemical digestion. Identify and count the taxa present. At this stage, each sample provides a reliable, albeit blunt snapshot of past vegetation integrated at the spatial scale of a town and the temporal scale of years to centuries, depending on the sedimentation rate and conditions. When two such snapshots are arranged in sequence, they tell a very short story. When that story is lengthened by other samples and embellished with other kinds of data from the same core—diatoms, isotopes, charcoal, mineral sediment, etc.—the result can be as rich as conventional history based on documents.[38]

Perhaps because the lake is so deep, there is no deep sediment core from Walden Pond. Using proper caution, however, paleo-limnologist Marjorie Winkler extrapolated records from nearby sites to Walden Pond in order to narrate its environmental history for the sixteen millennia following glaciation. The story begins with bare ground signified by the "clay pit" sediment at the base of all cores. Treeless tundra dominated by a mixture of sedge, grass, and hardy herbs such as *Hudsonia,* invaded by about 15,000 years ago, with arctic willow and stunted *(krumhholz)* spruce surviving in the most protected sites. It was a windy, cold, and open landscape, one almost certainly visited by mastodons, caribou, and their predators.[39]

By 14,000 years ago, spruce had expanded northward from refugia south of the ice sheet limit to create almost savannah-like parklands. Later, true boreal forests arrived, blending in complicated ways: spruce, jack pine, alder, and birch. By about 12,000 years ago, the spruce was permanently replaced by thick stands of white and jack pine, which had competitive advantage in a now warmer, drier climate, one in which summer sunlight was stronger than that of today. The result was the first genuine closed canopy forest, a northern evergreen forest with mixtures of white oak. By about 11,000 years ago, continued warming and drying, especially during summers, allowed pitch pine (which had replaced jack pine) and scrub oak to dominate in what could be called a "barren," with pockets of hemlock in shaded slopes and white cedar in lowlands swamps. Near the middle of our present interglaciation, between about 8,000–4,000 years ago, this savannah-like barren assemblage alternated with the more northerly northern influence of pine-hemlock-hardwood. The last several millennia (before the twentieth century) showed indications of cooling and increased moisture, with beech, birch, and hemlock rising in abundance.[40]

Stones and Bones

The indigenous *Homo sapiens*—Nipmuck, Massachusetts, Mohegan, Pequot, Wampanoag, Narragansett, Abnaki, and Passamaquoddy—were nearly extinct in Thoreau's day, owing to disease, genocide, migration, intermarriage, and cultural absorption. Though each of the original tribes had different historic territories and disparate customs, all spoke closely related languages derived from the Algonquin tongue, were genetically similar, and left an archaeology that merged back in time. Thoreau called all of these people "Indians," a misnomer, now being gradually dropped from vernacular English.

His sense of cultural place was infused with their archaeological remains. Though he wrote that these were "people who have left so few traces of themselves," he was nonetheless an avid and skilled collector of their remains. "By the time of his death," wrote biographer Walter Harding, "he had accumulated a collection of nearly a hundred pieces, including axes, pestles, gouges, mortars, chisels, spear points, ornaments, and a large number of arrow points of varied patterns and materials." When combined with those of Benjamin Smith, other private collectors, and more recent professional archaeological surveys, these physical remains demonstrate human continuity near Walden Pond for at least 11,000 years, from the Paleoindian Period to the present day. Construction of Concord's Mill Dam on top of a derelict fishing weir, and the exhumation of buried skeletons during the digging of village foundations, prove the rule that cultures replace each other in naturally strategic spots.[41]

In addition to his voluminous reading about Native Americans, Thoreau's own archaeological interpretations were quite keen for a discipline that had not yet coalesced. He reasoned that Concord's prehistoric peoples were far "older than the written character in Persia." He used the distribution and concentration of artifacts in time and space to document that different human groups had replaced each other. He deduced how form, material, and function related to one another, for example a "spherical stone probably" being "an implement of war." On artifact variety: "Arrowheads are of every color and various forms and materials. On the significance of tool concentration: "Chips which were made in their manufacture are also found in large numbers wherever a lodge stood for any length of time." And on material source: "And these slivers are the surest indication of Indian ground, since the geologists tell us that this stone is not to be found in this vicinity." He was very technical in his lithic analysis: "an Indian hoe . . . slate stone four or five eights of an inch thick, semicuculus, eight inches one way by four or more the other, chipped down on the edges." He was strategic in his reconnaissance surveys, waiting until "every stone is washed bright in the rain" on soil "now exposed by the plow." He developed hypotheses about site location, "a sandy spot on the top of the hill where I prophesized that I should find traces of the Indians." He correctly explained stonework in streams as the remains of fishing weirs, and was careful with his excavations and site restorations.[42]

Archaeologist Shirley Blancke and historian Brian Donahue have reconstructed the local prehistoric record of human culture near Walden Pond. There are nine sites in Walden Woods, though none at the pond itself. And except for the famous clamshell bank on the Concord River, the nearby record is quite sparse. Upland sites lack stratigraphic context, having been churned within the soil and moved downhill by creep. Most were temporary way stations anyway, places where someone stopped to shape a tool, lost something while spending the night, or perhaps inadvertently left a small cache. The dearth of stone artifacts near Walden says very little about whatever organic artifacts might have been present, because these would have been quickly disappeared in the acidic, oxygenated soils and waters of Walden Woods.[43]

The clamshell bank or Concord Shell Heap site is the closest thing prehistoric Concord has to a pharaoh's tomb, it being the only inland shell midden in New England. As early as 1832, Lemuel Shattuck had noticed "many hatchets, pipes, chisels, arrow-heads, and other rude specimens of their art, curiously wrought from stone, . . . an evidence of the existence and skill of the original inhabitants." Only there did Thoreau provide detailed information about his own findings. Combined with twentieth-century excavations, they suggest a kitchen waste dump where food residues were tossed by people who camped on the adjacent well-drained surface. The midden dates to 4,500 years ago, during an interval known by archaeologists as the Late Archaic Period, a time when the regional pollen record indicates a pine-oak forest similar to that of today. Clearly, the meadows of the Musketaquid—then and now—had far more to offer hungry people than the droughty upland sands of Walden Woods.[44]

This Late Archaic Period (6,000–2,500 years ago) was one of two intervals when natives seem to have been widely present in the region, and when sufficiently concentrated to make cemeteries. Its people traded widely, both the material used to make tools, and the ideas of how to work that material, yielding three recognized styles of projectile points that changed through time. The other interval was the Late Woodland Period, which began about the time of Viking exploration at about 1000 CE. Sites from this interval suggest larger populations supported by horticulture; corn, beans, and maize; and food storage containers made of crude stoneware ceramics. As with the late Archaic Period, Late Woodland Period settlement was largely restricted to the banks of the Musketaquid and its tributaries, with its marshes being the main draw.

The only indication of native peoples at Walden Pond claimed by Thoreau was its faint perimeter trail. Given its sterile soil and ascetic waters, there were few, if any, beaver, muskrat, mussels or fish. Potential food resources were lower than at nearby Flint's Pond, and far lower than those of the river. And given Walden's large surface area, waterfowl could evade hunters by resting away from shore. Walden's utility to Native America may have been little more than a beautiful drinking fountain between more useful places. During some centuries, human noises may have been completely absent from Walden Pond. During most, however, there were likely the occasional sounds of hunters passing through, women singing as they foraged, and children being children.

Colonial Outback

As much as Thoreau yearned for wilderness as an ideal, he preferred the settled country of post-colonial Concord. Returning from the Maine Woods in 1846, he found "it was a relief to get back to our smooth, but still varied landscape." Wilderness, he had learned, was but the "raw material of all our civilization . . . simple almost to barrenness." Conversion from that primeval condition was something he lauded up to a point. Concord's began in 1635 when the first settlers arrived to take advantage of the meadows, where freshwater hay grew in abundance and where the alluvial soils were fertile. Thoreau, who built his own house at the site of a woodchuck burrow, appreciated that the original colonists had done the same: "they burrow[ed] themselves into the earth for their first shelter under some hill-side."[45]

To him, these pioneers were "greater men than Homer or Chaucer, or Shakespeare, . . . rude and sturdy, experienced and wise men, keeping their castles, or teaming up their summer's wood, or chopping alone in the woods, . . . clearing, and burning, and scratching, and harrowing, and plowing, and subsoiling." Through such labor "he plants a town. He rudely bridged the stream and drove his team afield into the river meadows, cut the wild grass, and laid bare the homes of beaver, otter, muskrat, and with the whetting of his scythe scared off the deer and bear. He set up a mill and fields of English grain sprange in the virgin soil."[46]

Thoreau did not feel the same way about the farmers of his own era. Intensive commercial agriculture struck him as a blight: the "dust flies from . . . harrows across the field. The tearing, toothed harrow and the ponderous cylinder, which goes creaking and rumbling over the surface, heard afar, and vying with the sphere." Overgrazing by sheep reactivated ice-age dunes in a nineteenth-century version of the dust bowl. He resented the tradeoff of a "bare, ugly wall for an interesting grove." By the Walden years, some Concord lands had become so depleted that they were abandoned, leaving "well-preserved walls running straight through the midst of high and old woods, built, of course, when the soil was cultivated many years ago." He saw the regrowth of forest lands in Concord as a good thing, "a beautiful garden and boundless plantations of trees and shrubs" when compared to clear-cut Haverhill.[47]

The Walden years coincided with maximum Euro-American deforestation in Concord. Yet within this largely converted landscape, Walden Pond had somehow remained fairly unmolested, despite its having been named "as early as 1653, and apparently 1652." He understood why this "outback" was left largely alone: "there were no natural advantages,—no water privileges," and "the sterile soil" kept Walden from "low-land degeneracy." In contrast, the shores of more distant Flint's Pond were farmed because its soils were richer, being on silt-rich hardpan, rather than dry-washed pebbly sand. Historian Brian Donahue confirmed Walden's "outback" status with his exhaustive study of early settlement land records, adding that its "deposits are so coarse and droughty that they are virtually worthless for agriculture." For mainstream village culture, Walden offered little more than a place to hunt and fish, cut some fuel, or dig for clay. Yet even this was enough to dramatically impact the ecology, as Thoreau so clearly recorded in *Walden* as the extinction of bears and moose, deer and eagles. No longer was it "all alive with ducks and other water-fowl," as reported by "an old man who used to frequent this pond nearly sixty years ago, when it was dark with surrounding forests."[48]

Thoreau described three other outbacks: the "great wild tract . . . [of] . . . Easterbrooks Country? . . . A second great uninhabited place is that on the Marlborough road, . . . [and] the Great Fields." To these he added more than a dozen places still in their near-primitive states. The unnamed stream, a "recess apparently never frequented . . . [which] . . . flowed here a thousand years ago, and with exactly these environments. It is a few rods of primitive wood, such as the bear and the deer beheld." The swamps: "C. Mile's blueberry swamp, never cultivated by any"; and "Beck Stow's Swamp . . . deep and impenetrable" despite its proximity to "town or city!" Upriver, he found "retired natural meadows . . . nearly in their primitive state . . . how this country looked (in one of its aspects) a thousand years ago." To the east lay "Channing's moor."[49]

These four outbacks and the little-used corridors between them—riverbanks, wetlands, and moraine ridges—summed up to an "extensive 'common', [where] certain savage liberties still prevail." Where "I can easily walk ten, fifteen, twenty, any number of miles, commencing at my own door, without going by any house . . . Such solitude!" All this in "the oldest inland town in New England." "Perhaps I do not meet so many men as I should have met three centuries ago, when the Indian hunter roamed these woods." "It is surprising how much room there is in nature." One could still "enjoy the retirement and solitude of an early settler," provided one kept to "his proper path." Indeed, "it is a great art to saunter."[50]

That Thoreau's Walden was a fairly pristine outback was elegantly confirmed by Marjorie Winkler who went to Walden Pond with the goal of reconstructing the details of vegetation disturbance during the last few centuries. And because she did not need a deep core, she sampled under fair-weather conditions over the side of a boat using the freeze-core technique, which obtains a sample of organic sludge by freezing it to the outside of a tube. Her core spans six hundred years of the postglacial record, from roughly 1200 CE to the present, an interval postdating the onset of maize agriculture by the local Woodland tribes and Viking exploration of the New World. A more recent core by Drte Koster and others offers greater detail, but for the time period of interest here, its results are broadly comparable. With their better chronology, however, they were able to show that, for the period prior to Euro-settlement disturbance, the rate of pond sedimentation at Walden was slowing, suggesting it was becoming increasingly acetic.[51]

Eleven inches below the top of Winkler's core, the sediment archive shows that the pollen dusting the lake on an annual basis changed from its stable background pattern, signaling the arrival of European mixed husbandry. Winkler's evidence includes the abrupt rise in ragweed pollen

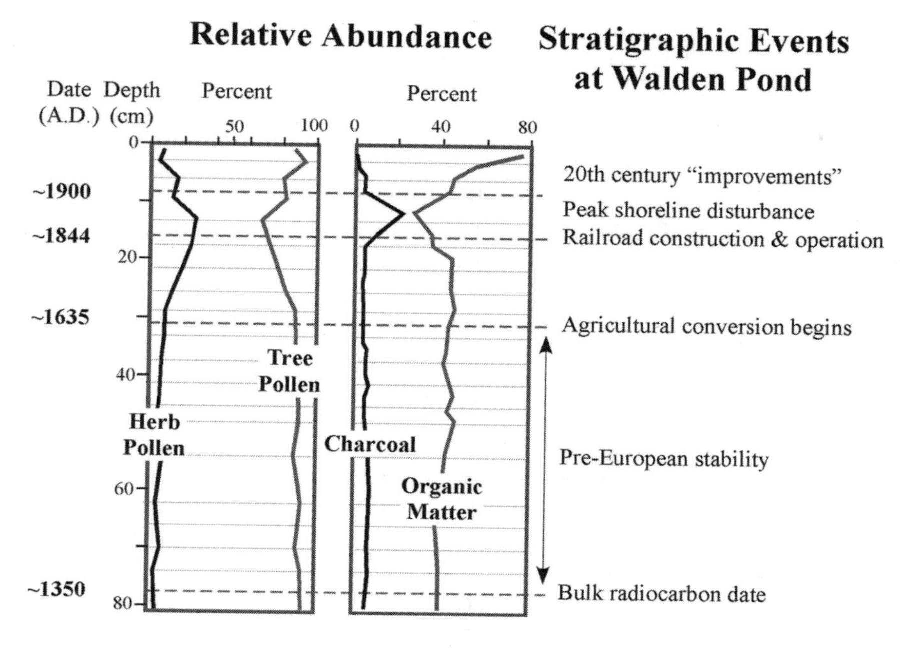

Figure 22. Recent sedimentation in Walden's deep hole. Recent changes in sedimentation at Walden are related to historical events. Top three dates defined by correlations between core stratigraphy and known events.
Redrawn from Figure 6 of Winkler, *Changes at Walden,* 206.

(which signaled land clearing and disturbance), a gradual rise in the abundance of herbaceous plants (pasture, hay, and grain) signaling "Farmers . . . rolling in their grain," and a decline in the relative abundance of tree pollen from the native woodland (deforestation). Everything else suggested a lake still "completely surrounded by thick and lofty pine and oak woods," which was Thoreau's description of the place when he "first paddled a boat" on its shores. This was the dreamscape lake of his boyhood, "one of the most ancient scenes stamped on the tablets of my memory . . . this recess among the pines, where almost sunshine and shadow were the only inhabitants that varied the scene."[52]

Six inches below the top of Winkler's core marks the beginning of the much more dramatic change that began in 1843 with construction and operation of the Fitchburg Railroad. This included the partial filling of one of its coves with loose sediment to create the causeway, likely "borrowed" from the fine-grained sediments of the Deep Cut. Some fraction of that grit became suspended in Walden's surface water, and some fraction of that fraction drifted out to the coring site to become part of an abrupt rise in mineral sedimentation. That same year Irish laborers built shanties along the shore. Their privy wastes and kitchen residues raised the nutrient load, shifting the planktonic sediment in recognizable ways. Simultaneous forest clearing, mostly for fuel wood, left stratigraphic markers of wash erosion such as charcoal and woody debris. The locomotive itself spewed ash and soot that became permanent components of the sedimentary archive in Walden's deep hole. Human beings had become the most important geological agent. The Anthropocene had arrived.

Philosopher Stanley Cavell insisted that we cannot "know what '*Walden*' means unless we know what Walden is." Thoreau made a serious commitment to learn how his lake behaved as a natural system. Those of us who appreciate his literature have a chance to follow his example by turning the page and exploring "what Walden is" in its natural state.[53]

INTERLUDE

6

THE WALDEN SYSTEM

Leo Marx's *The Machine in the Garden* is "the best book ever written about the place of nature in American literary thought." That was scholar Lawrence Buell's summary judgment of a title that elegantly captures the dialectic tension between industrial progress and the timeless beauty of nature: in this case between the Fitchburg Railroad and Walden Pond. Thoreau's pastoral remove there in 1845 took place midway between railroad construction in 1843 and his return to Concord village in 1847. During this four-year interval, the lake went from being an agricultural outback with little practical utility to a railhead where commodities like ice and timber were being extracted on an industrial scale and shipped to distant markets. The track cut straight steel lines through what had been a soft rolling landscape. The shrieking whistle of the engine and the clattering of cars put nature on a clockwork schedule. And the sediment archive in Walden's "deepest resort" began recording a dramatic pulse of anthropogenic change.[1]

Unfortunately, Marx's juxtaposition blinds us from seeing what Thoreau understood long before the iron horse came to his pond: that every object in nature is a "rivet in the machine of the universe," an *Organic Machine* with more power and stability than steaming locomotive. Conceptually, the nearest thing to an external machine intruding Thoreau's garden was the battery of factory-made tools tearing up his Bean Field: the steel axes to fell its trees, the steel chains to pull stumps, the steel plow to turn its earth, and the steel hoe Thoreau used to wage war against weeds.[2]

In contrast, breezy blue Walden and the fuming black engine were theoretically quite similar. Both were well-defined open systems and fine-tuned examples of steady state equilibrium: The lake idling in place and

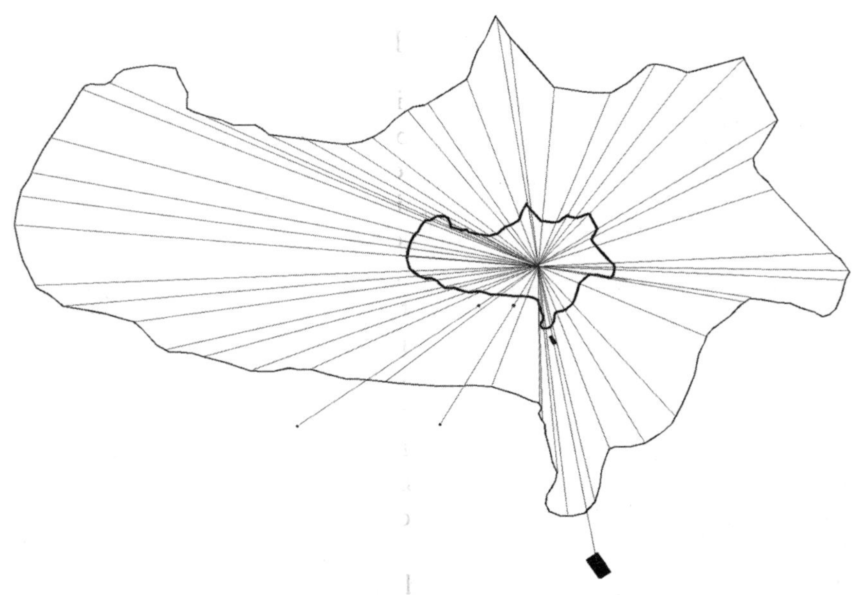

Figure 23. Radial plot used to draw Thoreau's "Reduced Plan" showing location of house (rectangle) and two local peaks.
Information adapted from Map 133c of online Thoreau Survey Collection, Special Collections, Concord Free Public Library.

the train rolling by at constant velocity. The cycle of the train matching the cycle of a day. The solar radiation powering them both, whether directly by radiation (sunlight) or indirectly by photosynthesis (firewood). The radial physical motion, whether the turn of a steel wheel on the track or by the circle of a nibbling minnow with the same silver sheen.

Author Bernhard Kuhn suggests that Thoreau's Walden was less a pastoral remove than "a site of engagement with the world." I adopt his approach, not because I am taking sides (I have exempted social history from my table of contents), but because the sole purpose of this chapter is to characterize the *site* of Thoreau's engagement before we move on to examine the *results* of that engagement in later chapters. Treating Lake Walden as I would any natural physical system, I begin by describing its physical boundaries, power supply, and spatial organization. After that, I identify its four main components of air, lake, land, and aquifer, take them apart, and put them back together again. I finish by examining the "rise and fall of the pond" as an integrated manifestation of steady state equilibrium controlled by interactions between these components.[3]

Conceptual View

At the most familiar level, the building blocks of Thoreau's Walden experiment—ant, pickerel, stone, and pine—are very different. Zooming in to the atomic level, however, they're nearly the same, all being composed of carbon, hydrogen, oxygen, nitrogen, phosphorous, and sulfur, straight from the periodic table of the elements. At this level, the iron in the black mica of the Andover Granite and in the blood of a black spider behave exactly the same. Zooming out to the scale of the lake, the same familiar objects—ant, pickerel, stone, pine—can be imagined as the "atoms" of something larger and more complex, here called the Walden System. Just as a stone is *one thing* composed of its minerals, so too is the Walden System *one thing* composed of its components, a discrete material entity at the scale of a mid-sized landform, in this case a bio-geo-chemical kettle. It may or may not have an ultimate purpose. If it does, Thoreau probably came closer to understanding it than anyone since. After all, he was the man who dipped his pen into ink and wrote, "The same sun which ripens my beans illumines at once a system of earths like ours."[4]

To understand the Walden System at the landform scale, we must first differentiate it from other systems associated with other Concord landforms. In April 1852, Thoreau recognized that the dry sandy stubble of "Peter's path" was "the only walk of the kind that we have in Concord." This is because it is the only significant patch of sand dunes in town, blown there by northwesterly winds from the shoreline of Glacial Lake Concord at a time when the buried ice at Walden was still melting down. Dune and kettle are separate systems. Ditto for ledge and swamp. Stream and cliff. Meadow and hill. Like different siblings in a family, all are discrete, yet all share the same glacial and tectonic inheritances.[5]

The glacial inheritance involves being covered and released by the same ice sheet on several separate occasions. The tectonic inheritance involves two aspects: one associated with making the rock (plate collision, rifting, tilting, and exhumation) and the other associated with making the climate (plate drifting to a specific latitude and longitude). At 42.43° north latitude, Concord has very distinct seasons. Winters cold enough to freeze Lake Walden toggle with summers hot enough to warm a lake to swimmable, though still bracing, temperatures. At 71.33° west longitude, Concord lies downwind from the prevailing westerlies and upwind from Atlantic moisture. Whirling gyres move generally eastward, whether high-pressure anticyclones from the north or low-pressure cyclones from the south. Chaotic interactions create day-to-day and week-to-week variations, whether the blue-sky invasions of dry arctic air or the drenching hurricanes of the oceanic subtropics. Transfers of heat between land and sea drive climate variations at scales up to a century in length.

All of Concord's discrete landform systems will eventually disappear, including Walden, because the long-term geological story on this passive continental margin is one of downward loss. Everything above sea level is being erased.

Returning to Thoreau's metaphor of the machine, he saw nature as an enchanted loom. At the scale of streambeds, its finer filigree cloth was the "woof and warp of ripples" running across the flow and streaked parallel to it, respectively. At the scale of Concord, its broader, coarser cloth was the perpendicular weave of the local trellis watershed, consisting of the woof of the cross-grain block faults and the warp of the main tectonic grain. At the middling scale of discrete landforms, the weave of Walden was like that of an orb-spinning spider. Its woof and warp were the rays and perimeters of Walden's western basin, a circular loom whose "fairy fingers" were busy turning phytoplankton into fish, rather than cotton into cloth. And seeds to songbirds. And scenery into literature.[6]

Powering the Walden System during Thoreau's residency were three energy sources. Most distant and ancient was the angular momentum of the local solar system, which pre-dates the ignition of our sun. The *Days of Henry Thoreau* were planetary rotations, his months lunar orbits, his years earthly revolutions and his geological cycles, Huttonian revolutions. Even the dust from his shooting stars followed the curved arcs of orbital gravity across the sky before nestling into Lake Walden, where "the water sleeps with stars in its bosom." The closest and most regular source of Walden's power is geothermal heat. It helped melt that stranded block of glacial ice about 0.00001 billion years ago to create its "deep and capacious" well. Even today, geothermal heat warms the lake's hypolimnion slightly during its winter hibernations and summer estivations.[7]

Walden's strongest source of power is solar radiation, which arrives after a seven-minute trip through the vacuum of space. It penetrates the atmosphere at visible and infrared wavelengths to illuminate the pond and adjacent land, respectively, and to warm it as well. Less directly, the stellar *astronomy* of the sun energizes Earth's atmosphere through differential absorption of radiant energy, causing differential heating. *Meteorology* then moves that heated air into and out of the Walden System, and within it as well: to flutter the leaves, ripple the lake, and import the gasses, from which the plants extract carbon, the animals oxygen, the soil water, and the waves physical momentum. *Hydrology* extracts liquid water from that air as condensed precipitation and then sends it to the soil via absorption, to the water table via infiltration, and to the open void of the lake via hydraulic pressure. *Limnology* then moves that standing water via sluggish thermal swirls, bottom-hugging density currents, and surface waves. *Ecology* then moves that water through food webs via the creation of organic matter by photosynthesis (catabolism), and its destruction by respiration (anabolism). All of this is a manifestation of solar power. Water is the key to the transfer between these components.[8]

When Walden was an infant lake about 13,000 years ago, the solar power was comparable to that of today, but the amount of that energy used by ecology was limited. This was due to the dearth of organic nutrients, which was due to the incipient state of soil formation, which was due to the still rigorous subarctic climate. By the time of the oldest known archaeological site, however, the climate had fully ameliorated, the soils had become well developed, the lake chemistry enriched, and the whole machine invigorated by life. By the time of Thoreau, the croaking bullfrogs, nibbling woodchucks, yelping foxes, and psalm-singing fishermen were all doing their thing using chemical fundamentals older than the Earth's crust.

In the air was the smell of dripping pine resin, the hum of flying insects, and the sight of dragonflies darting about. In the water these sights, smells, and sounds were muted and muffled: the ambush of a pickerel, the diving of a loon, or the rise of a turtle. At the shore, organic tissues from seemingly different worlds were blended together. Insect larvae squirming in the deep muck rose up by the millions each summer to pupate, fly, and mate. Caught in mid-air, they stuffed the bellies of swallows perching on the shoreline, which voided their waste nutrients down to the soil and thence back to the lake. "Part and parcel" of this system was Thoreau himself. Looking outward on the lake, he saw: "skater insects" use that solar energy to "produce the finest imaginable sparkle on it, or perchance, a duck plumes itself or . . . a swallow skims so low as to touch it . . . or a thistle-down" floats "on its surface, which the fishes dart at and so dimple again." Mechanistically, all of this activity is nothing but sunlight working with chemicals stewing in the kettle.[9]

The symmetry of Walden's organic machine is radial. Thoreau's experience began with the perfection of the sphere. Practically every molecule of lake water within it originated as a chemically pure, perfectly spherical droplet of liquid mist condensed from the emptiness of the sky. Only later, when enlarged to the size of a drop and falling through the atmosphere, did its purity change through chemical adsorption and its shape by friction. Place a single drop of that water on a glass microscope slide and it will collapse under its own weight to make a hemisphere. Add a few more drops, and that shape will flatten, except near the edge. Now invert that image in your mind. That radially symmetric microcosm crudely approximates the shape of Walden's western basin, being a flattened and truncated cone that, when rounded by time, became an imperfect hemisphere with rays and rings.

Thoreau used these rays to create one of the most potent visual images in all American literature, the undated "Reduced Plan" of his 1846 bathymetric survey. Beginning with a tracing of his larger map nearly two feet wide, he made a tiny pencil dot directly above the lake's "deepest resort," which lay above its focus. Outward in all directions from that geometric point of origin he drew fifty-one lines toward inflections on the shore. Next, he shortened each line to one quarter its original length, giving rise to fifty-one vectors radiating out from the center. Next, he connected the tips of these vectors—dot-to-dot—to make his map for the engraver. Finally, he plugged himself into that symmetry by adding his house and two of his favorite lookout points.[10]

Consider an ideal volcano like Japan's Mount Fuji, whose summit is the point of origin. Identical lines diverge outward and downward, their slopes diminishing with distance from the apex. Seen from every direction is a symmetrical pair of smooth, concave-skyward visual profiles that meet at the top where they are steepest. The opposite symmetry is true for the bottom of every ideal kettle. From each center, the profiles diverge outward and upward, their slopes increasing with distance toward the bank. When seen from any direction, a pair of concave-skyward profiles merge at the bottom to form a bowl. At some point above

Walden's shore the curvature of its bowl switches from being concave-upward to becoming convex-upward, thereafter flaring outward like the bell of a horned instrument until they reach the brink of the delta plain.

At a first approximation, all profiles drawn outward from Walden's western basin are similar in any compass direction. They are uniformly flat on the organic mud of the deep hole. They are equally flat in the pine-oak woods of the delta-plain at the top. Above the bottom, they steepen because the mud becomes sandier, making it stronger. Beyond that, the general pattern of steepening slopes followed by shallowing slopes is complicated by countless variations, making no two rays exactly alike.

Looking northward and moving clockwise, Wyman Meadow lies at 12:15. Thoreau's Bean Field is at 1:00. At 1:30 is a slope reaching nearly two hundred feet in height from the bottom of the deep to the top of the boulder-strewn moraine of Heywood's Peak, which crests at an elevation near 235 feet. Between 2:00 and 3:30 the rays extend beyond the western basin across the central and eastern basins to sweep the visitors parking area, the swimming beach, park headquarters, and the Thoreau Society's Shop at Walden Pond. At 5:00 the ray pointing toward Emerson's cliff crosses only a narrow remnant of the delta plain. At 8:00 is the railroad, where the tip of the cove was amputated by construction fill. At 10:00 lies the broadest expanse of the delta plain, the former site of the Lake Walden pavilion. At 11:30 is Thoreau's cove.[11]

Rings are the complements to rays. Most conspicuous are the cleanly washed stones of Walden's shore, where the combination of wave energy and intermittent submergence keep it mostly free of vegetation. Moving landward, there are "fluviatile trees next to the shore," which include willow and alder, followed by true pine-oak woods, the branches of which overhang the water. Next we encounter the steep bank, the upper convexity, and the terrace brink. Reversing direction from the shore is the wave-cut littoral zone of persistent shallow water where ripples mark the bed. Deeper still is the inner part of the littoral zone where leaves and organic detritus are being slowly decomposed by scavengers. Deeper still is a "leafy" ring of algae called *Nitella,* the "bright green weed" Thoreau "brought up on anchors even in midwinter," and which grows in water typically 20–40 feet deep. This lettuce-like alga interested him so much that he once consulted a specialist in Newburyport. Below this sunlit zone are slopes that shallow asymptotically down to the dark hole of the benthos.

In every direction are eight or more rings. Thoreau's compelling metaphor of the lake being "earth's eye" takes in six of them: pupil, iris, white, lid, lashes, and brow. These find their lacustrine equivalents in the open-water, littoral, and swash zones of the lake, and the terrestrial vegetation of the riparian bank, and brink. In Thoreau's language, these are the "deepest part," the "rippled margin," the "paved shore," the "fluvatile trees . . . shorn" by intermittent high water, the "bank," and the "woods."[12]

Four Realms

Given the complexity of the actual world, scientists have no choice but to break natural systems down into constituent pieces. Doing so with geographic places—cove, bank, terrace, beach—is comfortably familiar, but masks the flow of matter and energy moving between them. Doing so with academic disciplines crosses these physical boundaries more effectively, but at the cost of greater cognitive abstraction. Ecology, for example, is as aquatic as it is terrestrial: free to swim, fly, walk, wriggle, and burrow throughout the whole system. And though the largest organisms do favor certain habitats, their inputs and outputs cross all boundaries. Hydrology, of course, is ubiquitous, as is the case for its adjunct, microbiology. And there is no escaping chemistry and physics.

For the purpose of this book, I break the Walden System down into four material realms using two landforms, two fluids, and one solid. The two landforms are the *delta plain* and the *kettle* sunk into it. The two fluids are *air* and *water,* both of which seek the lowest level to which they can flow. Water, being denser, always displaces the air upward, whether in the larger void of the lake or the millions of voids in the aquifer. The only solid needed for defining the components is *earth,* the porous and permeable sand and gravel created from silica-rich rock by glacial crushing and meltwater washing.[13]

The horizontal boundary between air and water is easily seen inside the kettle as the flat surface of the lake. Above it is an enormous volume of air being temporarily held in the big pore of the basin, and being exchanged at the time scale of seconds to days, depending on the weather. Below it, and occupying the same big pore, is a much more conservative volume of liquid *water* that, like the air, is also being continuously exchanged, but at the scale of minutes to years. The air-water boundary inland from the edge of the kettle is the nearly horizontal water table. Below it, the billions of small pores are filled with water. Above it, simi-

GRAPHIC FORMULA

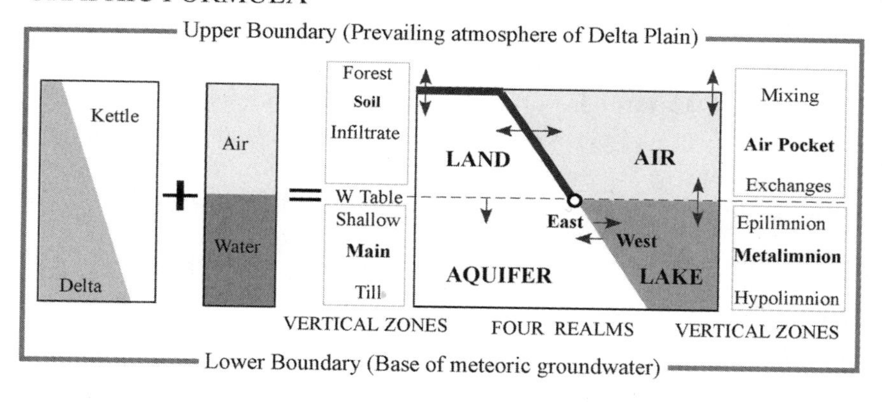

WORD EQUATION

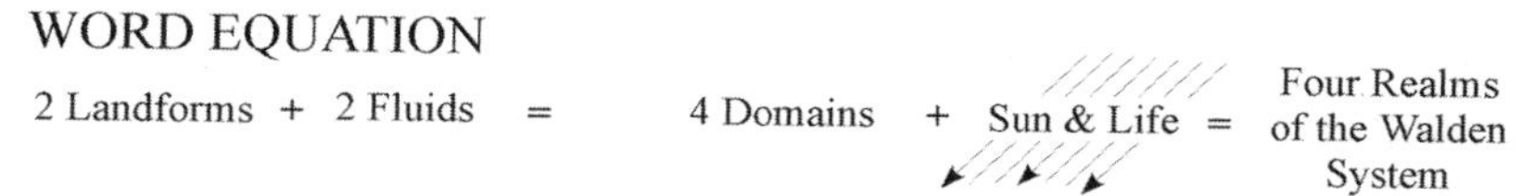

Figure 24. Graphic formula and word equation for the Walden System. In the idealized system of the western basin, each ray is identical across all four realms. See text for explanation.

lar pores are filled with air. Combining the horizontal gas-liquid boundary with the inclined boundary of the kettle wall partitions the Walden system into four physical *quadrants:* two below the surface of the water, and two above it. Add solar power, movement, and life, and these four quadrants enliven into four dynamic *realms.*

Easiest to define is the *lake,* the realm of standing water within the kettle, and the scientific domain of limnology. Above it is the realm of the *air,* Thoreau's "blue cauldron" and the domain of lake meteorology. The shared boundary between these realms is the "field of water" at the lake surface. Hardest to define is the realm called *land,* for lack of a better term. At the scale of the Walden system, it is the unsaturated *terra firma,* the three-dimensional volume where earth, soil, air, organic fiber, and transient liquids mingle together in complex ways, especially near the surface. Four scientific disciplines compete for this realm: ecology, hydrology, pedology (soils), and geomorphology (landforms). The final, simplest, and most conservative realm is the *aquifer,* which occupies the volume of earth below the water table, and is the scientific domain of groundwater hydrology. Little happens there except the slow, steady, invisible, and noiseless flow of infiltrated water through sandy pores toward and away from the kettle. Hydrologists call this the phreatic zone.[14]

The two dry realms of air and land share a common *top* boundary, the prevailing atmosphere of Concord and beyond. The two saturated realms of lake and aquifer share a common *bottom* boundary, which partitions the zone of seasonally recharged active groundwater flow (meteoric) from a lower zone (connate), where the flow cycles at the scale of centuries or longer, and therefore plays a negligible role in the system. Of the four realms, the lake is lowest and most central. And because Walden has an inward sloping bank from terrace tread to kettle bottom, the realms of the air and aquifer are larger in volume than those of lake and land for any vertical section.

Each realm has an optimal default condition. For the lake, it is the *metalimnion,* the mass of transitional water separating the cold, quiet, dark of the deep from the warm, wave-stirred, sunlit waters near the surface. For the air, it is the bulk of the air *pocket* above the complicated exchanges taking place near the lake surface and below the turbulent boundary with the prevailing atmosphere. For the land, the default condition is the soil *humus* separating the dominance of life processes above from the dominance of earthly processes below. For the aquifer, it is the *main mass* flowing toward the lake, traveling below transient pulses of recharge, yet above the transition to the connate zone.

Lake Walden lies in the namesake quadrant. Much of the radiation reaching it on a beautiful spring day is invisible to us, having arrived from the infrared part of the spectrum. When this wavelength strikes a molecule of water, the result is complete absorption and conversion into heat. To this is added the heat absorbed from visible light, especially in the red-orange-yellow range. Inevitably, the lake must warm in the upper foot or two where most of this energy conversion takes place. This warmed water, by virtue of its lower density, floats as a discrete layer above the colder water left over from the previous winter. As the heating season continues, wind mixes, but cannot destroy, this *epilimnion.* So instead it thickens to a depth of about twenty to thirty feet, and warms from a starting critical temperature of 39°F (4°C) to peak warmth up to about 77°F (25°C). When Thoreau mentioned that he had to travel elsewhere for his drinking water "almost every day in midsummer, when the

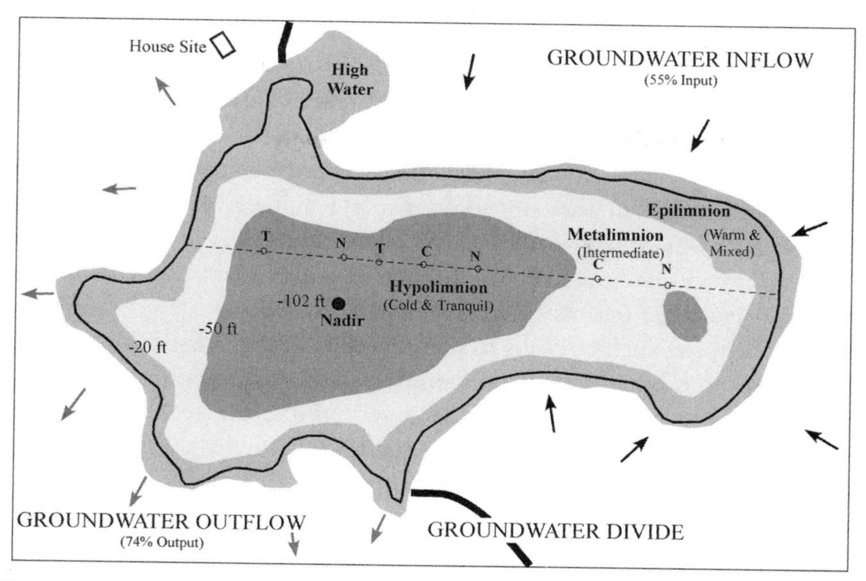

Figure 25. Hydrology and limnology of Walden Pond. Adapted from Colman and Friesz, *Geohydrology.* Epilimnion at high water shown by gray above normal stage, shown by black outline. The tops of the summer metalimnion and hypolimnion are approximated by the minus 20-foot and minus 50-foot contours, respectively. Black arrows (inflow) and gray arrows (outflow) show prevailing direction of groundwater flow on either side of a permanent divide (heavy black line). Dashed line shows in-lake segment of geological cross section. Circles show where it intersects the inner edges of submerged talus slopes (T), the deepest part of each basin on the line of section (N), and the crests between them (C).

pond was warmest," he was referring to the cove nearest his house, which is so shallow that it lies entirely within the upper layer. Meanwhile, at depth in the central basin, the cold water from the previous winter had scarcely warmed at all. This is still water of the *hypolimnion,* sent down the last time the surface water either cooled or warmed to its critical temperature.[15]

Thoreau described the freeze-up process during his first winter at the pond. "The north wind had already begun to cool the pond, though it took many weeks of steady blowing to accomplish it, it is so deep." That year, the enormous amount of heat gained during summer kept the pond surface warmer than the air until well into December. Only "when the nights get really cold, [do] the shadiest and shallowest coves" begin to skim over with ice "some days or even weeks before the general freezing."

Inevitably, the epilimnion reaches the critical temperature of 39° Fahrenheit, making it slightly denser than the stationary volume below it that went down the previous spring at the same temperature, but which has since been slightly warmed by geothermal, metabolic, frictional, and other sources of heat. At this threshold of buoyancy, the mass of cold, wind-stirred water at the top plunges to the bottom as a slow-motion continuous cascade known as the fall turnover. For at least a few days, all the water in his lake is swirled from top to bottom until everything reaches the critical temperature.[16]

Gradually, the surface chills to just above freezing (32–38°F), creating a thin layer that floats above the merely cold water at depth. Under calm conditions, the surface then glazes over with clear ice, which often breaks up or melts, only to reform again. But eventually, the new ice becomes an "icy shutter" too strong for the wind to break up, too massive to melt during a warm day, and thick enough to support snow. With

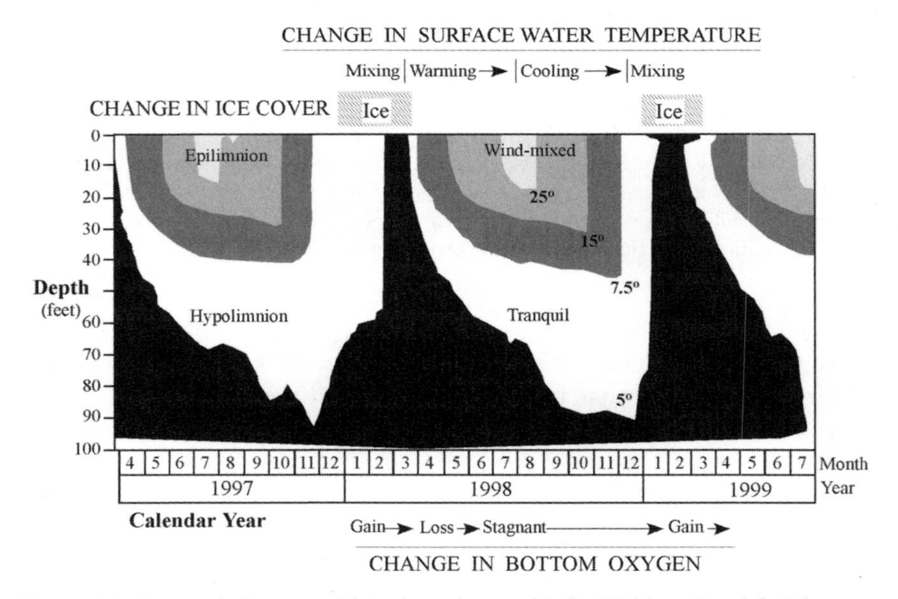

Figure 26. Seasonal changes within the volume of Lake Walden. Simplified from Figures 13 and 17 of Colman and Friesz, *Geohydrology,* 35, 41. Selected contours of water temperature through time show uniformly cold (< 5°C, black infill) conditions from top to bottom during times of mixing during late-fall freeze-up and late winter melt-off. Surface changes are dominated by warming, cooling, and thickening of upper mixed layer. Changes in tranquil bottom waters are dominated by a rise and fall of oxygen preceding 6–8 months of stagnation after midsummer.

the first snowfall, the lake "closes its eyelids for the winter," and darkens to gloom. Given the combination of thermal stability and near darkness, a floating layer of near-freezing water slowly thickens beneath the ice. The lake has become a thermal parfait of four layers of H_2O: snow, ice, near-freezing water, and merely cold water. From then on, the lake temperature remains virtually constant because additional cold is used to make ice. Of this, Thoreau wrote: "For four months in the year its water is as cold as it is pure at all times . . . I think that it is then as good as any, if not the best, in the town."[17]

With the coming warmth of spring, the snow melts to slush, and the ice thins from above. One day, the "eyelids" re-open. "This pond never breaks up so soon as the others in this neighborhood, on account both of its greater depth and its having no stream passing through it to melt or wear away the ice. . . . It commonly opens about the first of April, a week or ten days later than Flint's Pond and Fair Haven, beginning to melt on the north side and in the shallower parts where it began to freeze." The delay at Walden is due mainly to its proportionately greater mass of water relative to surface area, but also to the stability and shading within its deep, "embosomed" recess. Water melted by early thawing days becomes slush and usually quickly refreezes. But once the ice is gone, the sun and wind quickly work to warm the meltwater back to its critical temperature. Then, for the second time in a year, it descends as a continuous plume until the entire volume remixes to isothermal conditions. After this threshold, any additional heat begins to thicken the summer layer. The deep water is now isolated and motionless except when disturbed by life. Between the dynamic upper layer and the stagnant bottom water is a clearly defined zone of transition, primarily in temperature, but also of chemistry, appropriately the *metalimnion.*

Local complexities on this general pattern abound. The southern coves freeze first because they are most shaded. The northern ones thaw first because they are most sunlit. In spring, a moat of thawed water surrounds the entire north shore because its south-facing bank absorbs more radiation and emits it back as heat. Spring meltwater trapped above thinning ice can weigh it down until it finds a hole to seep downward into, creating radial channels of darker ice that collectively resemble starfish in shape.[18]

Though fairly small, wind-driven waves move over Walden's surface with great regularity in velocity, amplitude, wavelength, and period. In shallow water, these forward-propagating waves experience friction with the bottom, become slowed, bunch up, and increase in height as the wave energy is converted to the potential energy of taller waves and the kinetic energy of moving mass. At some critical height the water within them breaks with frothy turbulence. And if the wave propagation velocity (celerity) is fast enough, the waves curl to reveal the shape of the circular orbit just before it falls and swashes forward. Most of this energy is dissipated by turbulence, noise, and unseen heat. But some is used to grip, drag, and push particles of sand and stone back and forth, sorting them by size and dulling their edges with countless collisions.[19]

Chemically, North America's freshwater lakes run the full gamut from the deep, cold, and clear waters of Lake Superior to weed-choked, bird-crapped algal pools. Lake Walden ranks high on this scale, with its water chemistry more characteristic of a rocky highland lake than a sandy patch near Boston. Thoreau singled out Walden's "purity" for special praise. By this he meant its optical clarity. Lacking bordering wetlands, its water has a low concentration of the "tea" of humic acids that give rise to the "blackwater" streams draining bogs and swamps. Second, the input of suspended sediment is negligible, given the lake's stony shore and lack of an inlet. Third, the lake has a low concentration of the dissolved organic nutrient phosphorus, which keeps the phytoplankton count low, and therefore the turbidity.

Phosphorus, being in shortest supply, is Walden's limiting nutrient. It throttles back the metabolic engine of the lake toward what Thoreau called an "ascetic" condition and what limnologists call an *oligotrophic* status. Every other nutrient—carbon, nitrogen, potassium, and trace elements—is present in amounts that would otherwise push the pond toward a richer, murkier status. Walden's unusual ring of *Nitella* helps limit summer plankton growth by sequestering phosphorus in its biomass, which can then drift to the bottom for burial. Thoreau's observation that there were "few weeds" on the shore is consistent with its limited nutrient supply. But "though there are almost no weeds, there are frogs there," perhaps because the stones offer such good hiding places near a food supply concentrated by the shore.[20]

The lake's "schools of perch and shiners," reasoned Thoreau, "must be ascetic fish that find a subsistence there." They grow "cleaner, handsomer, and firmer fleshed than those in the river and most other ponds, as the water is purer. This was especially true for "its pickerel," which, "though abundant, are its chief boast." More generally, he remarked that Walden was "not very fertile in fish," which greatly understates the austerity of its waters, based on the 1832 report of Concord's first authoritative historian, Lemuel Shattuck. He claimed that "no fish were caught in it, till they were transplanted there from other waters." Even if this exaggerates the situation, his comment does suggest that colonial conditions may have been even more "ascetic" than what Thoreau witnessed. It is also consistent with radiocarbon dates showing a decline in Walden's organic sedimentation rate during the millennia prior to settlement. Thoreau's inventory of fish at Walden—"pickerel, . . . perch and pouts . . . , shiners, chivins or roach *(Leuciscus pulchellus),* a very few breams, and a couple of eels, one weighing four pounds" falls midway between what Shattuck reported and what Thoreau found in the nearby Concord River. Aside from fish and frogs, "there are also . . . tortoises, and a few mussels in it; muskrats and minks leave their traces about it, and occasionally a travelling mud-turtle visits it . . . These are all the animals of consequence that frequent it now."[21]

Though cold and completely dark, the seldom-seen soft bottom of Walden is vibrantly alive for much of the year. It is a nutritious cornucopia for whatever can survive this condition, typically the larvae of insects, a host of invertebrates, protozoans, bacteria, and bottom-feeding fish like pouts. Dominating inputs from above are the annual rain of leaves and seeds shed from shoreline trees, followed by pollen and spores carried in from great distance by the wind. Thoreau once remarked that "the sulphur-like pollen of the pitch pine soon covered the pond and the stones and rotten wood along the shore, so that you could have collected a barrel-ful." Pollen remains an important input for phosphorus today, second only to human urine due to the summer swimming season. To all this must be added the "ooze" of aquatic plankton that sinks slowly for days through the motionless water like manna from above. On the bottom, creatures churn their way through this bounty, passing much of it through their gastrointestinal tracks one or more times. Bacteria get involved too, coating things with gooey films or mats, and extracting what they can.[22]

At the bottom is a surplus of phosphorus. There, what limits life is the oxygen needed to consume all the food lying around. An infusion of dissolved oxygen arrives twice per year during turnovers: once during the onset of winter when surface water has been oxygenated by photosynthesis and atmospheric adsorption; and again during ice melt, this time mostly from oxygen mixed in by strong spring winds. Between these infusions, however, the supply of oxygen gradually diminishes, and is often used up by late summer and late winter. Under these chemically reducing conditions, phosphorous previously bound up in plant tissue is liberated back into the motionless water, which will mix back upward during the next turnover, recharging life at the surface from below.[23]

Air flows into and out of the top half of the kettle just as surely as water does so in its bottom half. Physically, air is mostly molecular nitrogen (N_2), which constitutes four-fifths of its mass. Light passes through it. Sounds propagate within it. When pushed or pulled, it ripples the water and makes waves run. Being more "fluid" than water, air can flow faster, though with less force because it is less dense. And with respect to heat, air is much more volatile because its heat capacity is so low. This means that for any given input of warmth, its temperature rises quickly, allowing rapid expansion, a concomitant change in pressure, and a lively breeze. Finally, the low thermal conductivity of air allows it to insulate losses and gains of heat like a fluffy blanket.

Ecologically, air is mostly about the oxygen (18 percent) and carbon dioxide (0.04 percent) within the final fifth of its mass. Organisms exchange these gasses back and forth. During the production of organic matter by photosynthesis, plants absorb the carbon and exhaust the oxygen. During its consumption by respiration, animals and microbes absorb the oxygen and exhaust the carbon. Round and round it goes. Also present within air are invisible trace quantities of life's vital ingredients. The most important of these is water vapor which, when condensed or crystallized, became the fogs, mists, clouds, snows and frosts that are so important to the prose-poetry of *Walden*. To this we can add the aerosols of dust, salt, soot, droplets, pollen, and "milkweed down."

Our language misleads us to think that air is a gas and water is a liquid. Air can be dissolved in water just as surely as water can be dissolved in air. Air can hold water in static suspension as tiny droplets, or as a steady state balance between falling drops and stormy updrafts. Drops of air rise upward from Walden's deep water just as surely as drops of water fall downward through its air. The physics is astonishingly similar, except for the algebraic sign. Conversely, turbulence at the shore traps air bubbles in water suspension, making whitecaps. And when that water is chemically softened, the surface tension diminishes to the point where foam develops, producing a solid emulsion composed mostly of air. Bubbles in Walden ice are held within a solid as if in suspended animation. Conversely, snow on the lake is mostly air, trapped by a skeletal framework of solid crystals.

Oxygen dissolves directly into the lake from the atmosphere, especially during stormy conditions. Photosynthesis releases it into the lake as a waste product. Diatoms do so invisibly. *Nitella* does so as tiny bubbles that "drip" upward before disappearing into solution. At the bottom, benthic processes release carbon dioxide and sulfur dioxide when there is oxygen available, switching to releases of hydrogen sulfide and methane when it is not. Typically these gasses remain dissolved in stagnant water near the bottom until turnover arrives. The consequent loss of pressure can cause them to exsolve—the opposite of dissolve—or fizz out like a slow-motion version of uncorked champagne.

Water vapor runs the visual "special effects" department at Walden. The slightest cooling or warming can create a visible haze that can either intensify into a mist or fade back to vapor in the span of seconds. A fog can transition to gentle rain, which can clear the air because larger drops grow at the expense of those remaining. Conversely, rain can evaporate before reaching the surface. Mists moving sideways can strengthen or fade as the temperature changes. Those moving upward can billow or flatten. Basically, any conceivable change can happen in any time sequence and in any spatial direction.

So volatile was the local meteorology in Walden's air pocket that, in lieu of mathematical equations, it is best understood by end-member scenarios. At one extreme, the regional wind is calm, the water cold, and the power of the sun diminished by the time of day, the date of the year, or by a thick stratus. Under such conditions, air in contact with water is chilled by conduction to a motionless state, except for the stirrings of life. Thoreau described passing through these cool strata, which he called "refrigerators, just as you would wade through a lake or at the bottom of a sea," with the higher, warmer strata being "comparatively lifeless or exhausted of its vitality." In a situation very much like lake turnover, stratified layers of air can de-stratify when a thermally driven density threshold is breached, changing the smell of the air in an instant. Or they can pour downhill over irregular topography, moving "from pillar to post, from wood-side to side-hill, like a dog that has lost its master." In another effect, "the coolness of evening" can "condense the haze of noon and make the air transparent and the outline of objects firm and distinct." In a related optical effect, the "greater coolness of the air over the pond" near his house relative to that over the darker rotten ice "condensed the vapor more there," allowing him to see "twice as far."[24]

At the other extreme, consider a storm blowing over the tree canopy of the delta plain. Initially, this lowers the air pressure near the edge of the kettle though the Venturi effect, allowing the air pocket to swirl upward, and if misty, to clarify. With slightly more wind, eddies will reach deep into this realm to mix it chaotically. With an even stronger wind, eddies reach all the way down to touch the lake surface and agitate it, especially when the air is cold and dense. Thoreau delighted watching this phenomenon take place on midwinter days, seeing the "very form of the wind" falling and splattering on the lake like "a mass of lead dropped upon an anvil." Finally, Walden has limited wave action, even during storms, because the deep hollow limits the frictional attachment of wind to rough water, and because the stony bank rapidly dissipates wave energy on resistant materials, like a breakwater.[25]

Thoreau recognized that "in summer, Walden never becomes so warm as most water which is exposed to the sun, on account of its depth." Indeed, the great amount of heat stored during summer offsets the daily and seasonal timing of air temperature changes, especially when the regional wind is slight. During those hours when the pond is warmer than the land—most commonly during the mornings of early fall—the air will rise above the pond, drawing breezes inland along rays. During hours when the pond is colder—for example during early spring or Indian summer—the rays travel in the reverse direction from pond to shore. During certain intervals of fall and spring, and under stable atmospheric conditions, these breezes can oscillate in and out like breathing.[26]

Slight variations in atmospheric pressure create standing waves that oscillate back and forth the full length of the lake, like the sloshing of a bathtub. Though generally invisible in the summer, they are easy to see beneath the ice, which acts as both a fixed point of visual reference and as a rigid membrane to contain the slosh and magnify slight differences in pressure. With his surveying instruments, Thoreau noticed what he had not been able see with the naked eye: "While I was surveying, the ice, which was sixteen inches thick, undulated under a slight wind like water . . . It was probably greater in the middle." He was spot-on with his last comment, especially for the circular western basin, which flexed up and down in the middle like a broad dimple, probably with a tempo of ten seconds or more.[27]

Life is present within the realm of the air just as surely as in the lake. Bacteria and microbes are everywhere suspended within it. Swarms of gnats and no-see-'ems resemble dust devils. Birds flit by day, bats by night. The sex life of plants involves copious releases of drifting pollen. Thoreau listed Walden's waterfowl: "Ducks and geese frequent it in the spring and fall, the white-bellied swallows (*Hirundo bicolor*) skim over it, and the peetweets *(Totanus macularius)* 'teeter' along its stony shores all summer. I have sometimes disturbed a fish hawk sitting on a white pine over the water; but I doubt if it is ever profaned by the wind of a gull, like Fair Haven. At most, it tolerates one annual loon."[28]

Embedded in Thoreau's famous parable of the loon in "Brute Neighbors" is an observation that indicates his recognition of the air pocket above Walden: "As I was paddling along the north shore one very calm October afternoon, for such days especially they [loons] settle on to the lakes, like the milkweed down." This down, having "floated high in air of hills and fields all day," finds the air pocket and then "sinks gently to the surface of the lake." Into this same pocket of still air, the much heavier loons drop down like stones from the stronger prevailing winds that keep them airborne above the forest canopy.[29]

"Land" is a hard word to define. It is a geographic place, a commodity to be bought and sold, a terrestrial environment, and a volume of material. Soil is part of land. Clay pits are part of land. Swamps, pools and springs are part of land. By mass, land is dominated by *earth,* which, in Walden's case, consists of unconsolidated silicate residues that were crushed and washed by glacial ice. By appearance, it is veneered by forest, an attribute that likely gave rise to the name "Walden," which originates from the Old English word *weald,* defined by the *Oxford English Dictionary* as either as a "heavily wooded area," or "a wild and uncultivated usually upland region."[30]

For our purposes here, the forest ecosystem at Walden is a thick membrane between air and earth, a terrestrial skin composed of organic fiber that extends from the tallest tree crown to the deepest tap root. It protects the underlying materials while at the same time modifies them. The trunks and crowns of the trees are like tufted hairs, which makes them a part of the land, even though the space is mainly air. They *buffer* the ambient meteorology, making the land a very different place from the lake under the same prevailing conditions. Over the lake, there is little friction to stop the wind, and any water that falls as rain simply adds to the mass already present. Over the land, however, the forest buffers the incoming flux of energy and mass. The fluttering of leaves and swaying of branches consumes wind energy through turbulence, making the forest floor much more tranquil than the treetops. What Thoreau called "roaring in the woods" was the sound of that energy being dissipated. During still conditions, the forest absorbs more energy on its dark canopy relative to the reflective lake. This creates slow updrafts that draw cool breezes from the water, concentrating them in middle layer above the canopy of brush but below the canopy of trees. Hydrologically, leaves and branches catch and hold substantial amounts of snow and rain. Much, if not all, of this evaporates before reaching the soil. Thermally, the shade of trees and the blanket of leaf litter keep the ground cool by minimizing heat absorption and maximizing insulation respectively, dampening the daily highs and lows that were so dominant during its tundra phase.

On clay soils, forest enhances the infiltration of water into the ground by creating pathways through which infiltrating water can flow. With coarse sandy soils like those at Walden, the forest is a net consumer of water. Indeed, there is a great sucking noise at Walden that is too faint to hear, the daily hydrologic work of billions of leaves evapo-transpirating water being drawn upward from the subsoil through tiny pores called stomata. Trees even drink from the aquifer below. Thoreau once traced the five-inch thick tap root of a pitch pine more than twelve feet "straight-downward into pure sand," all the way to the aquifer. A hundred gallons or more of water per tree per day can be exhaled. Sum that for every tree, and there is a whole lot less for the aquifer, and thus for the lake.[31]

During deglaciation, the land at Lake Walden resembled a vast, unvegetated construction site exposed to the elements. Today, the forest soil armors that same material, dissipating the brunt of raindrop impacts and the drag of flowing water. And beneath that armor lies a root *mesh,* which binds the earth together to a depth of at least a foot below the surface to create a thick and tenacious fabric under constant maintenance. In a lovely case of positive feedback, the growth and dieback of roots in the mesh is often the most important process driving soil creep. Yet without creep, the soils would quickly become sterile, diminishing the biotic beauty.

The most hidden component of the land's fibrous geo-membrane is the *humus* layer, otherwise known as topsoil or, in Thoreau's words, "vegetable mould." This is organic matter decomposed to the point where wood, seed fragments and coarse tissue are no longer visible. It is a dark, fine-textured solid that, in well-drained settings, is intermixed with the underlying silicate component of the soil, something like a moist mixture of espresso coffee grounds, cocoa powder, sawdust, and fine sand. In excessively drained acidic soils such as those at Walden, the humus is usually thin and relatively sterile, having "never fattened with manure," as have the more clay-rich soils of the deciduous woods. Importantly, humus retains far more moisture than an equivalent volume of silt or sand. This allows aqueous chemistry to operate continuously, giving infiltrated water the "bite" of carbonic acid that, when leached downward to the mineral soil, becomes an effective agent of weathering.[32]

Below the organic membrane of the Walden system, is the so-called subsoil. It's mostly permeable sandy gravel because the clay and silt were rinsed away from the delta plain to lake bottom settings. It is also relatively infertile because the source rock was mostly granite and associated high-grade metamorphic rocks whose dominant minerals are fairly resistant to chemical weathering. Under these circumstances, there is limited clay production in the soil. For this reason, ants, rather than "angleworms," are the main invertebrate mixing agents.[33]

If you look closely at the edges of excavations in the upland soil, you will see multiple horizons of different earth tones from the top down: the dun brown of the forest duff in which coarse organic fibers are waiting to be decomposed; the black-dark-brown of humus indicative of significant microbial decomposition; a few grains of ashy gray leached soil where the metal cations have been stripped away by organic acids moving downward with infiltrating water; the orange and vivid browns of the zone where iron being leached from above is precipitated back into the subsoil. And below all this are the "golden sands" of Concord, the light yellowish-brown sand of the trickle-down zone. Water infiltrating downward through this material carries little with it beyond some dissolved minerals, a trace of dissolved organics, and enough oxygen to precipitate iron as a yellowish stain and thereby minimize phosphorous transport.

The forest on the inclined banks of the lake behaves somewhat differently from the forest on the delta top. Being on a slope and more exposed to the wind, its trees must be more firmly rooted, and the foliage is thicker on the sunnier side. There is also more space to grow. A higher proportion of infiltrating water moves parallel to the slope, rather than straight down, making it richer for the longer journey. In a deep, steep kettle like

Walden, this richer forest of its rim is displayed like a billboard-in-the-round, a four-season tapestry best seen from the center of the lake. Every year the glorious colors of early autumn diminish as they fade to the twiggy deciduous browns and coniferous dark greens of late fall. In winter, gray trunks rise whisker-like through the snow around the stark white center. The icy breakup of early spring is followed by the darkening green spectrum leading to summer.

By and large, Walden hydrology simplifies to groundwater hydrology. The aquifer is both its largest and simplest realm. Thoreau knew this when he called the lake a "clear and deep green well . . . a perennial spring in the midst of pine and oak woods . . . without visible inlet or outlet . . . except by the clouds and evaporation," and "the water shed" to it "by the surrounding hills was insignificant in amount." Indeed, heavy precipitation—if not trapped in the canopy as interception, stored on the ground as snow, sopped up as soil moisture, or perched by some impermeable lens—seeps downward into the ground, especially on the Walden delta plain, which is so flat and dry it reminded Thoreau of the "prairies of the West." Locally, he was fascinated by "how rapidly our sandy soil dries up" relative to its reappearance in streams. "We walk dry-shod the day after a rain which raises the river three feet." Between these two visible manifestations of infiltration and stream flow lie the two great invisible sectors of groundwater geology: the trickle-down zone belonging to the land, where gravity pulls down vertically; and the aquifer zone where gravity operates mostly through fluid pressure and where the flow is dominantly sideways.[34]

In the land's trickle-down zone, grains of sand encountered along the way have first claim to any water moving downward. They are especially greedy during late winter owing to deficit of soil moisture caused by upward vapor transport toward frozen ground. This zone holds water against the tug of gravity by surface tension, which creates a film around each grain and connecting bridges between films. Only when the subsoil has absorbed all it can possibly hold, and only when the pore spaces are well connected, will water trickle down to reach the water table, beneath which all pore spaces are completely filled. We have reached the realm of the aquifer, where everything is saturated. If the pores are large and well connected—as they are at Walden—the water table is well defined, nearly flat, and has a very gentle slope that is sensitive to transient changes in pressure. At the water table, groundwater will flow down its slight slope from higher to lower elevations, threading its way through a maze of grains like someone moving through a crowd standing on a gently sloping hillside. Below that surface, however, the flow of every parcel of water is controlled by the sum of its elevation and pressure, something called its "potential," or "head."[35]

Upon reaching the water table, infusions of water spread out as irregular layers up to several feet thick. Near streams they thin fairly quickly because the local groundwater gradient is steepened by constant drainage. But on the broad divides between streams, this layer has nowhere to go until the water ahead of it is pushed out of the way by slight differences in pressure. Importantly, the sediments of the Walden kame delta are coarsest and most permeable at the top. This downward decrease in permeability causes the infusions of water to build up and drain away more slowly than if the strata coarsened downward. Snowmelt also helps smooth out annual flow variations in the aquifer because, as Thoreau said, "Nature's work has been delayed." With tongue in cheek, he wrote that this was because its snow "particles are less smooth and round," than water particles, "and do not find their level so soon."[36]

The average east-to-west gradient of the water table near Lake Walden is low, ranging from about ten to forty feet per mile. The pond surface, however, is horizontal. This requires that the eastern basin be a drain for incoming groundwater. There, Thoreau discovered inflowing cold springs, which he guessed were zones of enhanced permeability, likely layers of clean gravel formed in the deepest parts of stream channels. Conversely, Walden's western basin is mainly a spreading shelf of water that is slightly elevated relative to the nearby aquifer, which it recharges. There he saw a "leach hole" near its southwest side. Counterintuitively, Lake Walden's western basin drains outward along rays toward the tips of the coves. Beneath Thoreau's house site it flows northwesterly, not toward the pond, but away from it.

Heavy rain creates an interesting and transient complexity. The entire lake rises temporarily above the groundwater table. Temporarily, the top inch or two of the lake acts like a giant puddle leaking sideways into the soil in all directions.[37]

Finally, the water within Lake Walden at any time is not really being held. Rather, it is being exposed, like that of a deep pool in limestone cavern network. But instead of being surrounded by impermeable rock, the void is surrounded by permeable sand and gravel, especially in its wave-washed zone, which is porous like cheesecloth. And every molecule within that void's 113,411,286 cubic feet of space has its own history. A few molecules there today might have come from glacial meltwater that had been forced by pressure into a crack that recently broke. Millions of molecules might be a century old, having been trapped in the deep muck before being squeezed out by the weight of a sunken log. Untold billions of molecules may have entered as a rainstorm within the hour. On average, however, a molecule of water within Lake Walden remains in the basin for just under five years, its so-called residence time. Slightly more than half of those incoming molecules reach the pond by aquifer seepage from the east, with the remainder coming directly from rain or snow on the lake. In terms of departures, nearly three quarters of the outgoing molecules are lost by seepage toward the west. After leaving the pond and moving through an additional aquifer, they emerge as springs to the bed of the Sudbury River. And from there they flow onward to the Concord, the Merrimack, and finally to the sea.[38]

Steady State

What is Walden's shore? Instinctively, we think of the boundary between land and lake. But that same spot is equally the boundary between air and aquifer. Taking all four realms into account, we can more accurately visualize the shore as an italicized plus (*+*) sign marking the spot where the steeply sloping line of the kettle intersects the nearly horizontal line of saturation. This "X-marks-the-spot" junction is true for every ray drawn outward from the center. The sum of all such rays defines the shore, which has no beginning or end because it circles around on itself within homogenous material. This makes the shore a holistic place. Perhaps this is why, in a moment of reflective insight, Thoreau identified with it so strongly, writing, "I am its stony shore."

"The pond rises and falls," he wrote in *Walden,* "but whether regularly or not, and within what period, nobody knows, though, as usual, many pretend to know." Early in his career, he used this mysterious fluctuation as a symbol of transcendental correspondence between the height of lake stage and the height of a man's mood at any time in his life. But as he matured, Thoreau was increasingly drawn to the same phenomenon out of scientific curiosity. Nothing else can explain why his

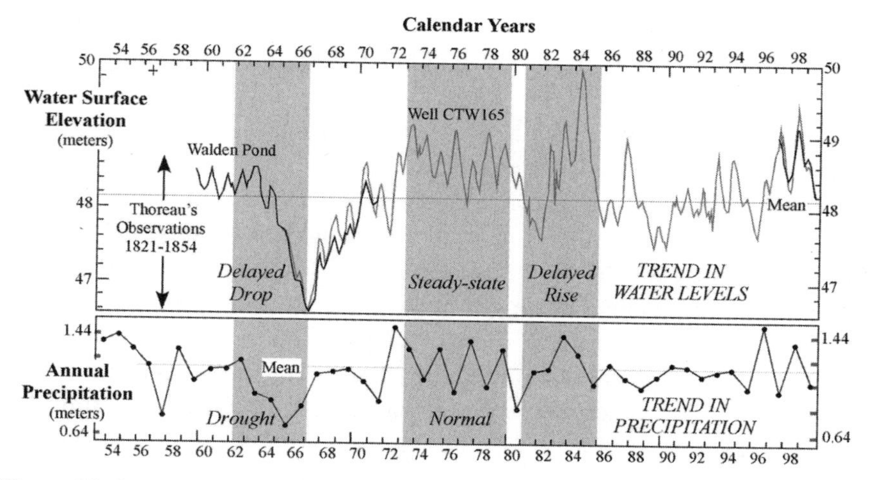

Figure 27. Long-term variation in Walden Pond hydrology. Data spans forty-six years (1953–1999) for three hydrologic parameters: Shoreline elevation (stage) of Walden Pond (top black line); the water table elevation in a nearby monitoring well (CTW165, top gray line), and the annual estimated precipitation (bottom black line). Modified from Figure 5 of Colman and Friesz, *Geohydrology,* 11. Selected intervals showing steady state, and time-delayed drops and rises of pond stage indicated by gray bars. Note that pond can fall when rainfall is increasing (1966 and 1991) and vice versa (1973 and 1981).

measurements of lake stage span a longer chronological interval, and were made more often, than any other type of observation in *Walden,* spanning three decades of data collection (1824–1854). Every glance at the pond during the time he lived there, and every visit back during his scientific sojourning stage, improved his understanding. By the time he published, he had enough data for three contiguous paragraphs.[39]

Thoreau's endless fascination with the ups and downs of the pond derives from the fact that lake stage is the most visible indicator of system-wide behavior. Using Thoreauvian language: "sky water" passes into the "earth," before moving down to a "perennial spring." From there it moves back into the volume of the pond, and then up to a "lake of light," where that same water "dissolves" back to the air to complete the circle-cycle. Four separate handoffs are involved. From air to land they involve the magnitude, duration, intensity, and phase of precipitation. From land to aquifer they are intercepted by organic matter, changes in soil moisture, evapotranspiration, and degree of wetness in the trickle-down zone. From aquifer to lake and vice versa they are the locations and gradients of groundwater surpluses and deficits. And from lake to air, the handoff involves every possible control on evaporation, including the presence or absence of ice.[40]

With respect to the pond rising and falling, Thoreau differentiated its "precessions and recessions, its cycles and epicycles." To these he added the more regular harmonics such as the "pulse of the pond" that swayed his ax, and the endless beat of small waves against its shoreline stones. He understood that the time-lagged seasonal component was caused by long-transmission times through the aquifer: "It is commonly higher in the winter and lower in the summer, though not corresponding to the general wet and dryness. And finally, he noticed non-cyclical secular variation: "I can remember when it was a foot or two lower, and also when it was at least five feet higher, than when I lived by it." His earliest record was from 1824 when he "helped boil a kettle of chowder, some six rods from the main shore," something that "has not been possible to do for twenty-five years."[41]

Based on continuous monitoring by the U.S. Geological Survey during the last half-century, Thoreau's observations and interpretations of pond stage have all been validated. The surface elevation of Lake Walden varies up and down with an average elevation of 158.36 feet (48.27 meters), and with a range of about 11.48 feet (3.5 meters). The height at any given moment reflects the instantaneous balance of gains and losses integrated around the perimeter by a suite of processes working on their own schedules within all four realms.[42]

Thoreau's longest time scale was climate change associated with Atlantic Multidecadal Oscillation, caused by variations in global oceanic circulation. Obviously, this cause of stage variability at Walden was extrinsic to its system boundaries. Within this long-term trend were the multi-year oscillations of El Niños and La Niñas, and their respective droughts, snowy years, and drenching subtropical storms. Thoreau recognized these long time scales "require many years for its accomplishment" because they involved massive changes in water storage within the enormous aquifer surrounding Walden. These long-term changes were town-wide, because other lakes like White Pond would usually "sympathize with Walden." Within these decadal and multi-year variations was the annual rhythm of the seasons, typically rising from a low stage in late winter to a high stage in late summer.[43]

These extrinsic rises and falls were rendered more complex by factors operating within the Walden system, beginning with the pattern of snow retention during the winter, and of heat-waves during the summer, both of which shift the balance toward evaporation. To this must be added the transit time for specific slugs of recharge passing along different rays within the same aquifer. Most instantaneous was the abrupt rise of the lake surface due to heavy rainfalls upon it.

More parsimoniously, five different time scales rode piggy-back on each other. Transient intrinsic responses like individual rainstorms rode on the steady-state annual average, which rode on the climatic mode (El Niño or La Niña), which rode on the climatic average, which rode on longer-term secular trends. Even more parsimoniously, a positive mass balance—due to any combination of these reasons at any one second—caused the lake to rise, and vice versa for a deficit.

Each time Thoreau visited the pond to monitor its stage he knew what to expect and what not to expect. He knew that the water level would be within a vertical range of about twelve feet. This is because the overall system is stabilized by internal negative feedbacks associated with water transfers between the realms. For example: the higher the stage relative to the aquifer, the faster Walden leaked out; and the lower the stage the faster it leaked in. Dozens of such processes worked together to buffer storm-to-storm, season-to-season, and year-to-year variability, thereby constraining pond behavior within predictable limits. Simultaneously, Thoreau could count on being surprised each time he visited. Chaotic interactions between more than a dozen variables kept the inputs, outputs, and internal transfers transiently imbalanced. This made the instantaneous stage of Lake Walden at any time—as well as the direction it was trending—probabilistic. This is a fancy word for a system that is predictable, but only statistically. Taken as a whole, the lake became a good analog for human life, a model system he used to help understand himself.[44]

At the much longer time-scale of geology, each of Walden's four realms contributed to the fixed properties of Thoreau's lake experience. In other words, they set the boundary conditions for whatever steady state system might result. Had Walden's average climate been less seasonal, the diminished thermal contrast between the realms of the air and the lake would have reduced the energy gradients. Such a Walden may not have frozen in winter nor stratified so deeply in summer, making it much more bland in its behavior than the one we know. In another example, if the realm of the land had not been vegetated, collapsing banks and gullies would have gummed up the works. If the realm of the aquifer had not been so large, significant runoff would have been the norm, creating what Thoreau feared most, a Walden with inlets and outlets linking it to the rest of the world, a lake whose story would have hardly been worth reading about.

In *Seeing New Worlds,* Laura Dassow Walls says this of *Walden:* "Thoreau's artistry creates a fiction of wholeness, roundness, an eternal cycle of withdrawal and reunion." I do not see the fiction. The attributes she names are all direct consequences of meltdown to the focus. In short, of the "Thaw Law." And now, with some trepidation, I step out onto the slippery slope between limnology and literature.[45]

II

THE BOOK OF THE PLACE

7

SENSING *WALDEN*

"This is a delicious evening, when the whole body is one sense, and imbibes delight through every pore. I go and come with a strange liberty in Nature, a part of herself." So begins "Solitude."[1]

Note that Thoreau's sensation is singular: as it was in the beginning nearly four billion years ago, a time before the Earth had continents. A time when its oldest microbes were evolving on some boiling volcanic vent, probably in the pitch dark below a global abyssal sea. Somewhere on the gradient between geothermal scald and oceanic cold was the ideal Goldilocks temperature of "just right." The microbes that found it survived best, thereby launching the evolution of human sensation. This touch of heat almost certainly preceded other aspects of touch such as pressure and texture.

Thoreau's choice of the word "delicious" was no accident. Taste may have arrived even before heat. For food, microbes exploited chemical reactions within metal brines. Somewhere on the gradient between "not-too-concentrated" and "not-too-dilute" was the ideal. From that salty sweet spot, our sense of taste would evolve. Smell, of course, is nothing more than taste in the air. But at this point in planetary history, life could not survive above the water owing to a sterilizing flux of ultraviolet radiation. This was a world without ozone because there was no free oxygen in the air with which to make it.

Sight began on the gradient between light and dark: a concentration not of heat or chemicals, but of photons. Microbes began to do what sunflowers and dandelions do so efficiently today, which is to turn and face the sun. Eyes would eventually follow.

Sound is as old as the planet. It is a pressure wave that moves through solids and fluids on a continuum with seismic waves. Because microbes

Figure 28. Mosaic of broken and refrozen ice on Walden Pond. January 2013.

do not have ears, paleontologists cannot tell when hearing originated. What they do know is that a barrage of loud noises occurred about 3.9 billion years ago when Earth experienced a staccato bombardment of asteroids. Were they heard?

Light and sight are not the same. The former is a real physical phenomenon, with no ambiguity or bias beyond what quantum physics requires. The latter is a biologically evolved sensation that creatures use to detect that physical phenomenon. Ditto for the physics of sound and the biology of hearing. And between pressure and touch, liquid chemistry and taste, and gaseous chemistry and smell.

Each of our senses is like a scientific instrument in the laboratory of our skulls. Lacking radiometers, we have eyes. Lacking seismographs, we have ears. Our skin is a thermometer, our nose a gas chromatograph, and our tongue a piece of compound litmus paper with strips for bitter, salty, and sweet. As with any array of lab instrumentation, each of our five senses evolved to their present degree of sophistication through the trial and error of natural selection. To feel as Thoreau felt, that "the whole body is one sense," is to let go of these more recently evolved distinctions, and to let ourselves time-travel back to the origin of life itself, when we were much more clearly "a part of herself." At that phylogenetic starting gate, life was little more than slime on rocks sensing a path toward awareness, and eventually intelligence.

Any talk of a more than five senses in humans is a matter of definition, clinical research, religion, and pure bunk. There is no doubt that some people have an uncanny "sense" of timing, others of direction. Thoreau claimed the latter. So do I, being constantly aware of the four cardinal directions whenever I am conscious of geographic place. Perhaps this comes from my being raised in the flatlands of North Dakota and northern Minnesota where the high latitude makes Polaris more significant, and where the Township and Range system is stamped onto the land by a checkerboard of fenced fields. Or perhaps it is my family's reverence for that original square of sodbuster homestead on an otherwise unmarked prairie. Or possibly it is my decade of work near the Arctic Circle, my half-century of surveying experience, and my delight in making maps.

Thoreau "saw as with a microscope, heard as with [an] ear-trumpet." This was Emerson's eulogy remark on the acuity of his friend's senses. Reginald Cook concluded that sensations, rather than thoughts, were "the chief means by which Thoreau achieved correspondence with nature," especially when he was sojourning: "The world of nature was resonant in his ears, sensitive to his touch, graphic to his eye, pungent to his nostrils, savory on his tongue." Victor Friesen pushed this concept even further in his book *The Spirit of the Huckleberry,* concluding that "sensuousness accounts for the essential Thoreau," making him more "Lockean" than Kantian in a philosophical sense. In Friesen's view, Thoreau's most ecstatic experiences fell beneath the threshold of perception, having come from a type of mysticism termed "super-sensuousness." Laura Dassow Walls refers to his "epistemology" of contact. Alan Hodder carried the idea to its rhetorical limit: "Notwithstanding Thoreau's reputation (partly deserved) as an abstemious prude, there is hardly a greater sensualist—in the strict sense of that term—among American canonical writers."[2]

This chapter characterizes Thoreau's scientific "sense of place" as preparation for the interpretations of *Walden* that follow in the subsequent chapters. By "sense of place" I refer not to the popular idiom of one's emotional geography, but to the neurobiology of his experience, the grand sum of signals arriving to his brain from sojourning space.

Scientific Sojourning

The very first word in Walden is a white lie. "When I wrote the following pages, or rather the bulk of them, I lived alone, in the woods." In *Revising Mythologies,* Stephen Adams and Donald Ross build on the previous work of Lyndon Shanley to document—in excruciating detail—that the bulk of the final version of *Walden* was written *after* Thoreau left the pond, with "most of the important changes occurring after 1852." They further contend that the book has two separate narratives, the first being the "false economics of Concord" leading to "spiritual anesthesia," and the second being the "basic poetic metaphor" of organicism. For my purposes, this is a fancy way of saying that *Walden*–Part I was early and has little to do with natural science. And that *Walden*–Part II is late and has everything to do with it.[3]

The essayist who wrote Part I lived at the pond and spent much of his time there writing *A Week* and other works with neoclassical and transcendental themes. The field scientist who wrote Part II lived with his parents in town, sojourned widely over miles of territory each day, and—in early 1852—decided to use his observations of natural phenomena to upgrade his dormant *Walden* manuscript, most of which was then social critique. During this later stage, Thoreau's modus operandi was to walk away from the village each afternoon, move outward into nature in whatever direction his inner compass suggested, and then drench himself in outdoor sensations until something caught his attention. At that point, his intellect would automatically engage, scientifically at first, and then poetically. On those days, Walden Pond was only one of many bright stars in the broader galaxy of his sojourning space. But on those nights and subsequent mornings—when sitting at his desk in his garret on Main Street—the pond became a potent literary black hole that drew everything inward and downward from his broader experience, concentrating the results.

This chapter surveys the outer regions of that experience with an emphasis on Thoreau's sensory input and his scientific thoughts about that input. "The true man of science," he once wrote, "will know nature better by his finer organization; he will smell, taste, see, hear, feel better than other men." Here he explicitly makes the connection between the acuity of his sensations and the quality of his science. Abstaining from alcohol, caffeine, hallucinogens, heavy foods, and sex helped keep his sensory instruments clean and well calibrated.

The scientific use of his senses is nicely revealed by an 1852 *Journal* passage written after a January stroll through Concord village.

> Where a path has been shoveled through drifts in the road, and the cakes of snow piled up, I see little azures, little heavens, in the crannies and crevices. The deeper they are, and the larger masses they are surrounded by, the darker-blue they are. Some are a very light blue with a tinge of green . . . the atmosphere must be in a particular state. Apparently the snow absorbs the other rays and reflects the blue. It has strained the air, and only the blue rays have passed through the sieve. Is then, the blue water of Walden snow-water? . . . the blue of my eye sympathizes with this blue in the snow.

Note that sight is the only sense being employed in this passage. There is no sound, smell, touch, or taste involved.[4]

On the surface, this is a typical journal entry, dense with sensory description and in need of editing. Below the surface, however, it illustrates a stepwise scientific engagement that Thoreau repeated like a fugue with countless variations. *Engage* the natural world in search of beauty, even on a gray winter day, in this case to discover "little azures." *Observe* details, here the objects he called crevices, and their attributes. *Associate* some of the details, in this case the color gradient relative to the depth and mass of the snow. *Hypothesize* about the association, here the ambient meteorological conditions causing color variations. *Import* an explanation from previous book knowledge, in this case the progressive adsorption of radiation in water. *Induce* a near-field scientific explanation, here the analogy between an optical filter and a mechanical sieve. *Induce* a far-field scientific explanation, in this case, a common color shared by the snow, the lake, and his own eyes. Only after all this does Thoreau reveal his true motive, the use of science to make meaning, in this case his declared affinity for pure snow. Motive, sensation, thought, and meaning are in their proper order. This was his sojourning technique.

This "little azures" passage was written near the beginning of Thoreau's self-declared "Year of Observation" (1852) when his daily sojourns were seldom complete without some element of hypothesis and experimentation. On several occasions he predicted the locations of archaeological sites and then tested his predictions with excavations or reconnaissance surveys. He wrote short memos to remind himself to validate some observation he had made, perhaps to return to some spot at another time, or to visit another spot to see if the same idea held true. Most famously, to "find the bottom of Walden Pond, and what inlet and outlet it might have." During a spring visit to the cliff in May 1853 he made a plan to return under different conditions "to observe carefully the direction and altitude of the mountains." In April 1852 he wrote parenthetically: "(Mem. Try this experiment again ; i.e. look not toward nor from the sun but athwart the line.)." In June that same year he reminded himself to search for other sites where the tender shoots of pitch pine might also face the sun. Having hypothesized that pond ice was more bubbly than river ice, he wrote: "There may be something to this. Let me look at the fresh ice of a pond that has a stream, and see if there are fewer bubbles under it." He returned to the same patch of Walden ice over-and-over again for repeat observations, being "curious to know what position my great bubbles occupied with regard to the new ice." By early 1854, and during the final push to complete *Walden,* his *Journal* at times, resembled a lab notebook: "Compare Three Ices: Cut this afternoon a cake of ice out of Walden and brought it home in a pail, another from the river, and got a third, a piece of last year's ice from Sam Barrett's Pond, at Brown's ice-house, and placed them side by side."[5]

The method-driven experimentalist I just described above was the same Thoreau who, in other moods, feared he was becoming too analytical. On those occasions he coached himself to "walk with free senses," as he had done as a child, when his life was "ecstasy," and when the "earth was the most glorious musical instrument, and I was audience to its strains." Even during his most scientific days, Thoreau often had to remind himself to "let my senses wander as my thoughts—my eyes see without looking." To learn, once again, by "direct intercourse and sympathy."[6]

I now turn to each of Thoreau's five sensory instruments in the order of their importance, based on their qualitative frequency in the *Journal.* In each case, I lead with a sensation. To it, I attach relevant scientific observations, and then explicate them, using Thoreau's language when possible. My primary purpose is to extract and separate the human sensations of this chapter from the *actual* physical reality of the previous chapter, and from the literary *perceptions* of the next. My secondary purpose is to bring readers—including those shy of "hard" science—to the level needed to appreciate geo-*Walden* as the joyous celebration of physics it is, and, to a lesser extent, of chemistry. Though I would love to explore the full range of details for each natural phenomenon, doing so would require a book in itself. For example, consider the work of physicist M. Minnaret, who spent all 362 pages of his delightful book, *Light & Color,* explaining only these two optical phenomena for lay audiences. I also avoid a detailed inventory of Thoreau's sensory experiences, because Victor Friesen has already covered that ground so well. Finally, I focus my discussion on those sensory phenomena most relevant to geo-*Walden,* singling out sight for special emphasis.[7]

Sight and Light

Above all else, Thoreau was a visual creature. "My only integral experience is in my vision," he wrote in a letter to H. G. O. Blake. "I see, perchance, with more integrity than I feel." Physically, each of his sights was a flurry of sensation being processed by his visual cortex. Darkness and brightness involved the concentration of photons, as measured in units of candlepower or lumens: the sum total of direct and reflected light from the sun, moon, stars, and human contrivances. Moonlight was mostly about reflections, fog about scattering, and shade about opacity. Seeing color was detecting which wavelengths were bounced toward the eye from some object; whether spread out by a rainbow, filtered toward red by a sunset, returned in full by a white cloud, or extinguished by a lump of coal. Whole scale reflections and refractions—for example the brilliant maples he saw above Walden's water line, or the seemingly bent oar he saw half in, and half out, of the pond, respectively—involved the bouncing and bending of parallel rays, respectively. In contrast, tiny droplets, dust, and molecules reflect and refract light in all directions, a phenomenon called scattering.[8]

It is no accident that Thoreau became a surveyor, a vocation literally based on lines of sight. Whereas touch and taste required direct contact, and whereas smell and sound required proximity, only sight could take him to the distant reaches of the earth and to the heavens. One

September night he looked up into the heavens through the "flashing clearness" of the sky to see "blue rays, more blue by far than any other portion of the sky . . . making a symmetrical figure like the divisions of a muskmelon, not very bright, yet distinct," separated by "bars or rays of nebulous light." These were shadows cast by the setting sun and diverging away from it, something known as anti-crepuscular rays, a word within Thoreau's vocabulary. Also on that same evening he saw the aurora borealis, or northern lights, caused by the channeling of electrically charged particles created by the solar wind into Earth's magnetosphere. Sight, like no other sense, allowed Thoreau to link Concord to the rest of his universe.[9]

Reading Thoreau's *Journal* can be a visually hyperbolic experience. Every sunset had to be the "grandest picture in the world," one in an endless series hung in nature's gallery, "painted and framed, held up for half an hour, in such lights as the Great Artist chooses." The "tints of the sunset sky are never purer and more ethereal" than through the "crystal clarity of the sky in January." Or, on second thought, was it February's "singularly crystalline, vitreous sky?" Adding strength to these visual metaphors was his expertise in optics, which he studied in college and which he thought about almost daily. From a Concord hilltop at exactly 7:30 p.m. on June 26, 1853, he differentiated eight separate optical effects during a single sunset:

> And now the sun sinks out of sight just on the north side of Watatic and the mountains, north and south, are at once a dark indigo blue," their "sierra edges all on fire. Three minutes after the sun is gone, there is a bright and memorable afterglow and a brighter and more glorious light falls on the clouds above the portal . . . When I invert my head these delicate salmon-colored clouds look like a celestial Sahara sloping gently upward, an inclined plane upward, to be travelled by caravans bound heavenward, with blue oases in it.

In those sentences are the effects of Earth's rotation, shading, refraction, power, reflection, symmetry, scattering, and critical thresholding.[10]

"Sunshine," Thoreau understood, "is nothing to be observed or described, but when it is seen in patches on the hillsides, or suddenly bursts forth with splendor at the end of a storm." Each ray leaves the sun's corona as a stream of nuclear-powered electromagnetic radiation. In the empty medium of space, the actual power of light, called its luminosity, diminishes with the square of the distance from the source. As with distant streetlights in a small town, most of space is dark, for one reason only: the lights are too far apart. Each night on Earth reminds us of this same fact. Though we face a million suns, they are too distant to shed much light. This inverse square law is strengthened when the medium is air, and even more so in water. This is why, when scuba divers reached the bottom of Walden at just below a hundred feet, it was dark. At that depth, the light has been scattered or absorbed. Limnologists say that it has become "extinct."[11]

On a clear day at noon, the sky appears white or light yellow near the sun because most of the incoming light passes easily through the six short miles of the lower atmosphere (troposphere). Farther away from the sun, however, the sky appears blue because the greater length of penetration preferentially scatters light with blue wavelengths. When the sun is low on the horizon, the spectrum must pass through more than a thousand miles of atmosphere. Then, only the longer and redder wavelengths are left to be scattered back. Similar optics apply when the moon rises orange under clear but humid skies. Like the sun, it whitens with further rise. Aesthetically, the best it can do is a creamy tinge of blue.

When a ray of light strikes an opaque object such as a rock or a tree, some of the energy is absorbed and converted to heat, while some is reflected back to the environment. Black is the color of complete absorption; white of complete reflection. Every other color we see is a partial sum bounced back to our eyes.

When a ray of light strikes a transparent medium, however, it penetrates downward and continues with a slower velocity. With increasing distance, much of the energy penetrating air and water is invisibly absorbed and converted to heat. But much is also internally reflected in myriad directions by molecular scattering, which brightens the media and colors it in three distinct ways: hue, saturation, and brightness. Hue refers to the true spectral color whether blue, green, yellow, or red. Saturation involves the purity of that color, or its vividness relative to a background of neutral gray. For example, one cup of blue tint in a pail of white is more vivid than a cup of the same in a pail of dark gray. Brightness refers to the magnitude of the experience, the actual photonic power on a scale between blinding bright and dull black. When Thoreau described

Walden water as "blue," "clear," and "dark," he was speaking of hue, saturation, and brightness, respectively.[12]

Thoreau was delighted and puzzled by the directionality of color in both water and air. Pure water is clear and colorless in the shallows because incoming light has not passed through enough mass to scatter back some color. Instead, we see reflections from the bottom of stone, sand, mud, and plant fragments. But beyond a depth of several feet, water begins to scatter back spectral colors from faint azure to dark emerald, with variations due to ray paths, chemistry and suspended materials. In one of his earliest entries, Thoreau noted the Sudbury River was "yellowish green" when "viewed from the bank above," a color that vanished on "a nearer approach." The Concord River offered him three distinct colors within an arc of only ninety degrees: "lighter" eastward, "very dark, deep blue," northeastward, and a "dark and angry flood" to the north. This likely had to do with the directionality of the sun at that moment relative to the incoming wave train. At other times, it was "blue ink," on one side of the boat only. Walden was "blue in the depths" when seen from afar on a typical day, but "green in the shallows," where the scattered blue of water combined with the reflected yellow of sand to yield a "vivid green."[13]

Looking straight down in the water from his boat, Thoreau noticed that "Walden is blue at one time and green at another, even from the same point of view." This was due to some combination of the light intensity and water clarity, which transmits green only under the clearest and most powerful conditions. Looking outward, "the farther off the water, the bluer it is." Standing in shallow water, the light changes from yellowish to a "light green, which gradually deepens to a uniform dark green." Only Walden Pond and White Pond could be green in the middle, and only under conditions of "serene weather." Other circumstances could produce that same color by different means: the "hollows about Walden, still bottomed with snow, are filled with greenish water like its own." Here, the green derived from lower luminosity amplified by scattering on submerged snow and reflected back through ultra-clear waters of winter snowmelt.[14]

Though perfectly clear, both Walden and White could become "dark, in the storm," because not enough light penetrated the clouds to reveal brighter colors. Lakes, as "liquid eyes of nature," could be "blue or black or even hazel . . . deep or shallow . . . clear or turbid" under different conditions regardless of look direction. Though the autumnal haze reduced the crispness of light, and made the shorelines of ponds more "indistinct," those same conditions blued "the distant hills" and made "the lakes look better."[15]

If light moves from one transparent medium to another at a high angle, it bends abruptly, for example from air into water or from warm water into colder water, and so forth. For most of the year, Walden water is multi-layered with less dense layers floating over more dense ones. At each boundary, the light within the lake is bent inward and downward, as with the lens of an optical instrument like the human eye. In muddy water, this doesn't matter much because the light is quickly absorbed or scattered. But in a transparent deep body like Walden, however, it allows the lake to appear almost luminescent, similar to the eye of a cat when hit by the beam of a torch light at night. For a lake, this effect is maximized when surrounded by dark forest and overlain by haze. This lensing effect is reversed for shallow water near the pond perimeter. There, light refracted inward and downward to the reflective sand travels back up through the water and is refracted outward toward the shoreline vegetation, intensifying its illumination with a spectral glow. Another aspect of refraction is the apparent distortion of objects seen partly above and partly below the water, for example the oar that looks bent but is not, or the "body of the bather . . . magnified and distorted" that "produces a monstrous effect, making fit studies for a Michael Angelo."[16]

At a more glancing blow on a surface, light is reflected, rather than refracted. In other words, it is mirrored, whether at the scale of crystal faces or the water surface. Sunlight effectively bounces off a lake surface like a tennis ball off a court, traveling forward (and usually upward) until it is either absorbed, scattered, or lost back to empty space. Under such conditions, the surface of the lake becomes a source of light like that of the moon, both of which reflect the sun. Seeing the water reflection of a moonlit boulder on the shore is detecting sunlight on its third bounce (moon, object, water), each of which changes the color, as well as other properties of light.

Any disturbance of the reflecting liquid creates ripples and waves that change the incident angle of sunlight. One side of a ripple, for example, might face the sun or moon; the other face away. At only one point on the sinusoidal surface will the incoming ray be exactly perpendicular to the water surface. At that point light moves straight inward to be scattered to whatever hue the depth demands. On either side of that point, however, the light will be refracted inward, its color slightly changed. Near the top of a ripple the angle between incident rays and the surface will reach the critical angle and be reflected to the eye or to something on the shore. Thus each ripple, each wave, and each disturbance changes the play of light on the surface of the water, and on its bottom, if clear enough, not only of color but also of reflections. In the process, "Lakes of Light" send off polychrome "flakes of light."[17]

Whole scale reflection on Walden's surface "multiplies the heavens" and brings them down to Earth. These were clearest when water was "blackness" and "smooth," features combining to produce to "the glassy surface of the lake." Reflections that reach our eyes as integrated patches of color are particularly vivid near the summer solstice when the intensity of light is strongest. Then, "the meadows are the freshest, the greenest green in the landscape," and "the river is a singularly deep living blue, the bluest blue, such as I rarely observe, and its shore is silvered with white maples, which show the undersides of their leaves."[18]

Scattered reflections from tiny clear droplets produce the white puffiness and wispiness of clouds. The same is true for frosted glass, which is perfectly clear under a microscope. When Thoreau remarked that "the dense fog came into my chamber early this morning, freighted with light and woke me," he was in the zone of maximum scattering near the edge of a low cloud at ground level, usually produced by the chilling of warm moist air by colder land or water. In contrast, scattering by the "haziness" of August "seems to confine and concentrate the sunlight, as if you lived in a halo." Clouds, when viewed from afar as objects on a clear day, always appear white because you are looking at scattering on their outsides. Inside those clouds, or beneath them when they are overhead, the colors are progressively darker shades of gray because the white light trying to penetrate them is being progressively dimmed by distance.[19]

Reflection combined with refraction produces mirages and other optical illusions, which fascinated Thoreau all his life. Looking down in the shallows, he enjoyed the net-like refraction shadows and bright spots caused by wave interference. He was also in the habit of inverting his head to see objects upside down. From the cliffs above Fairhaven Bay on a calm August day in 1852 he was awed by the reflected image of a small boat, "schooner-rigged, with three sails," whose location could be determined only by studying the "junction of the reflections." The "castles in the clouds" he loved so much are examples of what are called superior mirages because the image lies above the true line of sight. Patches of what look like shining water in the foreground of otherwise hot, dry fields were inferior mirages. Both are caused by the progressive refraction of light rays where the speed of light changes as a consequence of air temperature. Polarization, which occurs when light is filtered to travel in dominantly one plane, is also a consequence of differential reflection. Diffraction is yet another consequence of reflection. Points such as the top of a pebble or the crest of a wave collision can operate like a collection of pinhole cameras, each sparkling like miniature flash bulbs.[20]

Thoreau loved the ways refracted light interacted with his world: the "rainbow" from the misty prism of the sky; a "halo" from moonlight; or the "beautiful colors" of iridescent mussel shells in which tiny variations in the thickness of clear layers of aragonite spread out the spectrum. Refractions also brought the sparkle to stars, making each a "flower of light," and doubling the horns of the moon. Most curious was the optical fire he witnessed on the cliff in March 1854: "I find a place on the south side of this rocky hill, where the snow is melted and the bare gray rock appears covered with mosses and lichens . . . where I can sit, and an invisible flame and smoke seems to ascend from the leaves and the sun shines with genial warmth."[21]

By presenting an inclined land surface toward the sun, the cliff absorbed more radiation than elsewhere on an early spring day. And on those islands of extra heat were islands of even more concentrated heat, in this case, a patch of dark leaves absorbing more visible (short-wave) radiation than adjacent patches of snow. This created campfire-sized islands of heat (long wave radiation) that warm the air above them. The velocities of that rising air and changes in its refractive index were controlled by temperature gradients, creating the chaotic blend of laminar and turbulent flows responsible for invisible flames. One month earlier, he had enjoyed the "boiling of the air" beyond the blue smoke of a campfire, "through which you see objects confusedly." The details of his descriptions reveal his comprehensions.[22]

The "shimmering light" of the moon was special for Thoreau, a "creamy mystic medium," a "luminous, liquid, white light . . . which some might mistake for a haze," "silvery, as it were plated and polished smooth, with the slightest possible tinge of gold." These properties, he understood, arose from the reduced luminosity caused by reflections scattered and suffused by the dust of the moon. Only in such light would his "shadow" and the "moon" both bear a halo.[23]

On a warm midsummer night in 1852 Thoreau saw a distant river by moonlight flowing with a "certain glory," an "inward light like heaven on earth." There, the moonlight was likely being refracted downward into transparent, but pitch-black water, with some of it was being scattered back to his human eye. Something similar happened near the east shore of Walden during winter. Under just the right conditions when looking west a few moments after sunset, the lake become "more light than the sky,—a whiteness as of silver plating, while the sky is yellowish in the horizon and a dusky blue above." In late summer, and under conditions of thick stratus, a gentle rain, and tranquil conditions, it became "more heavenly and more full of light than the regions of the air above it."[24]

The maximum optical clarity of the atmosphere each year coincided with the minimum light intensity. "Though the colors are not brilliant, the sky is crystalline, on the coldest winter days." Only then was moonlight strong enough to illuminate the night sky to a "glorious blue" on "one of the most glorious nights I ever beheld." In other seasons, transient episodes of clarity occurred when the aerosols were rinsed from the sky during a storm, or when cold Canadian air blew them away. At such times, the White Mountains—which Thoreau called "Empurpled hills"—could be seen from atop the Bunker Hill monument. The daytime winter sun was too strong to paint them an alpenglow pink.[25]

We easily accept that no two snowflakes are alike, despite the fact that in every case the same mineral optics—crystalline, hexagonal, uniaxial, inorganic, transparent—applies. Even more true is that no two moments on the water can ever be the same owing to the astonishing complexity of the variables involved: the intensity of the light by the season or day; the balance between light sources, the instantaneous angle between water and sun, and the changing properties of the surface of the medium or the medium itself. A warm gentle rain on a cold-water surface, for example, will create yet another layer of water for an additional set of refractions and reflections. No two nanoseconds at the water's edge will ever be the same. There, the "nick of time" is always an instant.

Concord ice and snow varied in color as well. Walden ice could be "green as bottle glass . . . a vitreous glass green" in one direction, and an "unvaried white," in others. Lying in the street, a cake of ice cut from Walden was "emerald" green. The most complex colors were refracted from crystals before being reflected. "The sun being low, I see as I skate, reflected from the surface of the ice, flakes of rainbow somewhat like cobwebs. Colored ice was a phenomenon "to be put with the blue in the crevices of the snow," a phenomenon present only during some winters. For both blue snow and blue ice, and unlike air, red wavelengths are completely absorbed before they are scattered, which explains the impossibility of optical red snow relative to the commonness of optical red sky. True red snow is caused by dust and the life living on that dust.[26]

Natural Law

Like anyone else with an insatiable curiosity, Thoreau was constantly asking himself questions. In February 1851 he stumbled onto an intact raft of cranberry bog-peat about fifty feet long that had floated down the Sudbury River. Wanting to know how it got there, he asked himself five questions in a row: "Buoyed up (?), perchance. . . . Was there any pure ice under it? Had there been any above it? Will frozen meadow float? Had ice which originally supported it from above melted except about the edges?" At work here was the scientist within.[27]

Many non-scientists assume from their formal education that research scientists follow a logical, inevitable process toward the truth. This is laughable to any working professional who knows from their own experience that the tidiness of textbooks is an illusion, and that methodological rigor holds true only for data-collection protocols and final reports. The most important part of every scientist's job is not to collect data, get grants, follow a method, or be filmed for public television, but to ask questions that wander back and forth between taking nature apart and putting it back together again.

The taking-apart is called reductionism. It always begins with the question: What are the parts of the whole? In sequence, forest becomes tree, branch, leaf, lobe, cell, molecule, atom, and quark. In taxonomy, phyla are brought down to species. Thoreau's power of reductionism was as legendary as his skill at mensuration. One winter day, for example, he gives readers three pages of minutiae breaking down the search for wild honey, allegedly by unscientific men. When words failed, Thoreau took things apart by sketching their components, for example, parts of the halo of the moon. Each stroke of the pencil claimed a separate piece of the whole. Reasoning in this direction—from general principles to specific predictions—is called deduction. The more general practice of thinking this way is called analysis.[28]

My favorite example of Thoreau's reduction was his exacting descriptions of natural ice. One New Year's Eve he noted that the *slab* ice he had studied at Walden was only one of seven types of ice present on the Sudbury River, the others being: the *anchor* ice fastened to the shallows, *skimming* ice of the initial glaze, ice which consists of *individual crystals* formed from super-cooled water, *floe* ice which has floated intact downriver, *slush* consisting of multiple sources mixed together, and *refrozen slush*. In flowing water, the transition from pure liquid to wet solid formed a continuum of slush which, in its default condition, was "half ice, half water . . . the consistency of molasses or soft solder . . . a mass of crystals" flowing downstream, "looking like dark ripples in the twilight and grating against the edges of the firm ice." It was formed partly of new-formed crystals "shaped like bird's tracks" and partly by the residues of "thin trembling sheets," broken into pieces.[29]

At his next level of *reduction-deduction-analysis* were the various processes acting on each of his seven types of ice. Floe ice, for example could: change volume due to temperature, experience flexure from differential pressure, rupture, drift downstream, rise from buoyancy, become "dissolved," and many other processes. At the next level down, one of these six named processes (rupture) within one of seven types of natural ice (slab) could produce a host of unclassified sounds: "booming . . . as if it were a reasonable creature;" or "whooping like a cannon," or "a sort of belching . . . somewhat frog-like," or the ability to "mutter in a low voice," or to "thunder." These five named sounds were associated with five differ-

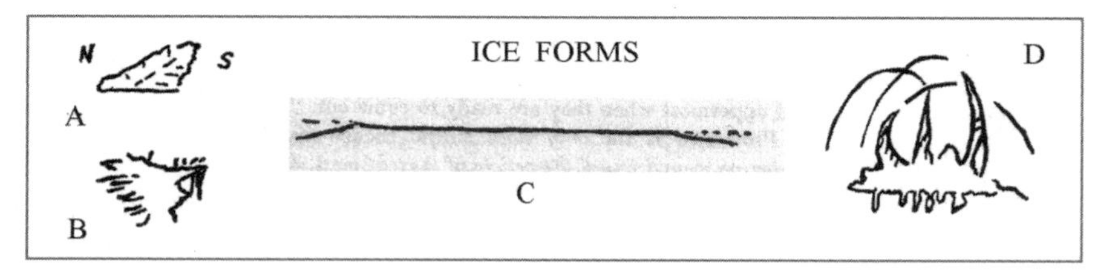

Figure 29. Thoreau's *Journal* sketches of ice. A, Fleets of flakes of ice in river, each resembling a sloop under sail (Feb 12, 1854); B, Organ-pipes on triangular projection at the cliff (Jan 11, 1854); C, River ice lifted by pressure from below and cracked downward near edges (Jan 14, 1854); D, Ice horns above frozen boulder at Clematis Brook (Jan 26, 1853).
Source: Thoreau, *Journal*, edited by Torrey and Allen, 1962.

ent kinds of ruptures, all of which were from one kind of process among six, acting on one of kind of river ice among many kinds of ice. To say "the sound of river ice" was simply not enough.[30]

The cracks associated with those noises were also complex. Those from slabs of lake ice are relatively simple because there is no mechanism for turbulent heating and no changes in fluid drag or pressure that might have influenced crystallization. Lake slabs had "cracks in the ice showing a white cleavage . . . somewhat like foliage, but too rectangular, like the characters of some Oriental language . . . somewhat rectangular with an irregular edge." More technically, they are "short, faint, flake-like perpendicular cleavages, an inch or two broad, or varying somewhat from the perpendicular." River slabs had a "white, leafy internal frostwork along the planes of the irregular flaring cleavages,—or call them deep conchoidal." To say that ice was cracked was simply not enough.[31]

Transformations of slab ice ice came from all directions. Tiny bubbles arriving from the bottom became part of its matrix; larger ones pushed down from the top became objects within that matrix. Holes in black ice of "various sizes, from a few inches to more than a foot in diameter . . . were perfectly circular" because "the circular wavelets so wore off the edges of the ice when once a hole was made." New layers added to old ice gave rise to special effects, for example ice that was "whitened and made partially opaque by heat." When covered by a transparent glaze from rain, it became "quicksilver on the back of a mirror." Floes on river and pond were "cracked coarsely and lay in different planes a rod or two in diameter." Marbling textures were caused by expansion and contraction "in a thousand directions" followed by the lensing effect of the cracks that "interferes with the rays of heat."[32]

Shoreline effects were fascinating too. Melting near land created moats with broad curves "answering somewhat to those of the shore, but more regular, sweeping entirely around the pond." Vigorous thrusting on the shore pushed up "ridges or ramparts," of sand "six or eight inches high, by a foot or two in width" with "stones as big as one's fist." When "heaved up," slabs of ice more than a foot thick creating pressure ridges in the "form of waves," where "the ice had crowded against the shore and forced it up." Though Thoreau didn't know this at the time, all of these compressional features were scale models of the same fundamental physical agencies that created the fold-and-thrust moraines of Martha's Vineyard, and the fold-and-thrust shearing the Nashoba Terrane.[33]

Above all kinds of river and lake ice was land ice. Freezing rain formed a "singular icy coat" that glazed the entire town. So "admirable" was one frozen waterfall that it took seven illustrations to capture it. Groundwater seeping from rocks on the Cliff created "great organ pipes . . . solid, pipe-like icicles" united to "form rows of pillars, or irregular colonnades" with "an annular appearance somewhat like the successive swells on the legs and tables and on bed-posts" draping the rocks. Splash ice created a "most splendid row of ice chandeliers, casters, hour-glasses (½) that I ever saw or imagined . . . prototypes of the ornaments of the copings and capitals."[34]

Reversing the direction of thinking gives us induction. Whereas deduction moves from general to specific, induction moves from specific to general. It always begins with a comparison between two objects or ideas, the detection of differences or similarities, and the asking of why. The answers that emerge connect previously separate domains in some way, for example the ripples of sand below air and those below water both result from fluid drag. In literature, the inductive link is metaphor. In science, it is law, ideally expressed as an equation. In the case of ripples, for example, the law is for fluid drag, one of several that apply. It is governed by at least four variables: the density, viscosity, and velocity of the fluid, and by a coefficient of boundary layer roughness. Just as deduction gives us the leaf as part of the tree, so too, does induction give us the forest as a collection of trees. Progressive induction leads to progressively more inclusive, or more general laws. Thoreau called these "higher laws," because they cover a broader variety of phenomena. I prefer to call them as "lower laws" because they are more fundamental, just as the foundation of the house is lower than anything that depends on it. What really matters is not the word used, but the absolute value of the induction relative to undiscovered reality. Novel inductions that unite seemingly disparate fields create what William Whewell dubbed "consilence," the jumping together of knowledge.[35]

Thoreau's habit of induction seems to have been as hard-wired as his habit of deduction. It seems he could not make an observation without comparing it to something else. Upon seeing an unripe walnut on the ground, his first thought was that of a green stone. Or realizing that "lakes without outlet are oceans, larger or smaller." Or wondering why rainbows in the morning were more dangerous to sailors than those of the evening. At greater distance, he compared the radial geometry of twinkling stars to sunlit flowers. Footprints in the snow became magnets for windblown leaves. Conversely, the tender green shoots of white pines became natural solar compasses. The shape of a beggar-tick seed suggested a solution for his loose shoelaces. Columns of rain-splashed sand beneath a protective pebble on the Marlborough road shared something in common with those of cooled basalt. One winter he asked, "Are not both sound and light condensed or contracted by cold?" It does not matter whether he was right or wrong. The point is that he instinctively linked ideas together in ways that could be tested with future observations.[36]

The intrinsic satisfaction of induction led him inevitably to more and more fundamental laws. Early in his "Year of Observation," for example, he became very excited while watching coarse foam develop within the frothy runoff of the Sudbury River. Tiny bubbles "making haste to expand and burst" created transient patterns of self-organization that reminded him "of the cells of a honeycomb." Instinctively, he mensurated, counting the edges of polygons, "four-sided, five-sided, but the most perfect, methinks, six-sided, but it is so difficult to count them, they are so restless and burst so soon." Here, Thoreau came close to inducing a general law of nature in which hexagonal cells originate under conditions of spatially uniform horizontal stress gradients, whether from packing, or from expansion-contraction. Examples include bubbles in foam, columns in cooling lava, cracks in desiccating mud, the boundaries of animal territories, or the comb in a honey hive.[37]

One July, Henry saw and sketched tablet-shaped clouds arranged like a line of tipped-over dominoes, likely two air masses creating an imbricate boundary layer. "Such uniformity on a large scale" he knew was caused by "the simplicity of the laws of their formation." If known, an "an experienced observer could discover the states of the upper atmosphere by studying the forms and characters of the clouds." Thoreau was so convinced that natural laws pervaded the universe that he even invoked one to explain his breakup with Emerson. "If I am too cold for human friendship, I trust I shall not soon be too cold for natural influences. It appears to be a law that you cannot have a deep sympathy with both man and nature."[38]

As late as 1849, Thoreau was still describing natural laws using the transcendental language of correspondence: "The laws of earth are for the feet, or inferior man; the laws of heaven are for the head, or superior man . . . Happy the man who observes the heavenly and terrestrial law in just proportion." Here, it is important to recognize that he is distinguishing two distinct sets of laws, those for the soul or spirit (heavenly), and those for nature or the body (terrestrial). Within the next ten years, and through a long and difficult process, Thoreau abolished the heavenly laws by pulling them down to earth. These he combined with terrestrial laws to create one unified set overseen by one ineffable "universal intelligence." For me, this marks the end of Thoreau's transcendental stage.[39]

Within his now-unified set of natural laws, Thoreau distinguished a polarity of higher and lower laws within his own psychology. These he featured in his "Higher Laws" chapter of *Walden.* One set of natural laws governed his "higher, or, as it is named, spiritual life." A "lower" set governed his "primitive rank and savage" life. Though he claimed to "reverence them both," he did concede that "primitive" and "inferior" laws gave rise to "spiritual" and "superior" laws as evolution took place. Modern science has since probed both of these categories and is seeking to unify them. Two book titles come to mind: the *Paleolithic Prescription,* dealing with the "brutish" needs of our modern bodies, and *The God Instinct,* dealing with the evolution of religion. Both books cover a range of subjects—diet, sex, violence, and politics—that Thoreau struggled with in *Walden*. Society continues to struggle with them today because both sets were potent drivers of human evolution or spandrels thereof. Both sets arose on planet earth, not in the heavens.[40]

This dichotomy of higher=spiritual and lower=primitive laws in that chapter of *Walden* must not be confused with the dichotomy of higher and lower laws described in the *Journal* during Thoreau's scientific sojourning stage. There, the words "higher" and "lower" were less about moral choices than stand-ins for poetry and science. Laws involving aesthetic beauty are *higher* than those of science because they are too complex to be governed by equations, and are instead controlled by ineffable aspects of mystery, meaning, beauty, and purpose. In other words, they are subjective. In contrast, the laws of science are *lower* because rules and formulas are possible, for example like those that govern the propagation of waves over the face of the water. They can be objective.

In the last of his many dichotomies, Thoreau divides the lower laws of science into higher and lower variants. Einstein's law of general relativity, for example is one of our lowermost laws. It is lower than Newton's law of gravity for the same reason the trunk of a tree is lower than its branches: the more general entity must predate and support the more particular. Similarly, Newton's branch of gravity is lower than the twig of Archimedes's buoyancy because the former explains the latter. In biology, the fundamental law of meiosis (nuclear cell division) gives rise to the laws of sexual selection involving peacock plumage and human fashion. Thus, each of Thoreau's scientific inductions was a downward step toward earlier, simpler, and more fundamental phenomena.

And to make matters even more confusing, Thoreau sometimes kept the dichotomies but stripped them of their labels and direction. "No good ever came of obeying a law which you had discovered." Here, I believe he is referring to laws revealed by scientific induction. Those that we should obey are the ones beyond science that are "continually being executed" by "the workman whose work we are." Similarly: "Obey the law which reveals, and not the law revealed." The former were the fruits of induction, those that help explain nature's "perfect government of the world." Resisting these deterministic laws were those of chaos, which involve complex, non-linear, dynamic interactions of otherwise straightforward, but low-ranking laws. The surface of Walden Pond, for example, was as chaotic as it was deterministic, depending on conditions. Thoreau revealed his understanding of this continuum between certainty and chance when he wrote "He who lives according to the highest law is in one sense lawless."[41]

One moonlit night in December 1853, the Sudbury River became laboratory for an inductive experiment with general wave theory. Rocking his boat, Thoreau watched waves propagate silently and effortlessly outward before breaking a thin layer of land-fast ice with noisy cacophony. "It is remarkable how much power I can exert through the undulations which I produce by rocking my boat in the middle of the river . . . The secret of this power appears to lie in the extreme mobility, or, as I may say, irritability, of this element. It is the principle of the roller, or of an immense weight moved by a child on balls, and the momentum is tremendous." Here, Thoreau demonstrates his understanding of wave theory: that they travel in circular orbits, and do so efficiently because the concentration of mass (density) is high, the fluid viscosity low, and therefore the conservation of energy is high. When slowed near the bank, the ultra-efficient propagation of wave energy in the "irritable" medium was converted to kinetic energy, which was dissipated to heat and disorder by friction, work, and noise on a surface. The following spring, Thoreau drew a moonlit sketch of waves on the Sudbury River in his *Journal* that matches his text: The "extremities were retarded by the friction of the banks."[42]

Before his boat-rocking experiment, Thoreau had induced the link between waves in air and water: "The surface of the snow in the fields is that of pretty large waves of a sea over which a summer breeze is sweeping." Specifically: "A slight, fine snow . . . [having] . . . drifted before the wind . . . has distributed over the ice as [to] show equal spaces of bare ice and of snow at pretty regular distances. I have seen the same phenomenon on the surface of snow in fields." Here is the "the track of the wind, the impress which it makes on flowing materials." Erroneously thinking that "the same law that makes waves of water" is involved, he correctly speculated on the cause of the latter: "Perhaps it may have to do with the manner in, or the angle at, which the wind strikes the earth." One year earlier, he asked himself, "Would not snow-drifts be a good study," continuing, "Are they not worthy of a chapter?" Indeed, they "are a sort of ripple-marks which the atmospheric sea makes on the snow-covered bottom." His final question on that day—"Are they always *built,* or not rather *carved* out of the heaps of snow?"—reveals his understanding of the equilibrium mass balance between deposition and erosion, respectively. He brought this concept to fluid boundary conditions between atmospheric layers, notably the fleecy imbricate clouds he had seen a year earlier, which rippled and drifted like the snow.[43]

Within a month, Thoreau was "struck by the gracefully curving and fantastic shore of a small island inside of Hull," Massachusetts, which he visited and sketched in July 1851. Knowing that Hog Island's continued forward movement—a "gently lapsing into futurity"—was required to maintain its shape, he suggested that the "inhabitants should bear a ripple for device on their coat of arms." Thoreau found it "refreshing" that lake and ocean waves were not only rippled alike, but also broke at the shoreline in similar ways. At Flint's Pond he specified the wavelength and amplitude of the waves as "ocean-like." And finally, moving to dry sand, he noted that the dunes of Cape Cod were "of every variety of form, like snow-drifts, or Arab tents." From this, he concluded that "water and sand also assume the same form under the influence of wind."

Through these, and other inductions, Thoreau taught himself that wave form and wave propagation were related, regardless of medium or material. At the next level, he induced that the boundaries between moving fluids and the materials they were flowing over could create ripples, dunes, and ridges, regardless of whether they were clouds, water, sand, or snow. Sojourn by sojourn, he gradually simplified his world through incremental induction.[44]

Four Other Senses

It is pleasant to work out of doors . . . my voice is echoed from the woods, if an oar drops I am fain to drop it again, if my cane strikes a stone, I strike it again—Such are the acoustics of my workshop." Thoreau enjoyed sound so much that he named a chapter of *Walden* after the phenomenon. Relative to the other senses, hearing brought him most often to a mystical, poetic state, and was most likely to make him weep.[45]

Thoreau understood sound as a wave propagated through the medium of air, "which our voices shake still further than our oars the water." He knew why the "ringing of the church bell" could be heard so far away. Beyond being deliberately manufactured to resonate with a pure and steady pitch, they were hung in high belfries to radiate that sound directly into clear air. The January bells of Bedford reached Thoreau's ear both at Walden and in the village because the air was typically dense and still in that month. When the Concord River was in flood during the cold of November, it offered a smooth sheet of water that allowed sounds to be heard from as far away as Lowell, in this case, a cannon fired in celebration of a Whig political victory. When in the village he wrote: "The loudest sound produced by man that I hear now is that of a train of cars passing through the town. The evening air is so favorable to the conveyance of sound that a sudden whistle or scream of the engine just startled me as much as it does near at hand, though I am nearly two miles distant from it." That particular sound penetrates the air sharply because its high pitch was designed to do so. The same is true for the piccolo in an orchestra, which is easier to locate with your eyes closed than the bassoon or string bass.[46]

The media through which sound travels can itself make noise when sufficiently agitated by turbulence. The "din," "roar," and "tempest" of windstorms gave Thoreau exquisite pleasure. Having heard the relentless "static" of the wind, he decided that the sound of "water falling on rocks and of air falling on trees are very much alike." This comparison provides an ideal case of separate sensations being unified by inductive perception.[47]

Echoes, he understood, were the reflections of sounds transformed into something new by the properties of the acoustic mirror they reflected from, and of the medium they traveled through. From echoes, he intuited the speed of sound based on known distances and vice versa: the distance to certain points based on the known speed of sound. Echoes circled as a function of distance, yielding a "rounder voice," circumnavigating the edge of Fairhaven Bay "with a sort of sweeping intonation half round a vast circle, *ore rotundo*." He linked the clarity of transmission to the difference in frequency: "You had to choose the right key or pitch, else the woods would not echo it with any spirit."[48]

Thoreau was particularly struck by the quality of echoes at Walden. Its woods would "ring" when gunshots were fired. "When, as was commonly the case, I had none to commune with, I used to raise the echoes by striking with a paddle on the side of my boat, filling the surrounding woods with circling and dilating sound, stirring them up as the keeper of a menagerie his wild beasts, until I elicited a growl from every wooded vale and hillside." Here, the acoustic roughness of the trees filtered out the high pitches, returning the lower ones that sounded like growls. On quiet summer evenings, the returns traveled efficiently within the discrete layer of cool air above the water. The symmetrical shape of Walden's western basin resonated like the main body of an acoustic guitar, the echoes from equidistant headlands reinforcing each other. Meanwhile, sounds entering the coves attenuated into rumbles. Echo, the pagan god of this lake, spoke in a low voice. Alan Hodder wrote that Thoreau's statements about the distinctiveness of the returning sound is a "curious fiction . . . that flouts the facts of physics." I disagree.[49]

Each bird made its own kind of music for Thoreau: the robin its joyous song, the sparrow its jingle, the bluebird its warble, the blackbirds their *conqueree,* and the larks their "insufficiently musical" performance. To this avian chamber orchestra Thoreau added the "long-drawn unearthly howl" of the loon, the "wildest sound" he "ever heard," and "probably more like a wolf than any other bird."[50]

In mid-January 1854, Thoreau heard a sound coming from the earth itself, a "singular buzzing sound from ground, exactly like that of a large fly or bee in a spider's web." Curious as usual, he listened carefully and then experimented with his walking stick, using it as a probe until he discovered the source. Air confined below the surface was "escaping upward through the water by a hollow grass stem," likely when air beneath frozen ground was being compressed by progressive down-freezing. In this case, the acoustics were comparable to that of a double-reed instrument like an oboe or bassoon. Another mysterious sound for Thoreau was the "telegraph harp," which whined a "clear tone" whenever there was enough breeze to move the wire. This is a challenging noise to explain except by analogy: something like an incredibly long banjo string that reverberates with constant pitch after being plucked into a standing wave by the wind. The constancy of the telegraph harp under breezy conditions matched the constant insect hum of bug season under muggy conditions. Near its peak on June 1852, "the air in this pitch pine wood is filled with the hum of gnats, flies, and mosquitoes." At other times of year, it was too dry, too exposed, or too breezy.[51]

Thoreau understood music as "the crystallization of sound." Its pitch, tone, rhythm, and resonance were as ordered through physical theory as the size, arrangement, spacing, and vibration of atoms are within a mineral. Pythagoras, he knew, had figured this out nearly two millennia earlier. Music, being "analogous to the law of crystals," explained why "melody can be heard farther than noise, and the finest melody farther than the coarsest." An Italian organ grinder walking the streets of

Concord, for example, could easily be heard in Walden Woods, whereas the cacophony of shouts and the whinnies of horses were not. One was crystalline, the other amorphous.[52]

Thoreau's tactile touch seems to have been his least acute sense based on the frequency and intensity of *Journal* entries. In fact, he comes right out and says "I see" and "I hear" far more often than "I feel," whether as touch or a psychological modality. There is one important exception, however. Thoreau was constantly engaged in the thermal touch of sensible heat, a category used by physicists to differentiate it from latent or radiant heat. Sensible heat was the touch he called hot or cold, depending on whether the heat is flowing into his body or out from it. Warm and cool were matters of degree.

"I bask in the sun on the shores of Walden Pond." This was in March 1840, five years before his move there, and well before he dubbed its northeast shore the "fireside of the pond" because it faced the direction of maximum solar warmth. This was one of the few places at Walden that he frequented outside its western basin, mostly before the railroad gave him direct access to the western basin. The calendar timing of his *Journal* entries about basking indicates Thoreau understood the principles perfectly. In late fall and early spring the sun was high enough to bear down directly, yet the ambient air was cold enough to make it worthwhile. By placing himself directly in the path of the sun's rays, his clothes absorbed its radiant heat, which transformed it into the sensible heat of his exterior, which raised his inner body temperature through conduction. "This is a glorious winter afternoon," he wrote on in February 1854, when "you feel a direct heat from the sun," so much better than "the relenting of a thaw with a southerly wind." Dead leaves melted downward into Walden ice demonstrate the same principle. The extra radiation they absorbed relative to the white reflecting surface allowed them to "burn" themselves more than an inch below the surface.[53]

The differential temperatures of groundwater springs fascinated Thoreau because they offered clues to the complexities of groundwater flow. Knowing that the bulk of groundwater went down into the "bowels of the earth, the cellar of the earth" as snowmelt and cold spring rains, he inferred that the "few really cold springs" indicated unusual subterranean situations in which early recharge was channeled into deeper, longer routes. Conversely, the presence of warmer springs nearer the surface explains why, in settings near large aquifers, the soils froze so shallowly.[54]

Aside from basking in radiant heat, Thoreau also used his body like a lizard to sop up sensible heat through conduction. On one cold night in May, a boss of rock had previously soaked up enough heat from the afternoon sun to keep him warm several hours after dark. In August he wrote that "the sand is cool on the surface but warm two or three inches beneath, and the rocks are quite warm to the hand." Here he highlights the link between heat flux and thermal conductivity. He knew that the depth of frozen ground increased with greater exposure to the elements, which rose with elevation. And he knew that the surest way to cool down on a hot summer day was to plunge into the bracing cold water of a pond or river. This experience involves enveloping flow of sensible heat through an excellent conductor with a high heat capacity.[55]

The annual budget of heat in Concord interested him nearly as much as its annual budget of water. Between late February and April, a great deal of nature's work throughout town involved melting all the snow: "twenty-one inches of ice on the ponds," and thawing "from one to two and a half or three feet of frozen ground." Given such a large sink for latent heat, actual warming of the soils was delayed. Lake Walden, he realized, was "a good thermometer of the annual heats, because, having no outlet nor inlet on the surface, it has no stream to wear it [the ice] away more or less rapidly or early as the water may be higher or lower, and also being so deep, it is not warmed through by a transient change in temperature." Though cumbersome, this passage illustrates Thoreau's clear understanding of every term in the lake's annual heat budget, which governed his heat budget as well.[56]

Smell was Thoreau's closest daily approach to chemistry. Though it was rumored that "no hound could scent better," this sense enters his journal far less often than for sights and sounds, and usually only for the most delicious or repugnant smells. Early autumn was one of the richest times for pervasive smells, when the leaves, fruits, and nuts of production were returning to the earth through decomposition. In September 1853, he attributed a three-week-long "strong musty smell" to fungi decaying in the woods. A month later the fermentation of wild grapes crossed his threshold of notice, and a month after that, a "traveller's pipe very strongly a third of a mile distant." In *Walden* he writes of a "dead horse in the hollow by the path to my house, which compelled me sometimes to go out of my way, especially in the night when the air was heavy." Ever the optimist, this assured Thoreau of the "strong appetite and inviolable health of Nature." The carrion smell of a dead cow "borne down over the meadow on the damp air" provided a link between the propagation of smell and the medium in which it diffused. Differences in smell also revealed to him the stratification of air under calm conditions. In the heat of late June 1853, he walked up and down through the hills and hollows of Walden Woods, noting "the fresher and cooler in the hollows, laden with the condensed fragrance of plants, as it were distilled in dews; and yet the warmer veins in a cool evening like this do not fail to be agreeable."[57]

Flooding was associated with particularly interesting smells. When on the causeway across the Great Meadow in mid-May 1852 he noticed "the sweetest fragrance I have perceived this season, blown from the newly flooded meadows . . . It is ambrosially, nectareally, fine and subtile . . . Is it not all water-plants combined? . . . I would give something to know of it." Quite likely, he was smelling two sources: the ambient freshness of spring growth mixed with aromatic gasses being expelled upward from the alluvial humus as floodwater infiltrated downward through the meadows. Something similar may have been the cause for an untraceable odor Thoreau described by the western shore of Walden a few weeks later in June. Perhaps water rising in the pond spilled outward to saturate adjacent soils, ventilating their gasses. Conversely, when water receded in the meadow, the smell reminded him of salt marshes, likely because the same hydrogen sulfide gasses were involved, the rotten egg smell created by anaerobic bacteria when oxygen was used up beneath the inundation.[58]

"Sugar is not so sweet to the palate, as sound to the healthy ear," wrote Thoreau on New Year's Eve, 1853–1854. Taste, it seems, was his least activated biological sense. This is no surprise for a man who prided himself on plain food: rice, hasty pudding, berries, hoecake, pickerel, wild apples, and the occasional woodchuck. He was, however, a great connoisseur of water. And he insisted that his own handpicked huckleberries tasted better than those handpicked by others. Though this suggests a discriminating palate, surely it is more symbolic of his extremist position on self-reliance combined with his animal instinct for foraging.[59]

Thoreau's scientific sojourns were the raw materials he used to create the nature writing in *Walden*. That is where I turn next.

8

WRITING *WALDEN*

My first reading of *Walden* in 1970 was a transforming experience. At the time, I was an eighteen-year-old freshman with a low draft number monitoring the nightly news from Vietnam. Thoreau's name came up. The first Earth Day was also in the news and rapidly approaching. His name came up again. During some restless moment, I found a paperback copy of *Walden,* bicycled seven miles out to our family's lakeside cabin in the pine woods of northern Minnesota, and began to read. I slogged my way through "Economy" before giving up, having discovered what Stanley Cavell was discovering at about the same time: that "*Walden* sometimes seems an enormously long and boring book." On my second try, I thumbed ahead, read "The Ponds," returned to "Solitude," and became hooked for life. Using one of Thoreau's words, I remember that "creamy" moonlit night at the lake with the clarity of yesterday.[1]

During the next several decades, I intermittently revisited parts of *Walden* for inspiration but never read it cover to cover. That didn't happen until my late forties, when I skimmed the entire text as a research task when writing a book on New England's historic fieldstone walls. My first truly careful read of the sort that Thoreau asks his readers to do in "Reading" took place when I was team-teaching a college course on *Walden* in my mid-fifties. After several careful reads and lots of student discussion, I remained convinced that Thoreau thought Walden Pond was "the most attractive, if not the most beautiful, of all our lakes, the gem of the woods."[2]

It was not. Those words refer to White Pond, which, as late as 1854, had not yet been profaned by "the wood-cutters, and the railroad," and, as he confessed, "I myself." Though he plainly says in *Walden* that White Pond was the true "gem," for some reason, I could not bring myself to

Figure 30. Thoreau's writing desk at Walden Pond. Replica of desk set up in its normal place in replica of house at Walden Pond State Reservation.
Props courtesy of Richard Smith. January 2013.

believe this until after I read the *Journal*. In it he reveals unambiguously that White Pond was the pedestal of purity from which Walden had fallen, a respite tucked into Concord's infrequently visited southwest corner, Thoreau's favorite direction. Being a stand-alone kettle, rather than a coalesced collection of several, White Pond had the advantage of being more radially symmetric. It was also a place of legend, having a mysterious upside-down tree and its own pod of Loch Ness monsters: "several pretty large logs may still be seen lying on the bottom, where, owing to the undulation of the surface, they look like huge water snakes in motion." Such logs, when soaked to neutral buoyancy and tipped at low angles, can magnify internal seiche waves through the levering effect of their lengths. This explanation is likely Nessie's authentic persona as well.[3]

The true twin of White Pond is Walden's western basin, also "about forty acres." Within Concord's transcendental "lake district," only these two qualified as "great crystals on the surface of the earth, Lakes of Light . . . too pure to have market value." Only these two were "crystal wells" that "plainly can never be spoiled by the wood chopper." Why just these two? I suggest it is because they were most closely matched in geologic origin: both sunk into the same materials within the same kame delta at nearly the same distance from the same former ice front position at the same time. They were like two peas in the same sandy pod. When Thoreau remarked that "they are so much alike that you would say they must be connected underground," he might have had these nearly identical origins in mind.[4]

After studying the *Journal* for this project, I reached an even more startling conclusion. That Thoreau's favorite place was not a pond at all, but an outcrop of the Andover Granite. I refer to the southwest-facing cliff above Fairhaven Bay, located about three-quarters of a mile southwest of Walden Pond, his favorite direction. Abbreviated the "Cliff," this was the largest exposure the Earth's crust on his side of the Sudbury River, located midway between his two favorite transportation corridors: the river and railroad. "In all my rambles" he wrote in 1850, "I have seen no landscape which can make me forget Fair Haven. I still sit on its Cliff in a new spring day, and look over the awakening woods and the river." He did this at all hours in all seasons: before, during, and after his famous sojourn at Walden. Members of his family had brought him there as a child. As a young college student in Cambridge, he fondly remembered the Cliff in an 1835 essay. When living unhappily on Staten Island in 1843, this is where he longed to be. It was there that he received his moonlit Arctic vision in 1852. Ironically, the frequency and aesthetic intensity of Thoreau's visits to the Cliff seemed to peak in 1853–1854 when he was hardest at work on *Walden*. Perhaps toying with the readers, Thoreau begins his pivotal chapter "The Ponds" with a sunset view from "Fair Haven Hill."[5]

From this rough granite outcrop, his spirit was fed by the crystalline beauty and authenticity of granite. The glint of feldspar. He was also fed by the richness and connectedness of the landscape mosaic. Distant sunsets. Mountain vistas. Breezy updrafts. An escarpment overlooking cultivated fields that sustained human beings and their communities. A lake vista similar in outward appearance to Walden, but with a meandering river running through it, one redolent with the smell of mud, rich enough for a carpet of mussels, and broad enough to tack a sailboat upon. This was a corridor for spawning fish, a flyway for geese and gulls, a fertile chain of marshes for inquisitive herons, and a priceless metaphor for the flow of time, the "stream I go a-fishing in."[6]

Literary Black Hole

All of these images from his sojourning country—seen and visited over a thirty-two year span—got sucked into the literary black hole of *Walden*. Thoreau not only let this happen, he encouraged it to happen in what I judge to be his most important compositional technique, the first of three reviewed in this chapter. In short, his masterpiece is a condensed memoir of a composite place. All of this time and space was drawn inward to Walden's epicenter and then downward to its focus on granite, Earth's most "fundamental rock."[7]

"Walden is a perfect forest mirror, set round with stones . . . a mirror which no stone can crack." The history of this essential metaphor illustrates how Thoreau's space and time became concentrated in *Walden*. The basic idea was published in the October 1843 *Dial*. The details were added a decade later in September 1853 after being imported from the sometimes-murky Concord River: "The river nowadays is a permanent mirror stretching without end . . . There it lies, a mirror uncracked, unsoiled." By my reading, the linear component from the river description—"stretching without end"—was replaced by the radial component of Walden's western basin—"set round with stones"—and then exported to the *Walden* manuscript sometime during the next six months.[8]

By the time Thoreau submitted his *Walden* manuscript in the early spring of 1854, he had visited Walden Pond regularly for more than twenty years, "almost daily" during the last ten. "One of his earliest recollections of Concord," wrote Frank Sanborn, a personal friend and nineteenth-century biographer, "was of driving in a chaise with his grandmother along the shore of Walden Pond, perhaps on the way to visit her relatives in Weston, and thinking, as she said afterward, that he should like to live there." Thoreau's first memory of the pond dates to 1822 or 1823 when he was a five-year-old child. Brought there by his parents on a trip out from Boston, he remembered it as "one of the most ancient scenes stamped on the tablets of my memory," its woodland habitat "another name for the extended world at the time." Early meaningful visits as a young man fresh out of college were recorded in 1837, the first year of his *Journal*. By December 1840, he had already made up his mind to live there when circumstances permitted, and he had already solidified his perceptions of the place. By May 1841, he was penning *Journal* phrases that eventually found their way into *Walden's* final text practically untouched, for example the one about charming the perch with his flute and seeing the moonlit bottom "strewn with the wrecks of the forest."[9]

Two years later in 1843, Thoreau featured Walden in *A Winter Walk*, although he kept its identity "concealed as a poet might conceal the name of his mistress," according to Perry Miller. Walter Harding concluded that this essay, "derived chiefly from the *Journal* for 1841," was "one of Thoreau's best . . . Indeed the spirit of the whole essay is close to some of the best pages of nature description in *Walden*. The later book at times even echoes words and phrases from it." One could go farther and say that *Walden* even "reprints" words and phrases from *A Winter Walk*. In it, he writes of a "woodland lake . . . in a hollow of the hills" containing "the expressed juice" of the forest and "without outlet or inlet," "Earth's liquid eye; a mirror in the breast of nature." The "woods rise abruptly," forming "an amphitheater." Its "fresh surface is constantly welling up," its impurities swept away in spring after the "plain sheet of snow" had been "swept down to bare ice in places."[10]

Indeed, some of *Walden*'s most lyrical metaphors had been imagined, shared with lecture audiences, and published more than a decade before the lake was profaned by railroad construction. These idealized visual remembrances from 1841 may have been the ones Thoreau wrote about in 1854, when he claimed that the pond had remained unchanged in spite of the human assaults thrown against it, namely when its shores were approaching the devastation of clear-cut conditions, and day-tripping tourists from Boston were already taking the train to picnic and swim at the pond.

Thoreau's old memories of Walden from 1841 also got mixed up with much more recent observations from White Pond. "Nature has woven a natural selvage . . . There are few traces of man's hand to be seen. The water laves the shore as it did a thousand years ago." This passage from the published text of *Walden* was lifted almost *verbatim* from a *Journal* description of White Pond entered in June 1853: "It is a natural selvage. It is comparatively unaffected by man. The water laves the shore as it did a thousand years ago." Another significant import was *Walden*'s famously triumphant clause: "frost coming out of the ground." In this case, the import went the other direction, from the debris-rich slabs of lake ice lying shattered along Walden's eastern shore to the Deep Cut of the railroad northwest of the pond. Sharon Cameron's stresses that "*Walden* and the *Journal* are autonomous writings." As projects they are. But with respect to its nature writing, they are hopelessly intermingled. With few exceptions, the latter was the mother lode from which the gold ingot of *Walden* was smelted. In fact Thoreau's main title, *Walden* is a space-time concentration of its subtitle, *Life in the Woods,* a purpose suggested by the preposition "or" between them.[11]

Making things disappear is another thing black holes do very well. I refer to those *Journal* observations regarding Lake Walden that disappeared before they reached the book manuscript. This made the Walden of *Walden* a prettier, cleaner, more innocent, and safer place than it actually was. *Walden*'s pickerel lying on the ice "possess a quite dazzling and transcendent beauty" like "precious stones . . . or crystals." Having done plenty of ice fishing myself, I strongly suspect that also on the ice that day were the blood and guts of fish and the urine stains of fishermen and their dogs. Also missing were *Journal* descriptions of Walden's ugly meltwater, a "dirty or grayish-brown foam . . . now like bowels overlying one another, now like tripe." This gray water still graces Walden today. Though I have done no analyses, I suspect it is some combination of chemicals excluded during the freezing process mixed with bacterial residues and air pollution from the nearby four-lane highway, U.S. 2. Also missing from the sanitized *Walden* is the true identity of what killed the pitch pines in front of his house. Though he convicted a suspect in *Walden,*—the gentle rise and fall of Walden's shore—he unambiguously identified the perpetrator in his *Journal* as the scissoring action of ice floes being thrust against the bank. What he does not say is that the "lips of the lake" there were not "licked" by water, but bitten by icy teeth.[12]

Also missing from *Walden* is the terror Thoreau must have felt when lightning struck his lakeshore. In the published version, he was merely "struck with awe" by the spiral black mark on a pitch pine where "a terrific and resistless bolt came down out of the harmless sky eight years ago." The source *Journal* passage is more ominous: "an invisible and intangible power, a thunderbolt." Even more ominous is the probable inspiration

Figure 31. Flowing gray foam above Walden ice. Similar to what Thoreau described as a "dirty or grayish brown foam" on Walden pond, produced by some combination of solutes excluded during freezing, aerosols, microbes, and other sources. Note central whirlpool and braided flow pattern.
January 2013.

for the source passage, a powerful lightning strike from June 1852. In his description of it, Thoreau used language reminiscent of Katahdin: "a thunderbolt, accompanied by a crashing sound . . . Jove's bolt . . . a Titanic force, some of that force which made and can unmake the world: . . . scorched . . . exploded . . . hurled . . . smashing. . . . splintered . . . broken . . . crashing." At each step toward *Walden* from the original inspiration, the language was softened and personalized. Terror, brutality, and savagery gave way to religious awe for a "sacred spot."[13]

In the real world of death and sadness and fright, *Walden* became Thoreau's secure retreat, a fallout shelter from the social dust of the world. A "snug harbor" and a "good port." Cupped by God's hands, and "embosomed" by protective hills, it was a "little world" of calm seclusion very unlike the oceanic shoreline of nearby Flint's Pond, which he exaggerated as a literary foil to highlight Walden. After overdosing on Walden's tranquility, Thoreau sometimes felt it "worth the while" to go to Flint's "if only to feel the wind blow on your cheek freely, and see the waves run" as a foaming surf above its broad littoral flats. He enjoyed seeing the broad sweep of its "sedgy shore" on windy fall days, feeling "the fresh spray blowing in my face," and finding "the mouldering wreck of a boat . . . amid the rushes; . . . as impressive a wreck as one could imagine on the seashore." At Flint's he could "admire the ripple marks on the sandy bottom, at the north end of this pond, made firm and hard to the feet of the wader by the pressure of the water, and the rushes which grew in Indian file, in waving lines, corresponding to these marks, rank behind rank, as if the waves had planted them." Using this and other literary foils, Thoreau set the Walden of *Walden* apart from the brute insensitivity and "inexhaustible vigor" of truly wild nature, manifest elsewhere as the "vast and titanic features, the sea-coast with its wrecks, the wilderness with its living and its decaying trees, the thunder-cloud, and the rain which lasts three weeks and produces freshets."[14]

In less salient ways, the Walden of *Walden* is a distortion of the *Journal*'s less scripted encounters. In the latter, we read of Thoreau's affectionate descriptions for Concord's hundreds of miles of fieldstone walls. In the former, we get hardly a peep. In the *Journal* we read: "All around Walden, both in the thickest wood and where the wood has been cut off, there can be traced a meandering narrow shelf on the steep hillside, the footpath worn by the feet of Indian hunters . . . And the same trail may be found encircling all our ponds." In *Walden* he reserves this observation for Walden alone, and paints the trail on the edge of sensory detection: so indistinct that a dusting of snow was needed to see it. In *The Maine Woods* we read of the true "lake-country of New England," the high, rocky "archipelago of lakes" in Maine and New Hampshire, where "boatmen, by short portages, or by none at all, pass easily from one to another." In contrast, the "lake district" anchored by *Walden* refers to a handful of disconnected, deeply inset ponds of vastly smaller scale.[15]

Threshing Floor

"From all the points of the compass," Thoreau once explained, "have come these inspirations and been entered duly in the order of their arrival in the journal. Thereafter, when the time arrived, they were winnowed into lectures, and again, in due time, from lectures into essays," and then into his books leading to *Walden*. The key word here is "winnowed," an agricultural technique used to concentrate cereal grain by letting the chaff blow away as the denser kernels fall downward. Such winnowing is the second and most important compositional technique reviewed in this chapter. Thoreau let the chaff of his scientific methodology blow away as the grains of literary truth fell to the threshing floor.[16]

At each step of refinement—notes, journal, lecture, essay, book—the scientific content of the text diminishes. In the *Journal* we encounter staggering detail and the sequential laying-out of arguments: "first . . . second . . . third," and so forth. In the final text of *Walden,* these are long gone, but never at the expense of a richly expressed summary. During final editing, Thoreau imagined having an "india-rubber" eraser to "rub out at once all that in my writing" he was so reluctant to "erase." By my reading, he had already used one vigorously to rub out hundreds of technical passages before blowing the editorial rubbings away from his manuscript leaves like chaff. Though the literary quality of geo-*Walden* reflects Thoreau's obvious gift for languages, his chief subject at college, the leaden density of its meanings derive chiefly from Thoreau's clear scientific understanding of how the Walden System actually worked.[17]

Poem 1494 by Emily Dickinson illustrates Thoreau's technique in the extreme. Of the heavens, she writes: "The competitions of the sky / Corrodeless ply." Of these seven words, only four are substantial. Yet that is all she needed to describe both the push-pull (centrifugal repulsion and gravitational attraction) forces that govern celestial motions, and the sterility (chemical inertness) of waterless space. Her four words of literary super-gravity abstract the whole of the astronomy course I took in college. Thoreau used a similar practice when he used the word "zephyr" in *Walden* to abstract a forty-two-word *Journal* sentence involving temperature, radiation, sensible heat, conduction, and heat capacity.[18]

Before I read the *Journal,* I had enjoyed geo-*Walden* because it resonated so accurately with my lifetime of experience with sandy kettle ponds, glacial aquifers and the dry soils of pine forests. Only after reading the *Journal* to the date of Walden's submission, extracting 841 passages from it, coding those passages using 49 attributes, and sorting its 41,209 data cells, did I discover why the lyricism of *Walden* captured the Walden System so well. It was based on good observations. Laura Dassow Walls correctly labels Thoreau as an "empirical holist" because he built his world together one technical detail at a time to erect the geological hydrological, meteorological, limnological, and ecological reality on which his evocative prose-poetry is based. Following Dickinson's example, I will start with four simple words.[19]

"A clear and deep green well." The image is lovely, in and of itself. This is also Thoreau's most condensed description of Walden. Setting aside the conjunction and the article, and being analytical, the remaining four words integrate four discrete physical phenomena, all of which were fully detailed in various *Journal* passages. Walden is not merely deep. It is the deepest lake in Massachusetts, allegedly too deep to fathom. Being so deep, it extends well into the aquifer, making it a natural well, albeit a very wide one. And because this well holds such a large volume of water relative to its surface area, is dug into permeable sandy soils, and is completely surrounded by forest, it is exceptionally clear, owing to the dearth of sediment and nutrient. And for water this pure, the extinction color of incident radiation is emerald green. When linked, these four words—clear, deep, green, and well—capture the holism of the Walden System. Each was carefully chosen to link natural fact to aesthetic expression. This "correspondence" is not transcendental. It is editorial.[20]

Thoreau's threshing technique parallels an aesthetic theory he once shared in a letter to the Emersons; one that Jeffrey Cramer used to help us understand the *Journal.* "It is the height of art that on the first perusal plain common sense should appear—on the second severe truth—and on a third beauty—and having these warrants for its depth and reality, we may then enjoy the beauty forever more." When applied to my four-word example above, the *common sense* involves the four observable natural facts of clear water, deep basin, green color, and groundwater source. The *severe truth* is the set of logical explanations I explicated above: low-nutrient, kettle sinkhole, optical scattering, aquifer properties. The *beauty* is the subjective feeling that emerges from the unification of these seemingly separate phenomena. To appreciate Walden, we need only read it. To understand it, however, we must work and wedge our way downward from Thoreau's final literary flourish to the original "warrant" of its physical entities. Illustrating this technique is the sole purpose of the remainder of this section. I begin with a few very specific examples before moving on to larger, more enriching themes.[21]

"The landscape radiated from me." Here, in the *first* sentence of the *first* paragraph of the *first* description of Lake Walden, Thoreau captures its most salient attribute: the radial symmetry of its western basin. Continuing: "I was seated by the shore of a small pond . . . but I was so low in the woods that the opposite shore; half a mile off, like the rest, covered with wood, was my most distant horizon." Here, in what reads like casual description, is an embedded understanding that the lake lies well below the general level of the terrain. "It was not always dry land where we dwell," he continues almost mystically. In fact, "I see far inland the banks which the stream anciently washed, before science began to record its freshets." This evocative passage captures and distills at least a dozen carefully documented geological inductions from sojourning space and recorded in his *Journal*. These we have already encountered in the geology narrative.[22]

The Flint's Pond of *Walden* was "more fertile in fish . . . comparatively shallow, and not remarkably pure." All three adjectives are physiochemically linked. Shallower translates to warmer because there is more surface area exposed to the summer sun per unit volume. And when this greater warmth combines with the higher nutrient content of Flint's more silt-rich (till) soils, the result is higher primary productivity for aquatic algae, and therefore of zooplankton, and therefore of fish. Thoreau had this figured out. He just spared us the details.[23]

In "Sounds," he writes: "Late in the evening I heard the *distant rumbling* of wagons over bridges—a sound heard farther than almost any other at night." This *Walden* passage is nearly verbatim from its source

Journal passage from 1845. Completely ignored are the eighty-three words of acoustical elaboration from his 1851 *Journal* entry. In this case, Thoreau the field scientist backs up Thoreau the nature poet with nine years worth of observation and analysis that he left out on purpose.[24]

Thoreau did not employ his winnowing technique in his essays predating his conversion to science. In *A Week*, for example, he went out of the way to show off what he had learned from books, writing extended, often uncredited, digressions. Consider its ten pages on fish anatomy, taxonomy, and behavior, and its five pages on the origin of potholes. *Walden,* in contrast, observed ecologist David Foster, is "surprisingly free of natural history observations." Ecocritic Rochelle Johnson noticed this too, concluding that: "most readers of *Walden* find it short on physical detail and long on philosophizing." This is because the informing science from the mature *Journal* has been deliberately winnowed away, leaving kernels of simile, analogy, metaphor, and allegory.[25]

One specialty tool in Thoreau's threshing toolbox was his use the words "perchance" or "perhaps" or the phrase "as if." These qualifiers are sometimes wild guesses. But most of them are shortcuts employed to circumvent the detailed arguments that would otherwise be necessary. Most are short for "I hypothesize that . . ."

Finally, we are primed to miss the science that informs *Walden* because its author introduces himself to us first as the social critic of "Economy" and then as the philosopher of "Where I Lived and What I Lived For." To meet the physical scientist, we have to wait until "The Ponds," which is midway through the book. Only then does he really begin to introduce himself as an observer of nature. This sequence of social critic to philosopher to field scientist derives from the *chronology* of his written drafts, not from the relative *importance* of his content.

One of *Walden*'s most knotty scientific phenomena is the lake's continuously changing vertical stage, because this integrates dozens of processes and time scales I have already described. "The pond rises and falls, but whether regularly or not, and within what period, nobody knows, though, as usual, many pretend to know." In this *Walden* passage, he emphasizes the mysteries because they make for more interesting reading than chains of logic and firm conclusions in the *Journal*. In the source passage, Thoreau is much more precise. "It is remarkable that the rise and fall of Walden, though unsteady, and whether periodical or merely occasional, are not completed but after many years. . . . It attains its maximum slowly and surely, though unsteadily." Winnowed from this more challenging source passage was the very thing that motivated Thoreau's return for his final monitoring observation: the long duration between rises and falls "many years" in the making. He also blew away the technical distinctions between three different temporal phenomena: "unsteady," which means chaotically random or stochastic; "occasional," which refers to an unpredictable external perturbation; and "regularity" which reflects periodic, if not harmonic fluctuation. And the last thing he let blow away before finishing this *Walden* passage was the idea that stage variations with higher frequencies are superimposed on those with lower ones, in this case transient spasms of recharge on the secular trends of rising or falling levels. Nitpicking these distinctions in *Walden* would have ruined it.[26]

In a few rare cases, Thoreau felt the need to hammer his reader with technical details that seem ends in themselves, most famously his multipage description of Walden's ice. In such cases, however, he had a different purpose: to share his enthusiasm for the beauty of small details, which is something Darwin did routinely. Even more rare are cases of scientific thinking in *Walden* that I could not find anywhere else. One humorous example was Thoreau's eleventh-hour anecdote about digging angleworms for bait, in which one's luck varied inversely with the square of distance from his house. This strikes me as a thinly veiled tease aimed at local "Men of Science" because he knew full well that the rule would not apply.[27]

Fluidity is present in all four of Walden's physical realms, but is most conspicuous at the boundary where air and water meet. Consider this example from "Where I Lived and What I Lived For": "As the sun arose, I saw it throwing off its nightly clothing of mist, and here and there, by degrees, its soft ripples on its smooth reflecting surface was revealed, while the mists, like ghosts, were stealthily withdrawing in every direction into the woods, as at the breaking up of some nocturnal conventicle. The very dew seemed to hang upon the trees later into the day than usual, as on the sides of mountains." Here, he addresses what can only be the microclimate of an isolated, forested, and rounded lake located within a mid-latitude macroclimate having strong contrasts between summer and winter. Otherwise the passage would not hang together.

The mists "stealthily withdrawing in every direction" require a disk-shaped pocket of air trapped below the prevailing atmosphere of the delta plain. Night mists form when the warm humid air above the lake is chilled by its cooler surface, causing condensation within the lowest layer. After daybreak, the lake "throws off this mist" in two ways. Initially, the slight increase in solar radiation shortly after dawn warms it, forcing the droplets to shrink and many to evaporate back into invisible vapor. Later, the whole surface layer becomes drawn outward "in every direction" as a consequence of the western basin's radial geometry.[28]

Under stable atmospheric conditions, a strengthening sun causes the air to warm more quickly above the dark woods than above the night-chilled surface of the pond. This creates a wide ring of lower pressure encircling a "donut-hole" of high pressure above the water. Given this energy gradient, conservation of mass requires that the air above the lake be pulled radially outward "into the woods," away from the pond center where air currents are the weakest because they lie furthest along the energy gradient causing the motion. This outward flow reveals the "smooth and reflecting" center of the pond and simultaneously creates "soft ripples" nearer the margin as a consequence of the strengthening breeze doing the revealing.

This outward flow also explains the heavy lakeside dew. On the forested bank surrounding the pond, droplets of mist collide with the leaves, needles, and twigs where they are captured by the high surface tension of water. This "cloud forest" mechanism—when combined with the slightly richer soils caused by the extra soil creep and the extra warmth of exposed slopes—explains why the woods are lusher on the rim of the kettle facing the lake than on flat tread of the delta plain. By the end of "Where I lived," Thoreau has extended his word imagery to practically every human concern: moral purity, mood, intimacy, solitude, enclosure, isolation, asceticism, and resilience. Undergirding these passages is the relevant physical science.

Continuing through his text to "The Ponds" and moving out on to the surface of the lake, Thoreau wrote of spending "many an hour, when I was younger, floating over its surface as the zephyr willed, having paddled my boat to the middle, and lying on my back across the seats, in a summer forenoon, dreaming awake, until I was aroused by the boat touching the sand, and I arose to see what shore my fates had impelled me to." As with the preceding explanation involving mist, this single sentence is also about radial micrometeorology, but with the viewer being *on* the lake rather than looking *at* it. Change one word of this lovely passage and it would not make physical sense. Winter is ruled out because the water must be in its liquid state. Late fall and early spring are ruled out because the radiation contrast is often insufficient to drive strong radial outflow and greater storminess overwhelms the thermal breezes. Late morning is required because the sun must be intense enough to drive the thermal engine, but the air above the lake surface must still be cool, which it is, thanks to the high heat capacity of water.[29]

Simultaneously, the leaves of the woods are darkest in full summer, and therefore most efficient at absorbing the energy. Energy not used for photosynthesis is radiated outward from leaf surfaces, warming the canopy air. Because air is such a good insulator, little heat can be conducted downward to the ground. And because it has such a low heat capacity, its temperature quickly rises, which lowers the air density. The result is the same ring of low pressure that caused the mists to withdraw, but now greatly exaggerated in velocity. Cooler lake air is drawn outward along a series of rays, accelerating toward the bank where the pressure differences reach maxima. This process shuts down automatically at midday when the top few inches of water become too warm. In the source *Journal* passages, Thoreau lets us know that if he paddled only part way out, he would blow right back to the same shore every time. Only in the sweet spot of the middle could the Fates send him in random directions. Only there could he surrender to nature's choice.[30]

Walden's sub-text of meteorological physics is further revealed by a multiplicity of *Journal* entries. "In the hottest day you can be comfortable in the shade on the open shore of a pond or river where a zephyr comes over the water, sensibly cooled by it; that is, if the water is deep enough to cool it." This language addresses seven separate terms in the heat budget that would require quantification in a numerical simulation model: the local temperature gradient, differential absorption of shade and sun, the boundary conditions of lake surface and shore, the conduction and advection of sensible heat, and seasonal thermal inertia. Thoreau was also aware that this mechanism switches on and off during the day, as the balance of heat changes: "He who passes over a lake at noon, when the waves run, little imagines it serene and placid beauty at evening." And that even when weakest, local air movements are revealed by distorted reflections: "The surface is not perfectly smooth, on account of the zephyr, and the reflections of the woods are a little indistinct and blurred." This daily cycle of local wind peaked in summer because by the time fall arrived, there was too little radiation for dramatic warming: "toward the evening of the day the lakes and streams are smooth, so in the fall, the evening of the year, the water are smoothed more perfectly than at any other season."[31]

"The shore is composed of a belt of smooth rounded white stones like paving-stones, excepting one or two short sand beaches, and is so steep that in many places a single leap will carry you into water over your head." Here—without any explanation—Thoreau correctly links the stoniness of the shore, the dearth of beach, and the steepness of the bank over most of its perimeter. Steep banks concentrate the wave energy of breakers into a narrow zone by creating what coastal engineers call a *reflective* beach profile. This contrasts with the *dissipative* beach profile of the shallows of Flint's Pond, where much of the incoming kinetic energy is consumed by friction with the bottom, a process that helps push sand upward toward the land. On Walden's shore, the reflective profile pulls sand outward into the water, which helps explain why Thoreau mentions only two "short" beaches in *Walden.*

Were the shore of Walden fixed in position by a spillway, its wave climate would produce only a narrow notch armored by a stone lag concentrate. But because the pond "rises and falls" up to twelve vertical feet, erosion is spread out over a broad range, creating a conspicuously "stony shore," a broad ring of granules, pebbles, cobbles, and boulders that Thoreau mentions six times by name in *Walden,* most of which lies beneath the water. "The stones extend a rod or two into the water," writes Thoreau, "and then the bottom is pure sand, except in the deepest parts, where there is usually a little sediment, probably from the decay of the leaves which have been wafted on to it so many successive falls."[32]

Another reason for the dearth of beaches is the limited wind on a surface so "low in the woods," in this case about two hundred feet below the terrace-top canopy. The prevailing wind at Walden comes from the west-northwest, nearly coincides with its east-west elongation, and is quasi-perpendicular to the most gracefully curved collapse slope of the coalesced system. This explains why natural and historic—not-engineered and recent—beaches were best developed at that end.

Wind data also explains why the sand bar fronting Thoreau's Cove is the only well-developed one in the lake. The strongest regional winds arrive from the south-southeast in March and April. Only during these months does the probability of winds capable of producing white caps on flat open water exceed 50 percent. This coincides with the season when the shoreline is least stable, having recently been freed from the winter ice, and when the water level is rising most quickly. Finally, before the twentieth shoreline makeover, the mouth of Thoreau's Cove lay immediately down the littoral drift direction from the tallest, steepest, and therefore most active creep slope above the pond, which would have maximized the delivery of sediment into the swash zone.[33]

At Walden, there were no "Neva marshes to be filled." And, he continues, "it is nowhere muddy, and a casual observer would say that there were no weeds at all in it . . . A closer scrutiny does not detect a flag nor a bulrush, nor even a lily, yellow or white, but only a few small heart-leaves and potamogetons, and perhaps a water-target or two." He understood that mud and marsh go together and why both are missing. Its banks are sandy and exceptionally stable. Any mud suspended in the water column by shoreline erosion usually enters during high pond stages and moved to the shallows. There it is re-suspended by weaker shoreline waves during the falling stage before being sequestered in the deep. Because falling stages are more protracted than rising ones, the odds of leaving mud on the shore quickly diminish to nil.[34]

The dearth of herbaceous and aquatic "weeds" on the shore is due to several factors. First is the chemical poverty of incoming precipitation, which, in Thoreau's time, was closer to that of distilled water. The remainder came from spring water that was mostly scrubbed free of biological nutrient, making it "acetic." Another factor was the unpredictability of long-term inundations, especially when combined with the mechanical potency of lake ice. Thoreau wrote, "This rise and fall of Walden at long intervals . . . kills the shrubs and trees which have sprung up about its edge since the last rise—pitch pines, birches, alders, aspens, and others—and, falling again, leaves an unobstructed shore." The roots of larger trees extend down to very near the average water level, weaving a selvage on the shore, which helps limit erosion by waves and ice gouging. Although the shore itself is generally clear of vegetation, the large trees overhanging the bank copiously feed the lake with acorns, pine cones, pollen, and deadfall limbs. I have been there in the fall when acorns were plopping into the water so frequently that it sounded like popcorn.[35]

Though ice is prominently featured in *Walden,* most of what its author knew about it was threshed out. My favorite example involves the conceptual distance between ice physics and flatulent dreams: "I also heard the whooping of the ice in the pond, my great bed-fellow in that part of Concord, as if it were restless in its bed and would fain turn over, were troubled with flatulency and had dreams." As with all sleeping partners, slab ice can make sudden, involuntary movements and unexpected noises. This includes flatulence, which is nothing more than the threshold release of compressive strain in tiny audible increments that are so high in frequency that the resulting noise is a brief stream of sound of variable pitch. Ditto for the release of stress within the ice and the incremental propagation of microcracks beneath a blanket of snow.

Thoreau asks why something as "large and cold and thick-skinned" as a slab of lake ice could "be so sensitive." He knew the answer, having already learned the "law" to which lake ice "thunders obedience" though not "exactly." Actually, he realized that multiple laws were involved. Strong elastic media such as thick lake ice respond to the buildup of mechanical stress by invisibly and silently changing their shapes via bending, expanding, shrinking, shearing, or stretching. When the strain becomes to great, however, it ruptures in a system of fractures that propagate until the stress is relieved. At Thoreau's Walden, the driving stresses were mostly horizontal and due to temperature-driven expansion and contraction. Vertical stresses included barometric pressure from above and hydrostatic pressures from below, set up by the weather and water level, respectively.[36]

Responses could be rapid. On Christmas Day 1853 Thoreau watched the "sun sunk behind a cloud." Almost immediately he heard the pond begin "to boom and whoop." Mechanically, this was the brittle rupture associated with elastic contraction under "clear, cold, and windy" conditions, those most likely to promote a quick response, especially when the ice was not insulated by snow. Acoustically, different sounds were caused by variations in the thickness of ice, the size and connectivity of fractures, their rate of propagation, the muffling effect of snow, and the distribution of that snow, if uneven. Groaning and moaning sounds came from slow propagation speeds, whether along one large fracture or the sequential tearing of micro-cracks. Those sounding like the rifle reports came from instantaneous, unmuffled breaks.[37]

Though Walden could not compete with other lakes in terms of wave processes, it outdid most in terms of ice thrusting. In *Walden*, Thoreau wrote: "On the side of the pond next my house a row of pitch pines, fifteen feet high, has been killed and tipped over as if by a lever, and thus a stop put to their encroachments." Here he quotes the source *Journal* passage of December 5, 1852, almost verbatim. At that point, he had not yet fingered the killer. That came two months later when he discovered the cause of the "lever" mechanism, writing "that the pitch pines by my shore were literally tipped or pried over by a force applied underneath," the same force that also "heaved up" large rocks and masses of sand frozen into the ice. The severity of ice thrusting near his "dock" made perfect sense, and likely increased its attractiveness by keeping it clear. In fact, some of the larger stones there are scraped today, presumably by ice action. Vertical changes in water level can fracture the ice, but cannot force it against the bank. Horizontal contraction during cold spells can pull it away from the bank, creating tension fractures that easily re-freeze. When warming returns, however, the resulting expansion presses strongly against Walden's sun-warmed shores. With colder ice to the south serving as a backstop, the expansion would have been sufficient to shear the trees from below. Alternatively trees frozen into the ice can be sheared if the water level rises before the outward push is relaxed.[38]

Walden's most severe ice thrusting took place at its eastern end, where the maximum length of thermal expansion would have occurred. There Thoreau saw frozen sand cast up in ridges "six or eight inches high, by a foot or two in width, . . . in the form of waves just breaking on the shore, as if the ice had crowded against the shore and forced it up." This makes perfect sense. Repeated thrusting against the bank and refreezing could happen with enough compressional shortening that "sometimes the ice itself, lying on the shore, was raised." Thoreau noticed a separate type of ice-thrust during the 1853 breakup. A "mass of ice in Walden of about an acre had cracked off from the main body and blown thirty or forty rods, crumbling up its edge against the eastern shore." He applied both of these ice-thrust mechanisms to explain the "rampart . . . about many ponds, in one place at Walden, but especially at Flint's Pond . . . several feet in height and containing large stones and trees," . . . "apparently by the action of the waves and the ice."[39]

Ice thrusting also provides a mechanism for paving the shore. Thoreau wrote of "a belt of smooth rounded white stones like paving-stones." Belt is surely the right word, given the pond's twelve-foot vertical range. And "some have been puzzled to tell how the shore became so regularly paved." Unfortunately, we can no longer see the regularity that so intrigued him, because twentieth-century changes have destroyed what was there, and the chances of getting a new pavement in our warming climate is nil. However, highly regular stone pavements are quite common to northern lakes and rivers. This process begins with an initial stone concentration created by wave action. When standing water above stones freezes, it grips the stones from above and incorporates them into the base of the thickening slab. When the slab contracts with the cold or buckles from drawdown, the slab will break in tension, and the gripped stones will stay put. But when rapid warming returns, the gripped stones can be sheared radially outward above unfrozen stones, pressing them flat like those of a garden patio, and filling significant gaps. Once such a pavement is created, the annual shearing and vertical loading maintain and improve it. Geologist Edward Hitchcock commented on this process in a section of his *Final Report* titled the "Power of ice in the removal of Bowlders in Ponds." Thoreau may have seen it before writing his passage.[40]

Compass Face

If *Walden* were a true literary black hole, it would have drawn in *all* of Thoreau's sojourning experience without discrimination of content. That this did not happen indicates that a third important compositional technique was involved, a two-stage process I call "culling." The first stage was subconscious. When sojourning, Thoreau's mind drifted more strongly toward certain perceptions than others, creating what Stephen Hawking calls "model-dependent realism." Thoreau knew this was taking place, as evidenced by his pithy summary: "The question is not what you look at, but what you see." The second stage of Thoreau's culling was fully conscious, and deliberate enough to be considered a creative technique. This was Thoreau's "imaginative perception," what Leo Marx called "the analogy-perceiving, metaphor-making, mythopoeic power of the human mind."[41]

Luckily, the *Journal*—from which *Walden* was culled—is a Fort Knox of perceptions. Unluckily, there is no order inside the two-million-word edifice. Rather it is a melange of topic, place, chronology, completeness, and mood; inconsistent from room to room. "With its circumlocutions and dead ends, its excesses and exhalations," wrote Sharon Cameron, it "exhibits a relationship which seems both unfamiliar and inaccessible," one "predicated not on connections but on the breaking of connections." (Maybe that is why I liked it so much.) Alan Hodder agrees, concluding it is "an ungainly and sprawling complex of documents, and resistant, after all, to easy distillations." Hence, to understand Thoreau's culling process, we need some way of sorting out and classifying all this raw material. After struggling with how to go about this for some time, I eventually stumbled on to a strategy that Thoreau employed for his own use. And athough flawed by qualitative judgments, it is an improvement over the global generalities of having no strategy at all: for example H. Daniel Peck's assertion that "The *Journal* is Thoreau's most characteristic work, against which *Walden* appears as an anomaly," or Sharon Cameron's contention that "*Walden* is splintered from the *Journal,* not the other way around."[42]

"Not till we are completely lost or turned round," wrote Thoreau in "The Village," "do we appreciate the vastness and strangeness of nature. Every man has once more to learn the points of compass as often as he awakes, whether from sleep or from any abstraction." Here, quite cleverly, he uses the straightforward physicality of his surveying compass as an anchoring metaphor to repoint and refocus his thinking after returning from the directionless world of awareness. This section adopts his compass metaphor as an organizing principle for his culling technique. But in place of his four quadrants of direction—northwest, southwest, southeast, northeast—I substitute four quadrants of perception—time, space, substance, and process.[43]

Within the quadrant of time Thoreau's perceptual compass shifted back and forth between four points. In his *Journal*, thoughts of story and sequence pointed him to *arrow time,* that endless, irreversible flow of history. In this direction lay his biography since birth, his heroes from Pythagoras to Darwin, and his paleontology from trilobites to mastodons. Competing for his attention in this quadrant was *cycle time,* driven either by harmonic oscillation or by negative feedback to steady-state. Again and again, this way of looking at time brought Thoreau back to wherever he had been before, whether at the scale of seconds or of Huttonian "revolutions." A third compass point in this quadrant involved *singularities,* those one-time perturbations, thresholds, or extremely rare events that seemed to come out of nowhere because they are never repeated. Lightning never strikes twice in the same place; so they say. Nor can the founding of Concord by English Puritans happen twice. Thoreau's final point of focus on time involved the *eternal.* Some things are truly timeless, for example the universal constants of physics such as velocity of light in a vacuum, or of chemistry such as the solubility of pure quartz in distilled water.[44]

Walden is dominated by perceptions within the quadrant of time. Specifically, Paul Friedrich noted that time was the "single most frequent key word" in *Walden,* "occurring no less than 167 times or an average of almost once per page." Biographers have structured Thoreau's *Walden* time as narrative: beginning in 1822, when he was a five-year-old boy remembering the pond for the first time, and ending in 1854, when he made his last change to the text. In contrast, what Thoreau did in *Walden* was to banish history altogether, as shown by his concluding parable of the Artist of Kouroo. Though discussions of geologic history in the *Journal* are intermittently prominent, in *Walden* he used the word "fossil" only once, and then only to negate geological history in his climactic proclamation that Earth is "not a fossil earth." His banishment of arrow time in *Walden* is consistent with Simon Schama's conclusion that Thoreau held a "fierce conviction" that history "was irreconcilable with nature." In such a view, the famous passage "Time is but the stream I go a-fishing in . . . Its thin current slides away, but eternity remains" makes no sense at all: my brain always hits this passage like a pothole on the road. Though lovely, this idea was imported from the Sudbury River where it actually applies.[45]

On July 9, 1852, Thoreau asked himself, "Do we perceive any decay in Nature?" This is another way of asking if disorder increases, which is another way of asking if entropy gains, which is another way of asking if time has an arrow at all. In the *Journal* he muddles ambiguously. In *Walden* he shouts: "No!" The pond is eternal, a "mirror which no stone can crack, whose quicksilver will never wear off, whose gilding Nature continually repairs; no storms, no dust, can dim its surface ever fresh." Paradoxically the eternal mirror of *Walden* is the meeting of two eternities: "past and future" hinging at "precisely the present moment," the very "nick of time."[46]

In place of time's arrow, *Walden* hyper-concentrates time's cycle at every conceivable scale. Observations from the decades before he lived at the pond "map" onto that of his residency and multiple revisions. His two years at the pond map onto one: "Thus was my first year's life in the woods completed; and the second year was similar to it." Even his reasons for coming and going to the pond map onto each other: "I left the woods for as good a reason as I went there. Perhaps it seemed to me that I had several more lives to live, and could not spare any more time for that one." Note that he does not refer to an individual *life* that might actually have history. Rather, it is his "lives" that he imagines cycling one after another. So central was time's cycle to *Walden*'s focus that the final manuscript carried a Hindu epigraph that was deleted by the printer without his permission: "The clouds, wind, moon, sun and sky . . . all revolve for thy sake."[47]

Like musical notes within phrases within movements within a concerto, the beauty of *Walden* lies in its embedded rhythms. At the scale of seconds, the tempo was set by a self-willed ax: "standing on its head with its helve erect and gently swaying to and fro with the pulse of the pond." At the scale of minutes the "pond fires its evening gun with great regularity." At the scale of hours, the ice "began to boom about an hour after sunrise, . . . which was kept up three or four hours. It took a short siesta at noon, and boomed once more toward night, as the sun was withdrawing his influence." On many days the pond cooled and warmed and froze and melted, respectively. Twice a year its waters turned over and mixed. Once a year it emerged from icy hibernation with the blink of a marmot's eye.[48]

Singularities were completely culled. What might appear at first glance to be a birth turns out to be a rebirth. Awakenings are re-awakenings. Not even something as momentous as the arrival of the railroad to the shores of Walden qualified as a singularity: "Though the woodchoppers have laid bare first this shore and then that, and the Irish have built their sties by it, and the railroad has infringed on its border, and the ice-men have skimmed it once, it is itself unchanged, the same water which my youthful eyes fell on; all the change is in me. It has not acquired one permanent wrinkle after all its ripples. It is perennially young."[49]

Within his compass quadrant of space, Thoreau's most instinctive point involved *network* geometry. The ponds of Concord, when linked by trails, became a network of places he called a "lake district." A more quantitative way of perceiving space involved *Cartesian* geometry, which he used when surveying. What mattered then were bearing, distance, and elevation. At map scale, it was latitude, longitude, and elevation, the XYZ of Walden's epicenter being at 42.4387° North, 71.3386° West, +158.3661 feet above sea level, respectively. Rectilinear landscape shapes were also part of his Cartesian perceptions, for example the joints of "nature's rockwork" and the strike, dip, and plunge of strata. Much more important were Thoreau's spatial perceptions involving the *radial* geometry of spheres, cones, circles, rays, and perimeters. His final compass point for space involved *fractal* geometry, whether of branching patterns such as those of hoarfrost crystals, or of embeddedness, the statistical self-similarity of bubbles within clusters of bubbles, within clusters of clusters. When surveying Tuttle Meadow Brook, for example, he noted a pattern of zigzags within zigzags, one that unified the landscape at multiple scales. What mattered were pattern and hierarchy, not size and location.[50]

Though the *Journal* delights in all four points within the quadrant of space, *Walden* steers hard away from what Lawrence Buell called the "cartographic imperialism" of Cartesian geometry. In such a world, the straight line of the railroad tracks and its singing telegraph are striking and recurrent anomalies. Squares and triangles are even more rare. Instead, *Walden* hyper-concentrates radial geometry. The "open sesame" to *Walden,* wrote Charles Anderson, "is a single magical word, 'circle.' " And "at the very heart of the book lies Walden Pond, the central circle image. All paths lead into this chapter, 'The Ponds'; all lines of meaning radiate from it." Having made this claim, however, Anderson puzzled over the discrepancy between the "oblong" shape of Walden and the circular imagery of Thoreau's prose. To explain it, he invoked its author's prerogative: "Since ponds are generally thought of as being round, Thoreau takes a poet's license and throws up a shower of circle imagery in its praise." I disagree. The coalesced system is indeed oblong, and for that matter, asymmetric and—let's face it—highly irregular. Anderson's *Magic Circle* was a product of his own imagination. Thoreau's circle was actually a circle, the western basin where he lived, where he returned most often when living in town, and where nearly all of his observations were made. The center of that basin was Thoreau's "one centre," from which all radii "can be drawn." He appreciated its geometric holism at every scale: from the "rule of two diameters" at the scale of a half-mile, to the tiny bubbles within the ice at the scale of one "eightieth" of an inch. At intermediate scale "trembling circles seek the shore."[51]

In *Walden,* the *circle of the book* surrounds its central chapter: "The Ponds," a lake district surrounding its central pond. In the *cycle of life,* Henry arrives, senses the beauty, holds it in his mind, and then departs, to return again and again, over and over throughout his life. The *cycle of the day* epitomizes the *cycle of the year* taking place on the *circle of the pond,* which is actually a series of concentric circles beyond and beneath its stony shore. These circles were created by the *circle of the ice,* the concentric edge of the Laurentide Ice Sheet. Its most recent expansion and contraction was the last of four produced by *cycles of the orbit,* which drove *cycles of climate.* These orbital oscillations are as timeless as the earth itself, created by *circle of the solar system* created when the original solar nebula experienced inward gravitational collapse and flattened into a spinning disk, allowing the accretion of four inner terrestrial planets and four outer gas giants. This circle of the solar system follows from the *circle of the universe* created by its explosive birth from the ultimate singularity. Since then, it has expanded outward as a sphere of matter giving rise to the circles of spiral galaxies, the spheres of stars, and the circles of planetary systems around them, respectively. When writing *Walden,* Thoreau did what surfers do. He stayed inside the circle of the orbiting wave as it moved forward through the cycle of time, at least until his energy dissipated. Only in that transient curl do the circle and cycle coincide.[52]

Superficially, two chapters of *Walden* appear to follow a network geometry: "The Ponds," which links lakes into a lake district; and "Former Inhabitants," which links cellar holes and abandoned houses into an archaeological community. In his broader cosmography, however, even these networks assume a radial shape. Like small cities around a larger one, Walden was the central and dominant node to which the others were connected by trails and roads: the village, the Sudbury River, the cliffs at Fairhaven, Nine Acre Corner, Brister's Hill, the Great Meadows, Pine Hill, etc. Every trip away from and back to Walden was an arc of "great-circle sailing" following "across-lot routes." In *Walden,* the locations of distant nodes were deliberately obfuscated in order to emphasize Walden's centrality, despite its trivial economic importance, a place that did not "much concern one who has not long frequented it or lived by its shore." Reinforcing this personal centrality and isolation was Walden's deeply inset topography: "I have my horizon bounded by woods all to myself . . . [and] . . . for the most part it is as solitary where I live as on the prairies." Welcome to *Walden*'s bubble universe, a conceptual place in which a chapter titled "Solitude" makes perfect sense. "I am no more lonely than the loon in the pond that laughs so loud, or than . . . Walden Pond itself. What company has that lonely lake, I pray?"[53]

Fractal geometry was generally, but not completely, culled from *Walden.* He reserved it for two important themes: the branching patterns of growth of the sand foliage at the Deep Cut used for *Walden*'s climax; and Thoreau's common statements of scale independence throughout the book, what Lawrence Buell calls "the aggrandizement of the minute and the conflation of near and remote." A drop, a pond, and an ocean were all the same to him. "Such was that part of creation where I had squatted," he wrote to illustrate that *Walden*'s *anywhere* was also its *everywhere.*[54]

Thoreau's material reality lay in the quadrant of substance. There, what physicists call "matter" occupies space, carries force, and changes through time. It is the basis of all action, whether a cloud blowing, a river flowing, a tree rustling, or a sun setting. Thoreau's most ubiquitous, most volatile, and most far-traveled substance was the *air* he breathed. Heavier and more visible was *water,* the medium responsible for creating his "lakes of light" decorated by sparkles, ripples, and waves. Heavier still was *earth,* the unconsolidated residues derived from silicate rock, the pebbles, sand, and silt below Walden Woods and the bulk of the soil. The fourth point in this quadrant was organic matter, the *fiber* of carbon-based life, the solid of leaf, needle, wood, rush, algae, grass, ant, woodchuck, pout (brown bullhead) and Henry himself.

In *Walden,* liquid water is the *sin qua non* substance. From seemingly out of nowhere, it "welled up from the bottom" of a "perennial spring in the midst of pine and oak woods to form a lake "so remarkable for its depth and purity as to merit a particular description," "so transparent that the bottom can easily be discerned at the depth of twenty-five or thirty feet," and so clear that, when floating and diving Thoreau "seemed to be floating through the air as in a balloon" above "transparent and seemingly bottomless water." Walden's water was the ultimate liquid crystal and religious symbol: window, mirror, bath, fountain, spring, and life-sustaining fluid. It cleansed impurities. It clarified minds. It invigorated bodies. It was strong enough to float a team of horses on the ice, and yet weak enough reveal "the spirit that is in the air." Its surface was a membrane sensitive to the tiniest insect on its surface, and yet resilient to all damage. As a volume, it was large enough "to give buoyancy to and float the earth," but small enough to be "God's drop." As a medium it was clear enough to scatter light from below to create a "lower heaven" that was "scarcely more dense" than the air. As a metaphor it was Earth's eye, composed of orbital fluid.[55]

Condensing, raining, snowing, sleeting, flowing, misting, melting, soaking, saturating, evaporating, splashing, trickling, dripping, and seeping: no other substance in *Walden* comes anywhere close to water in importance. When "God himself culminates in the present moment," this divine manifestation appears through liquid water: the "perpetual instilling and drenching of the reality that surrounds us." This made bathing "a religious exercise," a sacrament performed twice a day: a morning baptism of rebirth and an evening purification ritual to wash off the day. "Blue angels" hid "in the azure tint of its waters."

Locked away from the gates of *Walden*'s water temple was the unclean, but life-giving, water of dead streams and swamps, which he celebrated so compellingly in the *Journal* for their murkiness and stink. So pure was the water of *Walden* that Thoreau sometimes overdosed on its asceticism, writing, "I wish to bury myself amid reeds." If he stayed too long, he hungered for richer waters: "I pine for the luxuriant vegetation of the river-banks."[56]

In *Walden,* air was not culled as a substance, but was demoted into a supporting role to assist liquid water and, to a lesser extent other materials: zephyrs on the lake, fluttering leaves, and blowing dust. Nor was Earth culled as a material. Rather, with two critical exceptions, it was shunted to the limelight of the bean field, the clean sand of his cellar, the yellow sand beneath his rippled cove, and the dry dust of places where "man has broken ground." Those two exceptions were Walden's stony shore and the sand foliage of the Deep Cut. Fiber was rarely featured, for example his lichen-covered firewood and "vegetable mould."[57]

The closest Thoreau came to unifying all four substances of his perceptual compass occurred when he was lying on the thin glaze of new ice and looking downward to the sandy bottom of his cove through the two transparent media of air and water. Bubbles had floated up from the bottom to become trapped in the water beneath the ice. There, all three phases of H_2O were enjoined at the triple point where vapor, liquid, and solid coexist. And there on the bottom were "furrows" made by "some creature" strewn with the "cases of cadis worms made of minute grains of white quartz" sand (the larvae use sand to armor their bodies). There—beneath water vapor, water crystal, and water fluid—mineral grains had been wrapped around organic fiber before being dragged over earth to leave the mark of life.[58]

Thoreau's final quadrant of perception involved the dynamic processes taking place all around him. For brevity, I lump them into the familiar meta-processes called meteorology, hydrology, limnology, and ecology. Each was a cluster of physical and/or chemical agencies that operated on a substance, within a space, and through an interval of time. Since the nineteenth century, each of these has become a distinct "-ology," a bundle of scientific topics and laws considered a separate scientific discipline.

In sequence, everything begins with gaseous *meteorology,* the grand sum of all atmospheric phenomena ranging from the magnetic storms of the ionosphere to the shadier side of a stone. From that agency came *hydrology,* originating as raindrops, snowflakes, snowmelt, soil moisture, infiltration, and aquifer flow. Next in line was *limnology,* the sum of all processes in standing water driven mainly by sunlight: the flow of light and heat into the water, its stirring by winds, the resulting aquatic ecology, and density-driven turnovers. Last in line were processes associated with terrestrial *ecology,* which, for this book, involve the buffering, armoring, meshing, and enriching of the land by life processes. Thoreau was ninety-seven years ahead of the first use of the word "ecosystem," when he journaled, "In Nature the perfection of the whole is the perfection of the parts." His emphasis on the system, instead of on the individual, meant that there could be no death, only exchanges of currency in the grand "economy of nature."[59]

In the *Journal,* watershed hydrology and macro-meteorology dominate. In *Walden,* these were largely culled back into the limelight so that limnology could play its starring role. To Thoreau, the lake seemed alive, as if "it were all one active fish," its chemical essences flowing up from its deep liquid center, and then back down as the wet "dust" of aqueous life. The keys to Walden's acetic limnology include the sieve-like permeability of its gravel delta plain, its great depth beneath an otherwise richer world, and its steep rim of resistant stones. When cool air was cupped in the kettle, the lake surface resembled "glass congealed." When "small waves raised by the evening wind," reached the shore, one's "serenity is rippled but not ruffled." To Thoreau, the Walden system was a flawless organic machine featuring liquid water, a sunken garden in which "the perch swallows the grub-worm, the pickerel swallows the perch, and the fisherman swallows the pickerel: and so all the chinks in the scale of being are filled." And at the commencement of spring, this one great thing was reborn from below.[60]

In August 1854, after nine years of struggle, Thoreau finally published his "book of the place." A century and a half later, and over the next three chapters, I offer my personal interpretations of what it means.

9

INTERPRETING *WALDEN*

"I had three pieces of limestone on my desk, but I was terrified to find that they required to be dusted daily, when the furniture of my mind was all undusted still, and threw them out the window in disgust." No bachelor dusts daily, especially one living alone in the woods. So why the "disgust" from someone whose sister claimed he actually liked dust? There's a clue in one of his accounting tables in "Economy."[1]

Thoreau had only one complaint about the cost of building his house: the high price he paid for lime: "That was high," he entered opposite the $2.40 entry for two casks of manufactured lime, which translates to $75 in today's dollars. As someone who has purchased and mixed his own bags of cement for household projects, this strikes me as far too high for someone who wrote: "I might have got good limestone within a mile or two and burned it myself, if I had cared to do so." Earlier, he had made some lime by burning a few clams. And he was acutely aware that "thousands of tons of marble," called "limestone," had been burned in kilns in the north part of Concord, from which he may have obtained his desk specimens. Burning limestone produces quicklime. When mixed with water and coarse sand, it made a strong and waterproof mortar for his stone foundation and chimney. And when mixed with water, fine sand, and hair, it made plaster for his walls.[2]

A major purpose of *Walden*'s opening chapter was to illustrate Thoreau's self-reliance and Spartan household economy. He went out of his way to inform his readers where his building supplies came from and how he got them. Here was a man who cut down his own native pine for framing timber, hauled his own boulders for his foundation, and rowed across the lake to get his own sand for plastering. So why not make his own lime when he knew how to do it? Keeping costs down was also part of Thoreau's main message in "Economy." This is why he used second-hand materials—boards, shingles, nails and bricks—and perhaps forged his own door latch. It also explains why he used his lime so sparingly during construction, applying mortar "only in key positions such as the corners of the foundation and between boulders and brick." Despite this economy, he still needed two casks of overpriced lime for his tiny house. Surely it must have galled him to pay so much.[3]

I speculate that Thoreau regretted his decision to buy factory lime, instead of making it himself. Why else would he have multiple pieces of this potentially useful but aesthetically drab material on his desk, rather than in his mineral collection? And why would they evoke "disgust?" Laura Dassow Walls associates Thoreau's negative reaction to his ambivalence about housework or marrying a housewife. And this may very well be true. I suggest an alternative motive, that the unburned limestone reminded him of his failure to be self-reliant and buy lime at a reasonable price. My purpose here is not to argue this speculation, but to illustrate the main thesis of this book: paying attention to Thoreau's rocks and minerals raises new possibilities for interpretation.[4]

Ephemera

Thoreau asks us who would have learned more: "the boy who had made his own jack-knife from the ore which he had dug and smelted, reading as much as would be necessary for this,—or the boy who had attended the lectures on metallurgy at the institute." This question is autobiographical. Massachusetts geologist Eugene Walker reminded us that "Thoreau mentions how as a boy, he and his friends gathered with excitement on the rare occasions when a scow laden with red iron ore was seen drifting down the river to the smelter in Chelmsford." Thoreau's *Walden* passage on this topic may have been a wistful reflection by a man who studied metallurgy in college, and who may have hammered some bog iron himself. But above all, it is Thoreau the exteacher expressing his personal preference for hands-on and life-long education.[5]

The soil of Thoreau's Bean Field was, in his own words, a "light and sandy soil" a "dry soil, . . . which for the most part is lean and effete," a "dewey and crumbling sand" in a "yellow gravelly upland." This description matches the attributes of the unit officially mapped there by the U.S. Soil Conservation Service: the "Windsor Soil Series." This is the

Connecticut State Soil, a prized medium for growing vegetable crops, responsible for making the Connecticut River Valley between Middletown and Enfield the breadbasket of colonial New England. It's still prized today for growing the state's famous shade-grown tobacco for wrapping cigars. This type of soil, as historian Brian Donahue pointed out, was ideally suited for the plow, hardly the "briar patch" Thoreau's friend Ellery Channing made fun of. Thoreau's choice of this field reveals his "nose" for good soil.[6]

"The Boiling Spring . . . What a treasure it is . . . Who divined it?" Thoreau gives no answer, although I'm pretty sure he knew its origin, based on descriptions in *Walden*. In midwinter, he noted: "the temperature of the Boiling Spring . . . was 45°, or the warmest of any water tried, though it is the coldest that I know of in summer, when, beside, shallow and stagnant surface water is not mingled with it." This conservative cold temperature indicates its snowmelt source—combined with cold spring rains—and a deep, artesian pathway set up by the some unknown configuration of impermeable layers at depth. Of course there was no true boiling, because neither heat nor gas was involved. It may have been a simple gurgle, a low-pressure fountain. More likely, it was a fluidized mixture of fine sand that bubbled like a thin porridge. For this type of "boiling" to happen, the texture of the sediment must be just right: coarse-grained enough to settle, which allows excess pressure to rebuild, but fine-grained enough to intermittently cap and release the slight artesian pressure. In all my travels about New England, I have seen only two such places, both of which "boil" like the mud pots of Yellowstone National Park. Both of those places were in comparable geological settings.[7]

Thoreau dug his "cellar in the side of hill sloping to the south, where a woodchuck had formerly dug his burrow, down through . . . the lowest stain of vegetation, six feet square by seven deep, to a fine sand . . . The sides were left shelving, and not stoned; but the sun having never shone on them, the sand still keeps its place." Everything about this description—the location, depth, sediment texture, and stability—indicates that Thoreau recognized he had dug through the bio-turbated and creeping surface layer to reach the fine-textured, and fairly cohesive lacustrine sand of the original delta lobes. This made it the most "solid and honest" part of his house because, unlike every other component, it was *in situ,* a stand in for rock reality. To encounter this material at such a shallow depth proves that he had excavated through the zone of kettle collapse.

This observation may have contributed to his induction that Walden's was, in fact, "sunk" below the delta plain.[8]

During the winter of 1846, Thoreau took a short tour with the "ice-men," or "cutters." Southwest of Walden's shore they showed him "what they thought was a 'leach hole,' through which the pond leaked out under a hill into a neighboring meadow . . . It was a small cavity under ten feet of water." This anecdote reveals Thoreau's technique of linking description to explanation while leaving the mystery open. The location he describes coincides with the base of an unusually steep reach of measured hydraulic gradient, a drop of fourteen feet in a distance of about seven hundred feet flowing parallel to the direction of bedrock faulting at depth. Though this is scarcely a two percent slope, it is vastly steeper than the remainder of Walden's surrounding aquifer. And though there is no manifestation of bedrock on the surface, it still aligns the outlet drainage at depth.[9]

"All sound heard at the greatest possible distance produces one and the same effect, a vibration of the universal lyre . . . There came to me in this case a melody which the air had strained, and which had conversed with every leaf and needle of the wood, that portion of the sound which the elements had taken up and modulated and echoed from vale to vale." Here is Thoreau's explicit statement that high frequencies are acoustically attenuated and that low frequencies survive with increasing distance from the source. Since his time, seismologists have assured us that the whole earth chimes like a bell for days after being struck by large earthquakes, but at frequencies too low to be audible. Perhaps Thoreau's ear was attuned to *infrasound* frequencies below what the rest of us can hear. "Let him step to the music which he hears, however measured or far away."[10]

With these ephemera behind me, I now dive like the loon deep into the heart of *Walden* to explore two of its most important passages.

The Pond Survey

"As I was desirous to recover the long lost bottom of Walden Pond, I surveyed it carefully, before the ice broke up, early in '46, with compass and chain and sounding line." Desirous: yes. But why? Was it mere curiosity? I think not. Psychologically, measurement offered him the satisfaction of knowing something to be true that was not compromised by interpretation or comparison. Patriotically, his survey actualized the Manifest Destiny impulse to chart the uncharted. Artistically, it emulated one of his role models, the eighteenth-century British naturalist Gilbert White, who hand-surveyed his own Wolmer Pond and inserted the resulting map into *The Natural History and Antiquities of Selborne*. Finally, as Laura Zebuhr noted, fathoming provided a literary device, "a place to throw down anchor; it literally grounds the text." All of these actual reasons likely came into play, rather than the one he gave his readers: to dispel local superstition that such ponds were "bottomless."[11]

Thoreau's survey required three separate tasks. Taking them in sequence: "I fathomed it easily with a cod-line and a stone." This was actually his most difficult task. It required the cutting of more than a hundred holes through approximately sixteen inches of ice, probably with an ice chisel, though he used an axe as well, if only to keep the holes clear. From each one, he let drop and hauled back up a "cod-line and a stone weighing about a pound and a half, and could tell accurately when the stone left the bottom, by having to pull so much harder before the water got underneath to help me." Before moving to another hole, he recorded the depths in a notebook, likely with a pencil he manufactured himself. He also indicated whether the bottom felt hard (sand/gravel) or soft (mud/muck). After completing his soundings and bottom detections on his main transects, he repeated the process with closer spacing across three of the coves and the pond nadir. The total haulage up-and-down was easily more than ten thousand feet of cold, wet line chilled by the February air, likely with bare hands. At this point, Thoreau had committed himself to an extensive scientific research project that would continue intermittently for years.[12]

Last, he wrote: "I mapped the pond by the scale of ten rods to an inch." This was an indoor, cartographic exercise done to convert raw numerical data from his notebook into a graphic image, likely done with a table, pencil, straight edge, drawing compass, protractor, and rule. A version of this map might have been drawn before he chopped any holes because he needed something to guide his fathoming transects, which bisect the pond.[13]

What Thoreau does not describe is the technique he used to make the map. Instead of following the typical procedures of land surveying—the metes and bounds of compass and chain—he created an image of the pond using his surveying compass as if it were a human eye. Our binocular vision depends on a technique called triangulation, which links three lines by three angles. The first line is the optical baseline. On the human forehead, this is the distance between the pupils of both eyes. A second line is the ray of light reflected back from some object of interest to the left eye on the baseline. The third line is the same for the right eye. At the object's location is an acute angle of convergence between left and right rays. The smaller the angle of convergence, the farther away is the object. Our brain quickly computes the distance.

Thoreau used this same technique to "image" the shape of Walden Pond. First, he set up a fixed optical baseline on the ice between two surveying stations, treating them as if they were separate eyes nearly a thousand feet apart. Because he could not occupy both surveying stations at once, he drew a series of rays toward prominent places from one station, and then repeated the process for the same places from the other. This produced a population of penciled triangles whose acute apexes were points on the shore marking some important inflection, usually the tip of a bluff or the recess of a cove. With the triangles drawn, all he had to do to "see" the pond was to connect the dots of each apex, thereby revealing the hidden image. Unless someone had already floated high over the lake with a hot-air balloon, this would have been the first time anyone saw its true, and fairly ungainly shape.[14]

Some readers may find my distinction between surveying, fathoming, mapping, and imaging to be parsing Thoreau's text too closely. Yet, this is what is necessary in order to isolate and extract the physical science from his deliberately roundabout descriptions, out-of-sequence narrative, anecdotal allusions, and intentionally unscientific reporting. It is also my way to letting readers know that one of the most famous images in American literature was "seen" using a pair of "owl-eye" stations on the ice. This puts Thoreau's metaphor of "earth's eye" in a very different light. His map reflects the binocular vision of earth's "eyes."

Physically, Thoreau's fathoming project sent him down to a depth of 102 or 107 feet, depending on lake stage. He converted this relative depth below the ice surface to its actual topographic elevation by referencing it to a local datum or benchmark, in this case the surveyed railroad causeway. This put Walden's allegedly bottomless bottom at fifty-six feet above mean sea level. Is this topographic fifty-six the source for his allegorical "fifty-six" in *Walden?* The latter refers to the actual weight in pounds of the large cannonball he imagined sinking to Walden's bottom? Believing otherwise requires an exact coincidence for an unpopular number. Regardless of potential symbolism, this elevation was his lowest encounter in town, approximately forty feet deeper than the base of its deepest cellar and probably twenty feet below the deepest part of the Concord River channel where it was constricted at bridges. This revealed to Thoreau that in order to go "down to the pond," he had to first go up to the level of the kame delta-plain at about 200 feet. From there it was forty feet down to reach water, and then at least a hundred more to reach its "deepest resort."

Inductively, Thoreau's pond survey sent him deeper than anything he had done before. By my count, he walked down a methodological stairway through seven stepwise inductions. Reaching this depth requires that we sort the anecdotal *Walden* text for discrete tasks, work out their chronology using the *Journal,* and reconstruct his sequence of thoughts. Doing so reveals that beneath Thoreau's literary finesse is an extraordinary piece of scientific work unmatched by any of his transcendental peers. For each of seven downward steps, he worked from a clearly stated purpose; set up a hypothesis (often not explicitly); devised a method to test that hypothesis; collected data (and/or observations); drew conclusions relevant to the hypothesis; discussed the conclusion; and then induced a general "law."[15]

Thoreau's initial purpose, as stated in *Walden,* was to prove that Walden had a reasonably tight bottom. More explicitly, he invoked the null hypothesis that: Walden Pond was deeper than could be measured. To test this would have required no more than a half hour in a boat. He could have simply dropped a weighted fishing line over the gunnels often enough to verify that it always went slack. To locate the deepest point would have been only slightly more complex: a similar series of trials that moved in the direction of deeper water until a maximum was attained. To measure this depth would have required nothing more than to lay some unit of linear measure against his cod line. And to document its location would have required no more than a freehand sketch, or perhaps a triangulation from shore using his pocket compass.

Thoreau's methods for this research project are clear from the *Walden* narration. In sequence, he planned ahead by scheduling the project for late winter when accurate coordinates could be obtained from an ice cover thick enough to be safe. This meant that the work would be both cold and physically demanding. Next, he collected the triangulation data he used to make a base map, from which he laid out bathymetric transects parallel to the length and width of the lake. After marking off regular distances on these lines with his chain, the only thing left was to fathom the depths.[16]

His results are famous. Though this was the first complex surveying task of Thoreau's life, his results were astonishingly accurate, even by the modern standard of acoustic sonar and global positioning systems (GPS). However, calling his "Reduced Plan" a "three-dimensional pond map" is a stretch. Rather, it is a traced outline of three of Walden's four basins with most of the soundings on two lines. Only in the western basin are the control points dense enough to show the shape of the lake away from the lines. The Wyman Meadow basin was left unmapped, and the bulk of the central and eastern basins were left unfathomed.

If map-making was Thoreau's motive, he would have quit at this stage. Instead, he used the map as a source of information: "The greatest depth was exactly one hundred and two feet; to which may be added the five feet which it has risen since, making one hundred and seven." Having negated his original null hypothesis, he now did what scientists typically do, which was to discuss his conclusion in some broader context, looking for implications, potential problems, and new directions. More specifically, he wrote: "This is a remarkable depth for so small an area. . . . A factory-owner, hearing what depth I had found, thought that it could not be true, for, judging from his acquaintance with dams, sand would not lie at so steep an angle."[17]

This prompted Thoreau to move on to his second stage. Now his purpose was to show that the angle was not as steep as presumed. Framing this as a hypothesis, he used the width and depth data from his pond survey to test this idea, drawing the result in cross section, and concluding that Walden, "which is so unusually deep for its area, appears in a vertical section through its centre not deeper than a shallow plate." To support his conclusions he discussed the depth of Loch Fyne in Scotland, which had been greatly exaggerated as a "horrid chasm" by landscape theorist William Gilpin. Thoreau showed that it was proportionately four times shallower than Walden. To this he added four more examples of distorted thinking regarding width/depth ratios: "a smiling valley with its stretching cornfields . . . the shores of primitive lakes in the low horizon hills . . . puddles after a shower," and finally the "depths of the ocean . . . compared with its breadth." At this point, he arrives at his *first induction:* that bodies of water are generally quite shallow relative to their surface area.[18]

When examining the bathymetry, Thoreau became "astonished at its general regularity." By this, he meant two distinct phenomena. First: "In the deepest part there are several acres more level than almost any field which is exposed to the sun wind and plough." Taking the "flatness" of this area as a general hypothesis, he tested this with survey lines "arbitrarily chosen," concluding that "the depth did not vary more than one foot in thirty rods, "and more specifically, on rays drawn out from the center, "I could calculate the variation for each one hundred feet in any direction beforehand within three or four inches." At this stage he had a predictive graphic model for the radial symmetry of the bottom. To explain this model, he made (or perhaps borrowed) his *second induction:* that the "effect of water under these circumstances is to level all inequalities."

Next, he noted "the regularity of the bottom and its conformity to the shores and the range of the neighboring hills." The match was "so perfect that a distant promontory betrayed itself in the soundings quite across the pond, and its direction could be determined by observing the opposite shore. Cape becomes bar, and plain shoal, and valley and gorge deep water and channel." This insight became the basis for a *third induction,* that the outline shape of the pond was congruent with the topography of the adjacent hills. This, we now know, is caused by the alignment of differential rock resistance as a consequence of Paleozoic thrusting, magmatic intrusion and block faulting.[19]

Just ahead was an even more striking insight, certainly the most significant of his surveying career.

> Having noticed that the number indicating the greatest depth was apparently in the centre of the map, I laid a rule on the map lengthwise, and then breadthwise, and found, to my surprise, that the line of greatest length intersected the line of greatest breadth exactly at the point of greatest depth, notwithstanding that the middle is so nearly level, the outline of the pond far from regular, and the extreme length and breadth were got by measuring into the coves.

In translation, his hypothesis that the nadir coincided with the exact middle of the pond was verified by accurate measurements of new lines of maximum length and width obtained from the paper map. From this observation, he inferred that the cause of the symmetry must be more substantial than that for the details of the coves, and must be older than the sediment fill in its bottom. This led to his *fourth induction,* that the shape of Walden was governed by a general law, the "rule of two diameters," which, for radial geometry, are identical by definition.[20]

His "rule of two diameters" for topographic depressions like Walden applied equally well to topographic eminences like mountains: "Is not this the rule also for the height of mountains, regarded as the opposite of valleys? We know that a hill is not highest at its narrowest part." Thoreau's favorite mountain "Monadnoc"—with its "buttresses on all sides"—was indeed a fair reciprocal for Walden and its coves. It was to these secondary features that he turned next. "Of five coves, three, or all which had been sounded, were observed to have a bar quite across their mouths and deeper water within so that the bay tended to be an expansion of water within the land not only horizontally but vertically, and to form a basin or independent pond."[21]

Almost automatically, he compared the measured results of cove shape to those of "every harbor on the sea-coast." This led to a hypothesis that cove geometry was "regular," regardless of location. Using data from his map, he tested and confirmed this hypothesis, finding that cove geometry was indeed regular (roughly flat-bottomed triangles) and had little to do with the radial symmetry of the main basin. "Given, then, the length and breadth of the cove, and the character of the surrounding shore, and you have almost elements enough to make out a formula for all cases." His "formula," what we might call a "Law of the Coves," was his *fifth induction.* This one demonstrates unusual methodological skill because his use of the word "formula" means he fell just short of a quantitative mathematical model amenable to refinement and replication.[22]

Building on this success, and in a *sixth induction,* he hypothesized a comparable "Law of Radial Depth" for ponds with a quasi-regular shape such as White Pond. In his own words: "In order to see how nearly I could guess, with this experience, at the deepest point in a pond, by observing the outlines of a surface and the character of its shores alone, I made a plan of White Pond." Using his "rule of two diameters" he located the lines of "greatest breadth" and "least breadth," where "opposite bays [coves] receded" and "two opposite capes [bluffs]" converged, respectively. In a quantitative *test* of his hypothesis, he discovered that "the deepest part was found to be within one hundred feet of this [intersection], still farther in the direction to which I had inclined, and was only one foot deeper, namely, sixty feet." The depth at his predicted intersection lay within two percent of its true value, and its location only about five percent short of where it should have been.[23]

The close uniformity between predicted and observed depths and locations for the nadirs of Walden's western basin and White Pond gave Henry the confidence to write his *seventh induction:* "If we knew all the laws of Nature, we should need only one fact, or the description of one actual phenomenon, to infer all the particular results at that point." Taking him at his word, he is advocating an extremist version of mechanistic determinism that René Descartes and his followers believed. More generally he is validating his core belief that nature makes sense, or will make sense when we probe deeper and deeper into its secrets.[24]

At this stage Thoreau has traveled seven downward inductive steps from the messy and superficial "particular results" at the surface to "one actual phenomenon" at depth that could explain them all, in this case, meltdown collapse to a focus. Thus, beneath Thoreau's quirky narrative and freighted allusions is a research project eight years in the making that sought general explanations for seemingly unrelated observations, and converted these explanations into predictive models that either were, or could be, tested quantitatively. This marks Thoreau as physical scientist with methodological rigor. There is no other way to explain what he achieved. "In the finished version of *Walden*," concluded Robert Sattelmeyer, "Thoreau *is* a scientist . . . one who believes that the results of his investigations into nature expressed actual and not merely 'poetic' truth."[25]

Only after reaching his deepest induction does Thoreau take the transcendental leap to correspondence: "What I have observed of the pond is no less true in ethics. It is the law of average. Such a rule of the two diameters." Though this sounds like yet another scientific induction drawn from his research project, it was, in fact, an old, unoriginal idea borrowed from one of his earliest transcendental essays, "The Service," which was never published: "The golden mean, in ethics as in physics," he wrote, "is the center of the system, and that about which all revolve." During this early stage of his thinking, *circa* 1840, Thoreau was parroting Emerson's transcendental idealism and seeking correspondence between man and nature. Twelve years later on March 11, 1852, however, he had become a scientist. So, rather than just accept the dictum of the "golden mean" in nature, he set out to test it at White Pond.

His success in linking the two ponds with the same geometric law gave him the confidence to predict, three weeks later, that humans, being "part and parcel" of nature, must also be governed by natural laws. This blatantly anti-transcendental statement is nothing less than the central thesis of sociobiology and evolutionary psychology.[26]

Thoreau quickly moves back to the discussion phase of his research project. "Now we know only a few laws, and our result is vitiated, not, of course, by any confusion or irregularity in Nature, but by our ignorance of essential elements in the calculation." Similar statements are routine in scientific papers being written today because their authors know *a priori* that measurements contain error, are incomplete, and that the real world always departs from ideal conditions. "Our notions of law and harmony are commonly confined to those instances which we detect; but the harmony which results from a far greater number of seemingly conflicting, but really concurring, laws, which we have not detected, is still more wonderful."[27]

Translation? The process of scientific investigation is more satisfying than the results it generates because the results always point the way toward more investigation. Sometimes they lead outward and upward as deductions about the particulars of any system. More importantly, they lead inward and downward as inductions. But in both cases, the seeking is always more important than the finding. Creating ignorance with new questions is preferable to filling in the blanks of old ones. This impulse is the diagnostic mark of a true scientist. Thoreau bears that mark.

Thoreau's Intransitive Mind

Thoreau practiced science but shunned the label "scientist." His first nineteenth-century biographer, Ellery Channing, called him a "Poet-Naturalist." His second, Frank Sanborn, reduced Channing's hyphenated phrase to "poet." His third, Englishman Henry S. Salt, went out of his way to argue that Thoreau was "not a man of science," at all because "he could not properly detach the mere external record of observation from the inner associations with which such facts were connected in his mind." Quoting John Burroughs, Salt claimed, that Thoreau "made no discoveries of importance in the scientific field because he looked *through* nature instead of *at* her." This sweeping generalization is flatly contradicted by Thoreau's seven stepwise inductions from the pond bathymetry. They prove that he looked *at* nature in order to see *through* her complexities to reach more fundamental natural laws.[28]

By 1942, Edward Deevey had become quite frustrated by the purblind insistence on labeling Thoreau. "These truly scientific observations, and many more of equally astonishing perception, were made by a 'poet-naturalist,' a disciple of Emerson, and an accepted member of the Concord coterie of transcendentalists. This paradox implies either that he was alternatively scientist and mystic, or that being both, he was neither." Unfortunately, Deevey's paradox is a false dichotomy because the word "both" need not oppose the word "neither." It is indisputable that Thoreau used "both" modes of engagement at different times when sojourning. H. Daniel Peck recognized this long ago: "In certain important respects science and poetry were indeed separate and discontinuous for Thoreau." He "characteristically divided the activities of observation . . . and reflection into separate experiences, rather than intermixing them," yielding what Peck called a "categorical imagination." Simpler to me is to say that Thoreau was bimodal, or bipolar, or even bicameral, alternating his style, mood, and cerebral hemispheres, respectively.[29]

In systems theory, Thoreau's mind could be classified as an "intransitive" system because it had two equally viable equilibrium states: the poetic and the scientific. During the summer of 1852 he was tottering on the threshold between these two states when he wrote his widely quoted statement that "every poet has trembled on the verge of science." The backdrop for this statement was not Thoreau the poet being seduced by science, but Thoreau the scientist being pushed to the brink of poetry. At the time, he was culminating a month-long investigation of three-dimensional meandering in stream channels. Each day, he had been getting up earlier and earlier to explore how rivers work and, in the process, experience that intellectual rush of outdoor discovery that office-bound writers like myself can only dream of. Eventually, he put what he learned to good use, writing: "He is the richest who has most use for nature as raw material of tropes and symbols with which to describe his life." Translation? The better the science, the better the poetry.[30]

Leo Marx approached Thoreau's intransitive mind from the opposite, transcendental perspective: "To be sure, he always wanted to get the facts absolutely straight . . . A rigorous fidelity to the facts was for Thoreau a religious commitment. For he assumed, having accepted Emerson's premises, that any natural fact, if properly perceived, becomes an equally reliable yet far more exhilarating, moral-aesthetic fact." From the scientific perspective, however, "rigorous fidelity" is also a prerequisite to good scientific induction, which yields a wholly different kind of meaning. Not symbolic truth, but actual truth: the inner satisfaction of knowing what something truly is and how it truly works. I will grant you that it was poetic engagement that pushed Thoreau out the door into sojourning space and kept him burning the midnight oil in his garret. But the fuel he burned to keep that engagement going was his daily practice of science. His sojourning life was a pendulum swinging between these two modes.[31]

Quoting Alfred North Whitehead, Robert Richardson asserted that "the scientist must have, a 'union of passionate interests in the detailed facts with equal devotion to abstract generalities.'" This was certainly Thoreau's situation. My favorite example of this union involves the lightning-struck tree he encountered on June 27, 1852. He began simply enough, with careful observation and description: "Saw a very large white ash tree, three and a half feet in diameter, in front of the house which White formerly owned, under this hill, which was struck by lightening on the 22nd, about 4 P.M." Already, he has given us a sequence of events; the size, measure, and species of the tree; its location relative to two fixed points; and the timing of the strike to within an hour. At this point, Emerson would have likely taken the leap into symbolic correspondence. Thoreau will eventually become poetic, but only after he's truly gotten a hold of the "thing in itself" as a particular part of nature.[32]

His narration continues with the location of the initial strike; the intensification and dissipation of its electrical power; the hot scorch of the first fifteen feet downward; the stripping of the bark as the strike enters the wood along a ragged furrow; the bifurcation of the bolt below a high limb to both sides of the tree; and an initial explosion at the lowest point of general branching, which tossed off the large limbs "butt foremost." After reaching the ground, the bolt exploded back up the tree, "rending the trunk into six segments" about fifteen feet long that were blasted outward like rays into the shape of a cone still attached at the tree's base. Once in the ground, the surge of electricity also radiated outward, but through the roots, "furrowing them as the branches . . . and making a furrow like a plow" in the earth. One of these furrows "passed through the cellar" of a house about thirty feet "distant, scorching the tin milk-pans and throwing dirt into a pail, and coming out the back side of the house in a furrow, splitting some planks there." Following this is another long paragraph of detailed after-effects. Only after his 1,326th word does Thoreau finally jump toward poetic expression by drawing a parallel with his Katahdin experience: "There was a Titanic force, some of that force which made and can unmake the world."[33]

Thoreau's meaning comes not from transcendental correspondence, but from sensing the awesome power that was communicated at high voltage between the grounded lithosphere and the ungrounded atmosphere. The Titans who wielded this power received it from their mother, Gaia.

Another of Thoreau's scientific traits was to appreciate the hard work fellow scientists did for him. "I am grateful to the man who introduces order among the clouds," he wrote during the winter of 1852 because this gave him the chance to "look up into the heavens so fancy free." He followed this up with the remark: "I am almost glad not to know any law for the winds." Here, he had slipped comfortably back into the poetic mode of his intransitive nature, knowing full well what the other mode felt like.[34]

In *Consilience,* E. O. Wilson wrote: "The love of complexity *without* reductionism makes art." Conversely, "The love of complexity *with* reductionism makes science." Perhaps the presence or absence of the impulse to "reduce" was the toggle switch between Thoreau's modes, respectively. When turned off, he could drink in his sensations without thinking. When turned on, he could use those same sensations as a springboard for analysis. In *Seeing New Worlds,* Laura Dassow Walls wrote with equal elegance about why both modes were vital to prevent "knowledge without relation," which took "two degenerate forms: 'the man of science' who turns nature into objects to be collected, pickled, and abstracted; the 'scholar' or aesthete who prefers dry and dusty volumes of poetry to actual life." The former keeps the toggle switch of reductionism on at all times. The latter keeps it switched off.[35]

Prior to his initial conversion to science in late 1850, Thoreau was fairly idealistic about the subject, treating it like one of his Homeric heroes: "What an admirable training is science for the more active warfare of life." "Science is always brave, or to know, is to know good; doubt and danger quail before her eye." Afterwards, he became much more practical about its value for lancing through mumbo jumbo. The case of a prehistoric human skeleton accidentally dug up in Concord comes to mind:

> The skeleton which at first sight excites only a shudder in all mortals becomes at last not only a pure but suggestive and pleasing object to science. The more we know of it, the less we associate it with any goblin of our imaginations. The longer we keep it, the less likely it is that any such will come to claim it. We discover that the only spirit which haunts it is a universal intelligence which has created in it harmony with all nature. Science never saw a ghost, nor does it look for any, but it sees everywhere the traces, and it is itself the agent, of a Universal Intelligence.[36]

By the summer of 1853—and following three years of scientific reading, fieldwork, and classification—Thoreau had emerged to become Concord's self-credentialed consultant in natural science. That is when his *Journal* reports the curious case of an unidentified animal skeleton found in the wall of a shop owned by a certain Mr. Adams, who consulted Mr. Pratt, who consulted Mr. Thoreau, who used anatomical measurement, microscopic examination, and vertebrate taxonomy to make the proper identification. "With full faith in this and in science, I told Pratt it was a muskrat and gave him my proofs; . . . and [he] was convinced, much to his and to my satisfaction and our confidence in science!"[37]

Finally, Thoreau defended the joy of basic research, done "for purely scientific ends. Do not hire a man," he wrote of town civil engineers, "who does your work for money, but him who does it for love, and pay him well." In this passage, Thoreau clearly bestows his blessing on the idea of being a paid, professional scientist.[38]

In opposition to Thoreau's celebration of science as a means of personal engagement was his harsh criticism of it. Bernhard Kuhn described the motive for Thoreau's criticism: the "growing estrangement of the arts and sciences, from the self-assured epistemic unity of Rousseau's time . . . to the splintering of disciplines into competing ways of knowing under the pressures of specialization and professionalization in mid-century America." So, in due fairness to Thoreau's poetic mode, I hereby offer selected examples of his criticism of the opposite mode written during the peak of his scientific sojourning stage between August 1851 and November 1853, presented in reverse chronological order.[39]

"The rooster's crow as freshly and bravely as ever, while poets go down the stream, degenerate into science and prose." "The boundaries of the actual are no more fixed and rigid than the elasticity of our imaginations." "I love best the unscientific man's knowledge; there is so much more humanity in it." "The facts of science, in comparison with poetry, are wont to be as vulgar as looking from the mountain with a telescope." "The use of the rainbow, who has described it?" "Science affirms too much. Science assumes to show why the lightning strikes a tree, but it does not show us the moral why any better than our instincts did." "The sun of poetry and of each new child born into the planet has never been astronomized." "The skill of the savage is just such a science, though referred sometimes to instinct." "I suspect that the child plucks its first flower with an insight into its beauty and significance which the subsequent botanist never retains." "What sort of science is that which enriches the understanding, but robs the imagination? Not merely robs Peter to pay Paul, but takes from Peter more than it ever gives to Paul." "Facts should only be as the frame to my pictures; they should be material to the mythology which I am writing." "The astronomer is as blind to the significant phenomena, or the significance of phenomena, as the wood-sawyer who wears glasses to defend his eyes from sawdust."[40]

From my perspective as a properly credentialed physical scientist with forty years of research experience, scholars of Thoreau have paid either too much attention to his one-time refusal to join the Association for the Advancement of Science (AAS), or too little to his lifelong refusal to join a comparable poetry or abolitionist organization. All three seem equally likely outcomes for someone who was not a joiner, someone who once wrote: "wherever a man goes, men will pursue and paw him with their dirty institutions, and, if they can, constrain him to belong to their desperate odd-fellow society." His refusal to join organizations that met far away was automatic, such as with the annual AAS meeting in Washington, D.C. On the contrary, Thoreau did join local organizations that offered one or more practical benefits: for example the Concord Lyceum for its lectures and the Boston Society of Natural History for its library.[41]

And let us not overlook the fact that Thoreau's written response declining president Spencer Baird's invitation to join the AAS in 1853 was as cordial as it was appreciative. In fact, Thoreau went out of his way to express an "interest in the Association itself," before declining on the logistical grounds that "I should not be able to attend the meetings." His responses to the accompanying questionnaire were informative, and contained nothing resembling a snide remark. Against the blank "*Occupation (Professional, or otherwise),*" Thoreau wrote: "Literary and Scientific, combined with Land-surveying." Here, in his own words, his two main modes of engagement—or equilibrium states of his intransitive mind—stand side by side in their correct order, both supported by his day job.[42]

Contrasting with Thoreau's courteous declination to president Baird was his snarky private *Journal* response nine months earlier. On the face of it, this may indicate nothing more than Thoreau's legendary mood inconsistency. We all have bad days. But speculations about his ulterior motives have been legion. Perry Miller concluded that Thoreau was "furious" with himself for letting the science side of his mental dipole squelch his poetic side. Historian Donald Worster wrote more broadly about the scientific culture that put Thoreau on the defensive: "The world was not to be studied through love or sympathy—indeed, could not be, for it was widely subscribed to by scientists that nature had to be cleansed of sentiment and so deliberately made unappealing to human feelings. Such had been the Baconian mission from the first. The quest for objectivity also meant that the outer, physical world was to be kept firmly separated from all religious experience."[43]

To join the AAS under these terms would have branded Thoreau as one who "studies nature as a dead language," rather than as someone who prays "for such inward experience as will make nature significant." His unwillingness to swear off nature-spirituality, Bradley Dean concludes, was the "only difference between Thoreau and more traditional scientists." I agree. He wanted no truck with a profession that was atomizing into specialties, each with its own coterie of experts and policy regulations. This made no sense to someone who saw beauty in the holism and fluidity of nature.[44]

Yet today, with rare exception, young scholarly professionals are required to make the "Sophie's Choice" that Thoreau refused to make. For the sake of management efficiency, oversight, funding, and disciplinary inertia, academia requires its initiates to enlist in one of the two intellectual sub-cultures that C. P. Snow highlighted half a century ago, and which continue to have great purchase in popular culture. I refer to scientists and humanists, reason and faith, empirical and romantic, sense and transcendence, rational and irrational, order and adornment, logical and illogical, prudence and passion, astronomy and astrology, Apollo and Dionysus, materialism and idealism, and so forth. Evolutionary psychologists have long demonstrated that this dichotomy is real and rooted in the evolved architecture of our brains. Importantly, our more primitive (i.e. initial) response is always the subjective one rising from the midbrain, followed by cognitive processing of objective reality using the more highly evolved neocortex. Our "inner fish" always speaks before our inner attorney, accountant, or engineer.

Music captured this great divide in Thoreau's understanding. During the grueling final wordsmithing of *Walden,* he wrote: "Nothing is so truly bounded and obedient to law as music, yet nothing so surely breaks all petty and narrow bonds." To his subjective ear, music was pure joy, a sea of emotional power pre-dating the first drum or rhythmic dance by pre-civilized humans. To his objective ear, acoustics was pure science, a sea of equations that relate the wavelengths, frequency, amplitude, attenuation, and resonance of sound through the biological aspects of hearing, chiefly detection and neural processing. To navigate a straight line through each sojourning cruise of "great-circle sailing," Thoreau had to keep his rudder properly trimmed between poetry to starboard, and science to leeward. Such "steering" was a lifelong struggle for him, perhaps because the ocean of his experience with Nature was broader than for most of us, or because his waves of feeling were stormier. The more Thoreau tried to be objective, the more he realized that this was a lost cause. And the more he drifted toward science, the stronger he pulled the tiller back in the direction of poetry to keep his course straight. Cybernetically, he toggled back and forth between practicing calculus some evenings and playing the flute on others. "The scientist and the artist-poet were equally congenial to him," concluded Robert Richardson, "and his own best work partook of both." Using Stephen Jay Gould's terminology, both *Walden* and the *Journal* toggle back and forth between a clever mix of Franciscan nature poetry and Galilean intellectual delight.[45]

Deep Cut

Thoreau's criticism of the Fitchburg Railroad was well deserved. It fumed through his Eden, browsed his woods, fouled the Boiling Spring, and it truncated one of Walden's coves. But it was also a blessing in disguise, especially after July 1848 when Henry moved back to the family's "Texas House" on the west side of Concord village. From there, the railroad alignment offered a straight, flat, well-drained link to Walden Pond and the Cliff at Fairhaven, two of his most common sojourning destinations. This shared link carried him through the railroad's deepest excavation, its canyon-like "Deep Cut." Walking through it exposed Thoreau to the "dermis" of earth beneath the "epidermis" of ecology. And what he witnessed there—the lava-like flowage of silty sand—became "an insight vital enough to use as the climatic point" of *Walden,* concluded Robert Richardson. Decades earlier, Perry Miller had concluded that "all *Walden* is an adroitly suspended anticipation of the climax of thawing sand and clay in the railroad cut." More recently, Lawrence Buell dubbed this the "end point in Thoreau's epic" because it "breathes life into the biblical formula of humankind's earthly origins" by assigning priority to the life-giving role of the mineral kingdom over the descendant dukedoms of plants and animals.[46]

Before the 1843 excavation of the Deep Cut, the landform it sliced through was a typical creep-rounded forested hill of Walden Woods immediately underlain by kame delta sand and gravel. Excavation there went physically deeper than the shovel-dug, cleanly washed, fine sands of Thoreau's cellar. Stratigraphically, it went even deeper, down into the muddier pro-deltaic sediments where the sand was more uniformly and finely textured and the silt and clay content was higher.

Before excavation, frost penetration on this hill would have been minimal, its melting would have been invisible, and no surface flow would have taken place. This is because everything was insulated, armored, and

Figure 32. Sand foliage at the Deep Cut by Herbert Gleason (1900). Original caption is "Sand Foliage from Deep Cut on r.r. [railroad], Concord, Mass."
Courtesy Concord Free Public Library, Robbins-Mills Collection, II.1900.4 Box 16 (March 17, 1900).

woven together by the surface litter and root mesh, respectively. During construction, however, these organic horizons were stripped away and the deep subsoil was nakedly exposed to the elements. Additionally, the cold air of midwinter was channeled through the artificial canyon from both directions, enhancing the loss of heat from the ground. Under these conditions, the moist silt, sand, and clay froze downward to a depth of several feet. During this process, moisture from the deeper subsoil was drawn back up to the freezing front by a capillary wicking action and by the strong vapor-pressure gradient. These mechanisms—in addition to transient surface additions of water—filled the pores with ice to create a strong and impermeable mass that reminded Thoreau of "yellow sandstone." Based on one of his measurements, it reached a thickness of twenty-nine inches. This rock-solid agglomeration would eventually melt from the top down, releasing liquid water at the top of an impermeable mass. But before this happened, cold surface winds would have sublimated the uppermost pore-ice to create a thin veneer of loose fine sand. Also present above, and overhanging these dusty, ice-cemented banks, were massive snowdrifts blown into the "wind-shadow" created by the excavated gap, and held in place by plants draped over the top by the root mesh. They provided an additional source of water for the flowage yet to come.[47]

When solar warming began, liquid water melted from the pore-spaces at the top of the ice-cemented zone was blocked at greater depth, forcing it to ooze parallel to the inclined bank. This produced a thickening film of water flowing through loose, uniformly weak, saturated, fine-textured granular material. Inevitably, this created fluidized slurries that spontaneously transformed into rivulets, ribbons, and channelized masses. Inevitably, these stiffened downslope, diverting the continuing flow to one side, a process that repeated itself to create a bifurcating pattern. Local variations in water content, sediment texture, and bank irregularities near the fluid threshold gave rise to a seemingly autonomous system of great complexity with respect to transport and deposition. Thoreau was mesmerized by streams of sediment giving rise to lobes, heaps, levees, and fans resembling foliage, leopard's paws, lichen thalli, bowels, and brains.

The transfer of sediment from the upper part of a steep bank to its bottom is a predictable consequence of exposure. Under normal conditions, excavated cuts, especially in fine sand, are very unstable. They last for only a few years because the gravitational transfers lower the slope angle, thereby diminishing the power driving the process, and fostering re-vegetation and stabilization. But at the Deep Cut, sedimentation at the base of the bank would have clogged the drainage ditch, and therefore would have been removed as part of routine maintenance. Each re-excavation would have rebooted the energetics of the system, keeping the bank exposed and the sand flowing, year after year.

"Few phenomena gave me more delight than to observe the forms which thawing sand and clay assume in flowing down the sides of a deep cut on the railroad through which I passed on my way to the village." So begins Thoreau's 2,500-word buildup to the climax of *Walden.* Flowing mixtures of water and sand exhibit "a sort of hybrid product, which obeys half way the law of currents, and half way that of vegetation. Flowing, it takes the forms of sappy leaves or vines." Scholars have traced this inspiration back to Goethe's *Italian Journey,* which Thoreau read and described in his 1837 *Journal.* By 1842, he had expanded Goethe's idea in *The Natural History of Massachusetts.*

> It struck me that these ghost leaves, and the green ones whose forms they assume, were the creatures of but one law; that in obedience to the same law the vegetable juices swell gradually into the perfect leaf, on the one hand, and the crystalline particles troop to their standard in the same order, on the other. . . . This foliate structure is common to the coral and the plumage of birds, and to how large a part of animate and inanimate nature.[48]

Though this sounds like a description of the sand foliage passage in *Walden,* it is actually a passage about hoarfrost from his early *Journal.* By the time Thoreau left the pond in September 1847, he had walked past the unstable Deep Cut countless times, had extended Goethe's idea to its flowing sand, and had incorporated this extension into Version I of *Walden,* albeit with very limited scope: "As I go back and forth over the rail-road through the deep cut I have seen where the clayey sand like lava had flowed down when it thawed and as it streamed it assumed for the forms of vegetation . . . unaccountably interesting and beautiful." This one sentence was all he wrote before dropping his *Walden* project for several years.[49]

By the summer of 1851, however, Thoreau was emerging as a field scientist, and had entered nearly a dozen additional passages about the Deep Cut into his *Journal*. They described a variety of phenomena: the noise of "humble bees" resting in holes in the bank; the beautiful colors of ochre-brown, reddish-yellow, and dusty white; the sharp-pointed and densely dissected badland gullies resembling the molds for stalactites; the paper-thin delicacy of the geologic strata; the layers of air (cold air below warm and fog below clear); and expansion of the sand flowage. By June 13 he had credited the artistry to the work of the universal potter, working with human clay.[50]

Was this new awareness the result of Thoreau's enhanced observation skills? Given the timing, was he modeling Darwin? Or had he seen a beautiful engraving of sand foliage from Edward Hitchcock's *Final Report,* which he read in conjunction with his October 1849 trip to Cape Cod? In a section titled "Anomalous Effects of Water," Hitchcock's illustrates the phenomenon of sand foliage from the bank of the Westfield River in the town of Russell. His "Figure 59" was accompanied by this comment: should the "whole be hardened into a rock, it might perhaps be mistaken for an organic relic." Here, Thoreau may have encountered a link between his sand foliage and the fossil record.[51]

In early November 1851, Thoreau sojourned with Ellery Channing to the head of the Concord River Valley near Framingham. While overlooking Long Pond in Cochituate he saw what was likely a deeper cut than the Deep Cut in a place where the "composition of the soil" was "familiar," no doubt because it was also kame delta sediment from Glacial Lake Sudbury. There, he witnessed sand flowage at a much larger and more powerful scale: "Glorious sandy banks far and near, caving and sliding,—far sandy slopes, the forts of the land." The following day he reflected in hindsight: "When I saw the bare sand at Cochituate I felt my relation to the soil. . . . When I see her sands exposed, thrown up from beneath the surface, it touches me inwardly, it reminds me of my origin; for I am such a plant, so native to New England, methinks, as springs from the sand cast up from below."[52]

Here, in an entry dated November 8, 1853, Thoreau makes his final link on the continuum from Goethe, to hoarfrost, to sand foliage, to fossil foliage, to the scale of the human beings. This may have prepared him for his final rush of Deep Cut observations seven weeks later when, during the last three consecutive days of December, he watched the "artist" at work in his laboratory with deeper insight than ever before:

> The earth I tread on is not a dead, inert mass. It is a body, has a spirit, is organic, and fluid to the influence of its spirit, and to whatever particle of that spirit is in me. She is not dead, but sleepeth . . . this fundamental fertility near to the principle of growth . . . So the poet's creative moment is when the frost is coming out in the spring . . . Even the solid globe is permeated by the living law. No doubt all creatures that live on its surface are but parasites.

Robert Richardson credits this final *Journal* entry as the inspiration for Thoreau's return to the *Walden* manuscript after his long hiatus.[53]

But Thoreau was not done yet, returning to the Deep Cut on March 1852 to write his most physically exacting observations of sand flowage, and to broaden the scale to the deltas of rivers. Motivating his return may have been the search for physical causes, because only now does he explore the fluid mechanics of what's taking place, detailing: the thermally driven phase change from ice to liquid; the capillary tension of water; granular liquefaction; the control of slurry viscosity in causing either lobation or channelization; its shear strength as a function of water content; the flow rate as driven by slope; and the conditions fostering meandering. The difference between his symbolic descriptions of early 1851 and the rheological descriptions of early 1852, for the same phenomenon in the same place, exemplify his continuing shift toward physical science during this crucial year of his transition.

Thoreau's *Journal* for early 1853 documents nothing about the sand foliage. Conditions during the thaw may not have been conducive, either due to differences in ground freezing or to railroad maintenance. Other factors may have been involved. Perhaps this hiatus explains why Thoreau was so enthusiastic on February 2, 1854, when the phenomenon returned with a vengeance.

> That sand foliage! It convinces me that Nature is still in her youth,—that florid fact about which mythology merely mutters,—that the very soil can fabulate as well as you or I. It stretches forth its baby fingers on every side. Fresh curls spring forth from its bald brow. There is nothing inorganic. The earth is not, then, a mere fragment of dead history, strata upon strata, like the leaves of a book, an object for a museum and an antiquarian, but living poetry, like the leaves of a tree,—not a fossil earth, but a living specimen . . . The very earth, as well as the institutions upon it, is plastic like potter's clay in the hands of the artist. These florid heaps lie on along the bank like slag of a furnace, showing that nature is in full blast within.[54]

"There is nothing inorganic." Surely, this is *Walden*'s most important line. Life springs from non-life. Walden Woods springs from the sand, which springs from the stone, which springs from the "slag" of Earth's crust. The English writer Robert Chambers had proposed this controversial idea as early as 1844 in his *Vestiges of Creation,* published just as Thoreau began his famous experiment in deliberate living. The idea that every living thing on Earth—including its human institutions—is the consequence of the planet's "great central life," its geothermal "furnace" running "full blast within," still running steadily after four billion years worth of heat-driven tectonism and life-driven evolution. Thoreau's insight regarding this origin of life and its corollary of continuous creation stood in direct diametric opposition to the "dead history" of catastrophism, then the prevailing paradigm. Earth itself, he saw, was a single "living specimen."[55]

A week later, on Feb 8, 1854, Thoreau linked the sand flowage to the rebirth of spring: "This is the frost coming out of the ground; this is spring. It precedes the green and flowery spring, as mythology does ordinary literature and poetry." Noticing on March 1 that "the sand foliage is now in its prime," he returned for one last look on March 2. That evening, he hurdled over the final technical details to reach the core of his philosophy of Nature: the spontaneous emergence of order from disorder, of cosmos from chaos, of life from non-life: "How rapidly and perfectly it organizes itself! . . . The atoms have already learned the law . . . No wonder that the earth expresses itself outwardly in leaves, which labors with the idea inwardly."[56]

Earth's crust, Thoreau now plainly saw, was not the residue of something that *has* happened. It is the ultimate raw material for everything that *is* happening in the present moment. Ancient rock must be destroyed so that new rock can rise again. This was James Hutton's central idea: that creation is a never-ending loop of construction and destruction. The pitch pines of Walden Woods may seem parasitic upon its mineral soil, but only in the sense of having come later in time. In the grand scheme of things, the rocks need the plants to make the residues needed to make new rock. Pines are players on par with minerals.

And the growth of every oak tree branching upward and outward from its acorn is analogous to the growth of the tree of life bifurcating from its first microbial ancestor to the amazing biodiversity on Earth today. During the passage of deep time, the branching was always toward a "higher and higher level of complexity, order, and information," concluded Sven Jorgensen in his review of evolutionary thermodynamics, a book I found especially helpful. Thoreau's merger of the ceaseless annual cycle of rebirth at the yearly scale with the ceaseless cycles of Huttonian revolutions at the billion-year-scale, turned his planet into living poetry. In every moment, its rocks are giving rise to unconsolidated earth, which is giving rise to primary producers (plants), which are supporting consumers (animals), which are supporting individual human lives and their societies. These last two cycle one after another, as every cemetery and archaeological site reveals. The same is true of the fossil record, as every fossiliferous outcrop shows. The whole Earth and everything it contains was, is, and will be forever coming and going.[57]

To write of the "ur-phenomenon of the leaf" is to say too little too late, because this common fractal branching pattern was manifest in mineral dendrites, hoarfrost, and river watersheds long before animals of the early Paleozoic seas got the idea, and the plants after them during the Silurian Period. Last in line were humans, their languages, and the literary link between lobes of mud and the sounds of relevant nouns.[58]

Thoreau worked on this sand foliage section of *Walden* from his first to last draft. In final form:

> There is nothing inorganic. These foliaceous heaps lie along the bank like the slag of a furnace, showing that nature is "in full blast" within. The earth is not, a mere fragment of dead history, stratum upon stratum, like the leaves of a book, to be studied by geologists and antiquaries chiefly, but living poetry, like the leaves of a tree, which precede flowers and fruit,—not a fossil earth, but a living earth; compared with whose great central life all animal and vegetable life is merely parasitic.[59]

In this passage, the new phrase "to be studied by geologists and antiquaries chiefly" replaces the earlier phrase, "an object for a museum and an antiquarian." Why insert the word "geologists," remove the word "museum," and move from singular to plural? This editorial revision wags a finger at geologists—with an emphasis on the plural—who would might mistakenly interpret Hutton's timeless earth for something else: a "mere fragment of dead history, stratum upon stratum." Surely this must target the catastrophists flanking Thoreau's Concord: Louis Agassiz to the east, who viewed earth as a paleontological improving ground where God committed serial extinctions; and Reverend Hitchcock to the west, who spent a lifetime aligning the stony facts of geology to the literature of biblical scripture. Both of these leading "Men of Science" saw a series of "dead" histories, "stratum upon stratum" before God's choice to culminate his plan with the likes of us.[60]

Emergence—when used as a technical term—refers to the seemingly spontaneous appearance of something new at a higher level of organization that arises from simpler antecedent components. In short, the whole is greater than the sum of its parts. In September 1850, Thoreau paused to watch birds flock in "compact and distinct masses," wondering if they "were not only animated by one spirit but actually held together by some invisible fluid or film." Here, his symbolic focus shifts our attention from the spirit of the bird to the spirit of the flock. This idea, wrote Donald Worster, is as old as organic philosophy itself, with roots tracing back to the "surviving influence of Romantic idealism." As handed down from Rousseau to Goethe to Herbert Spencer, and then to Lloyd Morgan in the early twentieth century. Morgan struggled to "locate the middle ground during the prolonged and bitter wrangles between the mechanists and vitalists, those who respectively believed that life was nothing more than physiology, and those who believed in an *elan vital,* something special about life that no other scientific system possessed."[61]

Thoreau's challenging question about the "invisible fluid" was answered by Ilya Prigogine, who won a Nobel Prize in Chemistry in 1977 for his ideas about the spontaneous emergence of order, seemingly from nowhere. In decreasing complexity, Thoreau's flock of birds, the sand foliage at the Deep Cut, and whirlpools in the Sudbury River were all what Prigogine called "dissipative structures," because they survive by dissipating the ambient energy field around them. More loosely, they are called "emergent phenomena." Conceptually, each is a local island of order within the greater sea of growing disorder that surrounds and nourishes them. Mathematically, each is a discrete, nonlinear, dynamical system that thresholds into and out of existence when the ambient energy flux falls within certain limits: its domain restricted in some way. A stream whirlpool, for example, is a local island of order drawing energy from the river's loss of gravitational potential energy, which is a manifestation of increasing disorder. The whirlpool disappears when the current becomes either too strong or too weak, a frothing rapid or a tranquil laminar flow, respectively. Thoreau's sand foliage at the Deep Cut was also driven by the loss of gravitational potential energy, and was present only when the slurry energetics were just right. The fern frond he must have seen in the ditch below the Deep Cut was also a dissipative structure, in this case drawing power from the sun, which is slowly burning up its fuel. This thermodynamic commonality links the actual foliage of the fern with the pseudo foliage of the sand. Energetically, the fern, being more complex, is further from the common equilibrium state to which both must eventually fall: to death on the one hand and to destruction on the other. Fundamentally, both are heading toward the same place of renewal.[62]

Much, much, much more complex than either of these physical emergences is the neurological emergence of beauty in the human mind. The eruption of beauty—for example the emergence of notes into music—has not yet been satisfactorily explained by scientists. Thoreau was on this path as a young man when he wrote: "Music is the crystallization of sound. There is something in the effect of a harmonious voice upon the disposition of its neighborhood analogous to the law of crystals; it centralizes itself and sounds like the published law of things." Understanding aesthetic beauty remains a Holy Grail for cognitive scientists: to develop an objective theory for subjective phenomena such as a mother's smile, a symphony, a flower, or a reflecting pond. I am not referring to a chemical answer based on the concentration of hormones, a sociobiological answer based on evolutionary fitness, or an anatomical answer based on functional nuclear magnetic resonance imaging, because we already achieved these successes. I am referring to a cause explained by neurobiochemistry. Ilya Prigogine's later research sounded a cautionary note that such searching might be futile, and perhaps this is for the best.[63]

By the time *Walden* was submitted, Thoreau was beginning to see the emergent properties of nature everywhere: "The free, bold touch of Nature," he called it. "Give any material, and Nature begins to work it up into pleasing forms, even the ugliness of gray scum on the ice." Writer Joyce Carol Oates once reflected on the emergence of the sand foliage at the Deep Cut. Through its "fantastical designs on the embankment we are led to see how mysticism is science, science mysticism, poetry merely common sense."[64]

At this point, we have fathomed to the bottom of Thoreau's bathymetric inductions to reach his ur-theory of nature, that it makes sense on its own terms. And at the Deep Cut, we have fathomed down through the hierarchical complexity of life to its inorganic geothermal origin, to the living rock. I now turn to the mythology that emerged from all this fathoming.

10

MYTHOLOGY

Christmas had passed in the Emerson household. Lidian reigned. Ralph Waldo was on tour in Europe. Thoreau was man of the house. On December 29, 1847, he began a letter to the absent husband and mentor: "My Dear Friend, . . . I am here still, & very glad to be here—and shall not trouble you with my complaints." But then he does complain. A sarcastic rant about a certain unnamed professor whom Frank Sanborn clearly identified as Louis Agassiz. Apparently, the great scientist had stood them up socially on three separate occasions.[1]

"Lidian & I have a standing quarrel," wrote Thoreau, "as to what is a suitable state of preparedness for a traveling Professor's visits—or for whomsoever else—but further than this we are not at war. We have made up a dinner—we have made up a bed—we have made up a party—& our own minds & mouths three several times for your Professor and he came not—Three several turkeys have died the death—which I myself carved, just as if he had been there, and the company too, convened and demeaned themselves accordingly—Everything was done up in good style, I assure you with only the part of the Professor omitted."[2]

Thoreau's biting humor highlights a mystery I have not yet solved. Why did he never mention Agassiz's seventeen-year-old glacial theory in writing before *Walden*'s publication? Based on circumstantial evidence, I conclude that he knew it would wreck his book.[3]

Christian Natural Theology

As with the Laurentide Ice Sheet, natural theology developed slowly, culminated powerfully, and withered quickly. In short: human cognition dawned, bringing ideas. Ideas about ideas dawned, bringing philosophy.

Figure 33. Collapsed cairn at the house site overlooking west side of Thoreau's Cove. The population of stones is likely representative of the local near-surface material, which was glacially milled and meltwater-washed.
September 2009.

Philosophy was extended to nature, bringing natural philosophy, later called science. Science was incorporated into Christian theology, bringing natural theology, which reached its American acme during the Walden years before being annihilated in 1859 by Darwin's *Origin of Species*.[4]

Henry D. Thoreau received a heavy dose of Christian natural theology at Harvard before falling under Emerson's transcendental spell. During his junior year he was assigned two of the subject's most important books: William Paley's *Evidences of Christianity* and Joseph Butler's *Analogy of Religion*. According to Sattelmeyer, both were "orthodox, rational responses to [Enlightenment] deism and eighteenth-century skepticism that strove to prove that revealed religion, specifically Christianity, was compatible with the 'natural religion' of the deists." More succinctly, the goal was to show that the proverbial "Good Book" of scripture was consistent with the metaphorical good book of nature. In his senior year, Thoreau wrote a term paper on Paley's *Natural Theology,* which "contained the classic 'argument from design,'" which he later taught to his own students at Concord Academy, and which he used in his own thinking during his transcendental phase. When he graduated, Harvard President Josiah Quincy certified that "Henry D. Thoreau, of Concord, in this State of Massachusetts, graduated at this seminary in August, 1837; that his rank was high as a scholar in all the branches, and his morals and general conduct unexceptionable and exemplary." Though "qualified as a minister," chided the Athenaeum in London, he is "presently a pencil manufacturer," a discrepancy that spoke volumes about his true religious calling.[5]

Thoreau's nature spirituality of the Walden years emerged during what historian Herbert Hovenkamp—*Science and Religion in America 1800–1860*—called the "watershed decade" of American natural theology. Earlier in the century, colleges began hiring theologians to teach science, and their arguments became "gradually more sophisticated, more responsive to the natural sciences. It was generally acclaimed by Christians of all persuasions as an effective way of demonstrating the existence and attributes of God." Reverend Edward Hitchcock fit the archetype of this profile. After being ordained as a Congregational (conservative) pastor in 1821, he resigned his pulpit to become a Professor of Chemistry and Natural History at Amherst College in 1825. Two decades later, he was still hard at work at the same job but with a more honest title, Professor of Natural Theology and Geology, apparently in that order. The Book of Genesis was his academic specialty.[6]

Reverend Hitchcock viewed "the environment as the result of a divine analysis of man's needs," insisting that "science always supports biblical testimony and often explains spiritual operations only partly described in the Bible." His "greatest work, *The Religion of Geology and its Connected Sciences* (1852) [*sic*]," was published to "'defend and illustrate' the truth of the 'Christian Religion.'" His Amherst students, Emily Dickinson among them, were "treated to Religious Lectures of Peculiar Phenomena in the Four Seasons." His "Elementary Geology" became *the* New England textbook standard, which is why his son kept it in print after the Reverend's death. With respect to the diluvium controversy, Hitchcock's "Hail Mary" plan—against which there could be no counter argument—was deeply buried in the playbook of his *Final Report.* That for any true "sacred historian," the potential glacial origin of diluvium was of "little or no importance, since it does not prove the non-occurrence of the Noachian deluge, even though no traces be now remaining upon the earth's surface of that great event." In other words, faith trumps science, even in the face of compelling negative evidence. In spite of this caveat, Hitchcock could not resist citing in his *Final Report* a personal letter from a missionary colleague in Turkey who described a meltwater flood at Mount Ararat. This anecdote linked his favored diluvial mechanism to the very spot where Noah's ark was alleged to have landed.[7]

"Truly there were catastrophists in those days," wrote George Merrill in his masterful summary *The First Hundred Years of American Geology,* published by the Smithsonian Institution in 1904. And they surrounded Thoreau on all sides during the Walden years. To the east was Louis Agassiz in Cambridge, the ardent anti-evolutionist who believed that serial mass extinctions were evidence that God was trying to get things right. To the southeast was Northern Baptist Seminary in Providence, now Brown University, which still staffs an endowed chair in natural theology. To the southwest was Benjamin Silliman Jr. at Yale, whose textbook on geology contained "a long essay on the 'Consistency of Geology with Sacred History,' complete with a chart illustrating the 'Coincidences between the Order of Events as Described in Genesis, and that unfolded by Geological Investigation.' " To the west was Reverend Hitchcock whose ideas about icy debacles were widely known. Northerly was Emerson's brother-in-law, Dr. Stephen Jackson, who saw the boulders on Mount Katahdin as proof of the greatest flood of all times, inconceivable except by the hand of a caring Christian God.[8]

Thoreau was so put off by the catastrophist zeitgeist that he invoked the issue of "diluvium" only once in *Walden,* and only then as a sarcastic adjectival prick against William Gilpin's "diluvian crash" in a Scottish loch. Thoreau reserved his written commentary about diluvium for his other works, especially *Cape Cod,* in which he lampooned Hitchcock for mapping almost the whole peninsula with the "d" word.[9]

English poet Samuel Taylor Coleridge—whom Thoreau greatly admired—railed at the absurdity of natural theology, which he said "confused nature with the supernatural; it confused science with religion and forced clergymen to lose sight of what they should really be doing; helping men to find faith in an unknowable world." Thoreau railed sympathetically: "There is more religion in men's science than there is science in their religion," he complained, wanting no part of their "Mosaic account" of deluge, debacle, diluvium, and drift, whose chronology was so short that the span from "Adam and Eve . . . down to the deluge" was "one sheer leap." Twisting the knife, he added that "without borrowing any years from the geologist, . . . the lives of but sixty old women, . . . say of a century each, strung together, are sufficient to . . . span the interval from Eve to my own mother. A respectable tea-party merely,—whose gossip would be Universal History." And in a final huff: "Father, Son, and Holy Ghost, and the like . . . in all my wanderings, I never came across the least vestige of authority for these things. They have not left so distinct a trace as the delicate flower of a remote geological period on the coal in my grate." These specific complaints, and more broadly those of the entire American transcendentalist school, were part of a global, century-scale movement dubbed "detheologization" by German philosopher Immanuel Kant. Thoreau, who resigned from Concord's Unitarian Church as a matter of public protest, was an ardent radical of this movement.[10]

In America, the "watershed decade" for Christian natural theology began in 1850 when "the University of Virginia invited fifteen distinguished scholars to present lectures on the 'evidences of Christianity.' " The local, albeit unofficial, response to the Virginia invitation came two years later in 1852, when Hitchcock published *The Religion of Geology and its Connected Sciences,* historically his "greatest work." During that same year, Thoreau was watching his first book, *A Week,* fail in the marketplace, in part because of its strident anti-Christian message. That Hitchcock's publication coincided with Thoreau's return to the *Walden* manuscript after a three-year hiatus was no accident, given the spiritual fumes in the air at the time.

Natural theology's watershed decade also coincided with what biographer Laura Dassow Walls calls "the decade of Humboldt," a time when "his popularity approached cult status." Both movements—Christian natural theology and Humboldtian holism—were driven by the same religious impulse to unify head and heart. Humboldt's "reflexivity of mind, society, and nature," seen by Walls as the "overriding argument in *Cosmos,*" "demands that painstaking observation be supplemented by the emotional responses, subjective impressions, aesthetic judgment, intuitive insight, and informed imagination of the observer." Such reflexivity is blatantly antiscientific because it can exist only in individual minds, whereas science, by definition, requires replication by others. Geographer John Pipkin notes that, although Humboldt is considered the founder of physical geography, this field quickly distanced itself from his "humanistic and idealistic facets," which they saw "as romantic and dispensable appendages to his specialized empirical work and his innovative techniques."[11]

Though elegantly presented, vast in scope, based on empirical observations, and intuitively appealing to scientific minds, the fusion of "order and adornment" in Humboldt's *Cosmos* was simply the latest, albeit highly sophisticated, attempt to erase the age-old duality between reason and superstition, atheism and theism, left brain and right brain, fact and myth, anterior prefrontal cortex and limbic system, science and art. Humboldt, as well as his predecessors and successors, was right to claim that we cannot think straight without ordering our thoughts, nor can we enjoy them without adorning them with meaning. The first half of this is science. The fusion is not.[12]

The lingering demise of Christian natural theology was put out of its misery by Darwin's *Origin of Species* in November 1859. Thereafter, concludes Hovencamp, natural science and Christian theology generally went their separate ways, happier for the divorce. Louis Agassiz and Edward Hitchcock, however, never consented. In their respective citadels of Cambridge and Amherst, they remained holdouts to the end, bracketing Thoreau's Concord with supernatural catastrophism.[13]

Uniformitarian Natural Theology

For the sake of argument, let's assume that philosopher Stanley Cavell is correct in identifying *Walden* as a "sacred text," and that "its words are revealed, received, and not merely mused." This raises the corollary question: Can *Walden* be considered an "alternative" natural theology? One arising from its author's "living earth," rather than from Bronze Age myth? One associated with the eternal circle as opposed to the arrow of biblical scripture, which is highlighted by singularities such as the Fall, the Resurrection, and Armageddon? A theology manifesting the timeless steady-state uniformitarianism of Hutton, Lyell, and Darwin, rather than Agassiz's serial catastrophism, or Hitchcock's great debacle? One inspired by the *Bhagavad Gita* where "the reader is nowhere raised into and sustained in a higher, purer, or rarer region of thought," rather than by the King James Bible, which Thoreau also appreciated solely for its literary content?[14]

Cavell continues: "The Bhagavad Gita is present in *Walden*—in name, and in moments of doctrine and structure" and like *Walden,* is "a scripture in eighteen parts." In fact, Melvin Lyon considered the *Gita* to be Thoreau's "bible." Charles Anderson concluded it was completely "absorbed into the text, or used imaginatively for taking off into an extended metaphor." For Robert Sattelmeyer, "passages from the Eastern scriptures sharpen Thoreau's critique . . . and "the idea of the yogin's life is fully integrated into the perspective of the narrator." According to Philip van Doren, the "sacred scriptures" of "poets and philosophers of Persia and India" were so important to Thoreau that he was "said to have the best library of them in the country." His closest sojourning friend, Ellery Channing claimed that "no one relished the Bhagvat Geeta [*sic*] better." Perry Miller saw *Walden* as "Thoreau's translation of Krishna's teaching into artistic terms." Paul Friedrich devoted an entire book to the subject, asking us to see the *Gita* within *Walden* both "in the most general and abstract terms as well as many specifics of image and trope." Thoreau was so smitten by the book that he read it "in three translations, two English and one French, and studied it not only during his two-year sojourn on the lake but off and on during the seven revisions of Walden." Finally, and most "significantly, the most resonant Eastern fable in *Walden,* the story of the artist of Kouroo, is apparently a product of Thoreau's own invention, demonstrating the extent to which he had internalized and absorbed this material."[15]

Less well appreciated is that *Walden* is equally suffused with the cyclical gradualism of Charles Lyell's *Principles,* which opens by drawing a parallel between geology and Hindu scripture. Thoreau had long been smitten by this link between science and spirit. Within *A Week* (1849), he recapitulated (without credit) Lyell's most important *Principles.* That Earth is so inconceivably old that, for all practical purposes, it can be considered eternal. That nothing under the sun is really new. That everything everywhere is always being recycled from a previous state of existence. That there is no such thing as progress. That the whole Earth manifests steady-state equilibrium through an endless series of creations and destructions via the rock cycle. That the biggest things happen in increments almost too small to be noticed, meaning that creation is ongoing and culminates in every moment.

Emerson—who is directly responsible for bringing both the *Gita* and the *Principles* to Thoreau's attention—was of the opinion that "human nature—the human mind—is and has been essentially the same in all ages and places . . . All ages are equal." Thoreau simply extended this big idea back further in time to the Earth. "As in geology," he wrote in *A Week*, "so in social institutions, we may discover the causes of all past change in the present invariable order of society." "The longer the lever the less perceptible its motion. It is the slowest pulsation which is the most vital. . . . All good abides with him who waiteth wisely; we shall sooner overtake the dawn by remaining here than by hurrying over the hills of the west." "The newest is but the oldest made visible to our senses." "In reality, history fluctuates as the face of the landscape from morning to evening. What is of moment is its hue and color. Time hides no treasures; we want not its then, but its now." "The eyes of the oldest fossil remains, they tell us, indicate that the same laws of light prevailed then as now. The gods are partial to no era, but steadily shines their light in the heavens, while the eye of the beholder is turned to stone . . . We will not be confined by historical, even geological periods . . . Men, will be succeeded by a day of equally progressive splendor; that, in the lapse of the divine periods, other divine agents and godlike men will assist to elevate the race as much above its present condition."[16]

Walden echoes these passages from *A Week* by extending Lyell's trinity of cyclical gradualism, steady-state equilibrium, and deep time. The cycle of time banishes its arrow, creation is continuous, and gradualism, rather than catastrophism, is the natural law.

Walden also echoes passages from the *Gita*. In the translation of Stephen Mitchell: "The presence that pervades the universe / is Imperishable, unchanging / . . . Birthless, primordial, it does not / Die when the body dies. / . . . If you know that one single day / Or one single night of Brahma / Lasts more than four billion years, / They are gathered back into my womb / At the end of the cosmic cycle— / A hundred fifty thousand / Billion of your earthly years— / And as a new cycle begins / I send them forth once again, / Pouring from my abundance / The myriad forms of life."[17]

In turn, the *Gita* echoes the beliefs and myths of other primitive religions, all of which trace their origin back to the earth and its cosmic origin. In his *Myth of the Eternal Return,* scholar Mircea Eliade noted "their revolt against concrete, historic time, their nostalgia for a periodical return to the mythical time of the beginning of things, to the 'Great Time.' " History is annulled by cyclicity. "No event is irreversible and no transformation is final." Reginald Cook recognized this primitive, "extra-Christian dimension" in *Walden,* as "ancient, ritualistic, and hieratic," the "reactualization of the archetypical gestures of archaic man—the gestures of baptism and planting and harvesting, or curative ceremonials and rebirth."[18]

Aside from its content, the timing of Thoreau's manuscript revisions, and his frequent use of the word "sacred" support the notion that *Walden*–Part II was Thoreau's alternative to the prevailing Christian theology dominating Concord churches at the time. Bernhard Kuhn concluded that Thoreau's "sense of wonder and aesthetic appreciation is in large part a sign of the continuing influence of [Christian] natural theology within the field of natural history." If true, Thoreau's version was heretical in the extreme, though not reactionary, arising not so much from his animus toward the Puritan tradition, but from the quality of his intimate immersion experience at Walden Pond. There he consulted Nature's oracle after being deeply steeped in both the *Principles* and the *Gita.*[19]

My interpretation is consistent with Joel Porte's summation that "Thoreau's solitary life at Walden Pond" was "the spiritual exercise of a man who has lost his interest in formal religion but not in the habits that it once engendered." And it certainly qualifies as one of William James's *Varieties of Religious Experience:* "the feelings, acts, and experiences of individual men in their solitude . . . in relation to whatever they may consider the divine." To bathe in Walden each morning was, for Thoreau, "a religious exercise, and one of the best things which I did." Its water was "as sacred as the Ganges at least." Within a year of *Walden*'s publication, Frank Sanborn acknowledged Thoreau as the "high priest" of the "Walden Pond Society," an informal religious movement widely recognized in Concord.[20]

Robert Sattelmeyer concludes that the final version of *Walden* is a "theory of nature, which, as Emerson had said in his first book, *Nature,* was the aim of all science." If so, then the tie between Lyellian cyclical uniformitarianism and Hindu cyclical spirituality is a "perfect correspondence of Nature to man." The most famous example of this correspondence is Thoreau's sand foliage passage at the Deep Cut, which provides a unifying metaphor for birth and rebirth in every domain of existence from flowing sand to linguistic evolution. All *crescunt* from simplicity to complexity. Thoreau understood that "order was not dictated rationally from above but emerged cooperatively from below . . . from the collective interactions of constituted individuals," whether from atoms or societies. This notion of continuous creation from within, rather than a supernatural creation from without, emerged to become what John Gatta calls Thoreau's "sacred cosmology." His purpose in *Walden* is to take us *down* all these branches of divergent complexity to the unified trunk of truth in order to realize that all aspects of nature are one, including human spirituality. Or, as Robert Sattelmeyer put it, the study of nature is ultimately the study of the self. Looking downward into "earth's eye," Thoreau could legitimately ask of the lake, "Walden, is it you?"[21]

In *Walden*'s "Conclusion," Thoreau invents the Artist of Kouroo, who creates a perfect staff as a symbol of a perfect world.

> Time kept out of his way, and only sighed at a distance because he could not overcome him . . . By the time he had smoothed and polished the staff Kalpa was no longer the pole-star; . . . [and] . . . Brahma had awoke and slumbered many times. . . . When the finishing stroke was put to his work, it suddenly expanded before the eyes of the astonished artist into the fairest of all the creations of Brahma. He had made a new system in making a staff, a world with full and fair proportions; in which, though the old cities and dynasties had passed away, fairer and more glorious ones had taken their places. And now he saw . . . the former lapse of time had been an illusion.[22]

This parable is as Lyellian as it is Bagavadian. The arrow of time stands aside while Earth cycles through its endless Huttonian "revolutions," a word chosen by the "father of geology" to mirror the celestial revolutions of Isaac Newton and Nicolaus Copernicus before him. In Concord, the latest rock-cycle revolution is still taking place, the creation of the Acadian Mountains and their continuing denudation. At a finer scale of time, revolution involves the making and unmaking of subsequent Waldens during each glacial cycle. Thoreau actually used Hutton's word "revolution" in the first draft of *Walden,* though he later edited it out, probably because, by Thoreau's time, the word had taken on negative, convulsive political connotations. But the basic idea remained central to the final text of *Walden,* and to that of his contemporary *Journal:* "Is not the world forever beginning and coming to an end, both to men and races?" That Thoreau believed this in the very core of his being is evidenced by his final days. He went calmly and happily to his death knowing that both his spirit and his "Saxon race" could live but "one world at a time."[23]

Glacial Dilemma

To assert, as Barksdale Maynard does, that "Thoreau and his contemporaries did not know about Concord's history of glaciation" is a convenient generalization. Ditto for John Pipkin's statement that "Thoreau did not work contemporary glaciology into his thinking," or Robert McGregor's comment that "the glacial origins of Concord's hills and ponds seemingly held little attraction . . . [because] . . . he makes almost no mention of the subject in his journals." A close reading of the facts shows otherwise. Shortly after reading Darwin's and Kane's descriptions of existing glaciers in 1851, he accurately envisioned all of New England being overwhelmed and buried by a continental ice sheet, and he seemed to know why.[24]

Was Thoreau's Arctic vision a genuine scientific induction? Or was it yet another example of his habit of collapsing spatial scales, in this case, a thick snowfall for an ice sheet? Was it a wishful projection of Elisha Kent Kane's well-publicized descriptions of Arctic phenomena to his home turf? Or was it an unattributed summary of Agassiz's glacial theory, which Thoreau had presumably read by that time? Whatever the sources of Thoreau's ideas about glaciation, not one tweet on the subject made it overtly into Walden. There, the closest he gets to glaciation is the species name for the common loon, *Colymbus glacialis.*[25]

William Howarth believes this deafening silence on a then-popular subject indicates that "Thoreau either forgot about glaciation or regarded it as an unproven theory." I disagree on both points. No naturalist of Thoreau's caliber could *forget* something that intuitively makes so much sense, a paradigm that has since brought consilience to disparate observations of the Concord landscape, and which would have given Thoreau the additional benefit of being a thorn in Edward Hitchcock's side. Thoreau's *Journal* shows that he greatly enjoyed speculating about unproven theories and teasing their promulgators, for example Robert Hunt's support for "Actinism," the misguided idea that sunlight caused rock weathering during the day but the effects were reversed at night. Having puzzled over the gap between what Thoreau knew and what he let on that he knew, the most parsimonious explanation is that, when revising *Walden,* Henry decided to keep the glacial theory at arm's length owing to its *sin qua non* of cosmic catastrophism in the American version, which was Agassiz's version.[26]

Art historian Rebecca Bedell convincingly demonstrated that "geology reigned through much of the nineteenth century as America's most 'fashionable' science." Setting the academic stage was the founding of a "gentleman's" geological society at Yale College in 1819 (the American Geological Society) to match the founding of the London Geological Society in 1807. "Between 1820 and 1870," she continued, "men and women across the country . . . avidly pursued an interest in the field." Geology had entered the "heroic period, or romantic period" that typifies a new science. We must keep in mind that it was a group of geologists who, in 1848, launched what would eventually become the most powerful scientific organization in the United States, the American Association for the Advancement of Science (AAAS). Its immediate parent was the Association of American Geologists and Naturalists (AAGN), founded in 1842. And its grandparent was the American Association of Geologists, founded in 1840. The dates of these name changes signify the critical importance and volatility of geology during the explosion of American science and its professionalization during Thoreau's early writing career.[27]

Walter Harding reported that when the Concord Lyceum opened in 1829 it offered "numerous lectures . . . on geology, botany, and ornithology," and that Thoreau became one of its most active supporters. In 1830, Lyell's *Principles* would begin to unite the field into a true paradigm. Ralph Waldo Emerson's tilt toward nature in 1832 followed an epiphany in the *Jardin des Plantes* in Paris. After seeing its mineral and rock collection, "amber containing perfect musquitoes, grand blocks of quartz, native gold in all its forms of crystallization,—threads, plates, crystals, dust; and silver, black as from fire," he wrote: "Ah! said I, this is philanthropy, wisdom, taste,—to form a cabinet of natural history. . . . I am moved by strange sympathies; I say continually 'I will be a naturalist.'" By 1835, the *New England Magazine* noted that Bostonians had gone "geologically mad" for what they considered the "queen of the sciences," second in dignity only to astronomy. In 1836, Emerson had followed through on his self-promise to become a naturalist, publishing *Nature*. In 1837 he gathered Thoreau under his wing, a recent college graduate who would go on to build his own "cabinet" of mineral specimens that included a few mundane, but important, rocks from Concord: "granite" and "mica schist."[28]

By 1840, both Emerson and Thoreau had read Lyell's *Principles*. In 1842, its author voyaged to America and "delivered a series of public lectures in Boston, New York, and Philadelphia," to audiences supposedly composed of as many as three thousand "persons of both sexes, of every station in society." Thoreau could conceivably have been in the crowd.[29]

Louis Agassiz's 1846 arrival to America was part of this same crescendo of "geology rising." Within a year he had toured the northeast and concluded that all of New England had been shaped by a great ice sheet flowing down from the north, grooving rock on the summits of mountains, creating Cape Cod as essentially one giant moraine, and sprinkling erratic boulders during its recession. All true. Within a year he had lectured on the subject to Silliman's Yale geology students, and to "thousands of Bostonians crowded into the Tremont Temple, on some evenings as many as five thousand of them, to hear the foreign professor." So great was the public interest that Agassiz repeated his lectures each day to a second audience. He became a wildly popular "rock star" throughout America; his "fame and fortune" arising from his "charm, erudition, and brilliance." New England's intellectual luminaries courted him as a friend. This cultural phenomenon culminated in 1847 when Agassiz was offered and accepted a newly created academic professorship at Harvard.[30]

By the summer of 1847 Thoreau was personally supporting Agassiz's research program as a paid collector of specimens. At some point that year, the two men met in the flesh, talked at the Marlborough Chapel, and had plans to meet again. By 1848, Thoreau was using Agassiz and Gould's *Principles of Zoology* as a reference, despite its catastrophist premise. By 1849 Thoreau and Agassiz were cordial correspondents. By October that year, Thoreau had been exposed to Agassiz's *Etudes* and facsimiles of its plates within a lengthy review prefacing Hitchcock's

Final Report. Finally, and most directly, Thoreau's personal knowledge of the link between the great Swiss scientist and his glacial theory is confirmed by a cryptic comment in a private 1853 letter to H. G. O. Blake in which he writes of a "Switzer on the edge of the glacier" who can be none other than Agassiz himself.[31]

At the time, Thoreau was also being exposed to other authoritative glacial texts. Elisha Kent Kane's *Personal Narrative* is full of accurate scientific descriptions of glaciers. It pervades Thoreau's *Journal* for February 1854, with some of its material entering *Walden*. The striking similarities between Kane's Arctic observations and Thoreau's winter observations that year may have been what prompted Henry to ask: "are not most of the arctic phenomena to be witnessed in our latitude on a smaller scale?" Kane wisely avoided Agassiz's geological leaps of cosmic significance. Instead he stuck to matters of natural fact, citing Professor James Forbes's physical observations of existing glaciation in six separate sections of his book. Kane, while reflecting on the moraines, giant boulders, streamlined rock knobs, drift, and till fronting modern ice caps in Greenland and Baffin Bay, wrote: "It was impossible not to have suggestions thrust upon me of their agency in modifying the geological disposition of the earth's surface." Thoreau almost certainly read this very line, and may have imagined something similar for Concord as well.[32]

Given Thoreau's fascination with arctic phenomena, Agassiz's looming presence at his *alma mater* in nearby Cambridge, and the wild popularity of geology at the time, I find it highly implausible that he was ignorant of the Switzer's views. Robert Sattelmeyer wrote matter-of-factly that Thoreau knew the "principal works of Louis Agassiz," and the *Etudes* is certainly the most important historically. For Thoreau to accept the glacial theory on Agassiz's terms, however, required that he also accept its supernatural catastrophism. That Walden Pond had originated from a solid version of Noah's Flood sent by God to reboot the planet prior to Man's arrival. To accept it would also have required that he support the views of a "rival," an archetype "Man of Science," a consummate authority figure with a swaggering ego, and one who spoke with a thick French accent to one of America's most ardent patriots. To accept Agassiz's theory would have been to endorse a man who killed, stuffed, and pickled hundreds of animals all in the name of science. Frank Sanborn, speaking of Agassiz, recalled that his friend Thoreau "avoided the man of science." I suggest he avoided his words as well.[33]

Contributing to Thoreau's deafening silence on Agassiz's glacial theory in *Walden* may have been the pettiness of perceived social offenses to the Emerson household. Or perhaps it was mistaken fealty to his new role model, Darwin, whose early conversion to the glacial theory in 1842 may have gone unnoticed by Thoreau, having been buried in a footnote of his 1846 copy of *Journal of Researches*. Or perhaps it was financial. During the *Walden* revisions, Thoreau was in serious debt owing to the commercial failure of *A Week*, and his literary career could not survive another disaster. By mid-1852, he was fully committed to a manuscript featuring a lake of eternal purity and endless cyclicity. Accepting Agassiz's hyperbolic catastrophism at this stage would have directly contradicted the book's central thesis. Regardless of motive, I conclude—but cannot prove—that Thoreau exercised his author's prerogative to ignore the validity of ice sheet glaciation until after his book was delivered to the publisher.

And when was that? Scholars agree that he sent batches to the printer during the interval late February to mid-March 1854. Philip Van Doren concluded that "typesetting went fast" because "Thoreau had evidently promised not to make many changes, allowing the book to go straight into page proofs," skipping the galley proof stage. March 13 was Thoreau's only known trip to Boston during this interval, as shown by that day's *Journal* entry: "To Boston." By March 16, he had signed an indenture (publishing contract) for *Walden* with Ticknor & Co. Given these dates only three days apart, it is reasonable to suggest that Thoreau hand-delivered the first part of his manuscript on the 13th and picked up the contract. The alternative would have been for this particular self-reliant author to trust its handling to someone else, a manuscript he had agonized over for nine years through seven drafts. After the contract was signed, *Walden* was typeset almost immediately. Page proofs were complete before June 7 when Mr. Fields was in the process of hand-carrying them to England before he aborted his trip in eastern Canada due to illness.[34]

There are two tidbits of circumstantial evidence supporting my assertion that Thoreau dropped off the first portion of the manuscript himself on March 13. On that day, he purchased an expensive collapsing military telescope that he confessed in his *Journal* he had been wanting for quite some time. Also on that same day—but without mentioning it in his *Journal*—he checked out the two most important books on ice sheet glaciation from two separate libraries in two separate cities: Agassiz's *Etudes* from Harvard in Cambridge, and James D. Forbes's more analytically rigorous *Travels through the Alps of Savoy . . . With Observations on the Phenomena of Glaciers* from the Boston Society of Natural History. Could it be that the telescope was a self-reward for delivering the manuscript? And that the pair of books on glaciation were now acceptable because the die for *Walden* had been cast?[35]

That Henry spent time reading these books during the next several weeks is indicated by the content of his *Journal* entries. Ice becomes a dominant theme. He drew four sketches between March 18 and 22, 1854. On March 28, the day after he wrote "Got first proof of 'Walden,'" Thoreau "had the experience of arctic voyagers amid the floe ice on a small scale," a subject covered by both books. And then—completely out of the blue on March 31—came a passage that scholars have puzzled over for years: "In criticizing your writing, trust your fine instinct. There are many things which we come very near questioning, but do not question. When I have sent off my manuscripts to the printer, certain objectionable sentences or expressions are sure to obtrude themselves on my attention with force, though I had not consciously suspected them before. My critical instinct then at once breaks the ice and comes to the surface."[36]

"Breaks the ice"? The glacial "ice"? Could it possibly be that the "thing" Thoreau came "very near questioning," but chose not to, was his decision to ignore the glacial theory? Did his recent exposure to this pair of authoritative books—*Etudes* and *Travels*—raise "certain objectionable sentences or expressions" that obtruded his "attention with force," even though he had not "consciously suspected" them before? Had he allowed his "critical instinct" for steady-state uniformitarianism to overrule the possibility of catastrophe? This, of course, is pure speculation on my part. But something important was clearly involved, something well beyond cosmetic wordsmithing.[37]

During the next month, Thoreau left compelling proof that he had examined both books before finishing his corrections to the *Walden* manuscript. He borrowed from Forbes's section on glaciology the claim that blue is "the color of pure water, whether liquid or solid." A *Fact Book* excerpt entered into the *Journal* on April 27 contains the following sentence: "Forbes says that the guides who crossed the Alps with him lost the skin of their faces,—apparently from the reflection from the snow." This incidental aside is immediately followed by the very last *Journal* passage that found its way into *Walden*—about the rise and fall of the ponds—and is immediately preceded by an unusually cryptic passage that begins: "No man ever had the opportunity to postpone a high calling to a disagreeable duty. Misfortunes occur only when a man is false to his Genius."[38]

Speculating again, could Thoreau have been reframing the editing crisis he had a month earlier? Could the doubts he overruled as a disagreeable duty have been associated with the glacial theory? Overruled in conscious deference to his literary genius? This guess is weakly supported by Thoreau's relatively quick return of *Etudes* to the Harvard Library on April 18, 1854, with what reads like a petty taunt. In the accompanying letter, he strikes the word *Etudes* from its title and replaces it with the word "Agassiz." This transforms the title from a "Studies of Glaciers" to "One Man's Opinions About Them." In contrast, he kept Forbes's book until May 9 of that year.[39]

At a minimum, these library patron records prove beyond reasonable doubt that Thoreau had examined Agassiz's book with three months left to spare before *Walden*'s publication and nine days before he made his final corrections to the page proofs. Had he wished to make any changes, he could have easily done so. At the very least he could have followed Hitchcock's example by notifying his readers of Agassiz's competing ideas, however ludicrous they may have struck him.

The banishment of Agassiz's theory from *Walden* is consistent with a separate and more clear-cut mystery. Why did Thoreau describe the glacial grooves and striations on the summit of Mount Monadnock so clearly *after* the book's publication in 1854 but not *before?* These features were—and still are—salient landforms on its bald summit, impossible to miss for an observer as keen as Thoreau, especially during his "midnight by moonlight" rambles, when the grooves would have glistened with polish like no other feature. Thoreau made two trips to its summit before *Walden*'s publication, in July 1844 and on September 7, 1852. The second trip post-dates his careful reading of Hitchcock's 1841 *Final Report,* which highlights the distribution and alignment of "diluvial" grooves and striations with a foldout map and with a two-page numerical table listing their locations and azimuths. Among these sites were outcrops along the Lexington Road in Concord, and on the summits of all three of Thoreau's favorite southern New England mountains: Monadnoc (now Monadnock), Wachusett, and Saddleback (now Greylock). Also in Hitchcock's report was his review of Agassiz's theory complete with original engravings of Monadnoc's striations and *roche moutonees*. In his main text, Hitchcock wrote that Monadnoc [*sic*] was "thoroughly grooved and scored" with common "diluvial" grooves and scratches . . . One groove measured fourteen feet in width, and two feet deep; and others are scarcely of less size. Their direction at the summit, by a mean of nearly thirty measurements with a compass, is nearly north and south." These lineaments are so obvious that even my own, marginally interested eleven-year-old son noticed them without being prompted during our first summit climb together.[40]

Though Thoreau seems to have ignored these lineaments before *Walden*'s publication, he described them enthusiastically on his first trip afterwards on June 3, 1858: "The rocks were remarkably smoothed, almost polished and rounded; and also scratched. The scratches run from about north-northwest to south-southeast. . . . The rocks were, indeed, singularly worn on a great scale. . . . as if they had been grooved with a tool of a corresponding edge."[41]

Setting aside my speculations about Thoreau's overt silence on glaciation in *Walden,* two curious facts remain as a matter of historic record. Before its publication, Thoreau checked out the two most authoritative works on the subject of ice sheet glaciation, examined them both, and cited one. Second, before its publication, he almost certainly beheld the most conspicuous and astonishing field evidence of ice-sheet glaciation in New England, and then wrote nothing about it.

Myths and Allegory

"It is very certain, at any rate, that once there was no pond here, and now there is one." Surely, this is *Walden*'s greatest tease. The author of that sentence knew very well that Walden had been created by a void sunk below an ancient river terrace associated with an earlier geological epoch when the landscape may have been buried by glacial ice. Overtly, we get none of this in *Walden.* Covertly, we get it all under the guise of the book's literary structure, creation myths, delicious puns, and clever allegories veiled with calculated obscurity.[42]

Thoreau begins with a myth borrowed without credit from a Walden-esque kettle located about fifty miles to the southwest:

> Some have been puzzled to tell how the shore became so regularly paved. My townsmen have all heard the tradition—the oldest people tell me that they heard it in their youth—that anciently the Indians were holding a pow-wow upon a hill here, which rose as high into the heavens as the pond now sinks deep into the earth, . . . and while they were thus engaged the hill shook and suddenly sank . . . It has been conjectured that when the hill shook these stones rolled down its side and became the present shore.

On his personal copy of *Walden,* Thoreau confessed in writing: "This is told of Alexanders Lake in Killingly CT, by Barber in his Con. Hist. Coll," short for Jonathan W. Barber's *Connecticut Historical Collections* (1835).[43]

That Thoreau borrowed this story of topographic inversion for Walden suggests he thought it applied there as well. Morphologically, he was justified. Both are similar sized kettle lakes sunk into delta plains. When the Walden ice block was first being detached as part of the debris-covered stagnant fringe, it would have indeed loomed up into the air. As it melted downward toward the elevation of the lakeshore, stones and debris would have been shed away from the highest places in all directions. But after the ice block was completely buried beneath the delta plain, further meltdown would have caused the stones to roll into the deepening cone of collapse, where they could later pave the shore. Also during deglaciation, the rapid loss of vertical weight, accompanied by the reduction in glacio-hydrostatic water pressure and the flexural tilting of the crust, would have led to an epoch of more intense earthquake activity than at present.[44]

Thoreau continues: "And this Indian fable does not in any respect conflict with the account of that ancient settler whom I have mentioned." Here, he alludes to the pioneering English Puritans who settled Concord in 1635, and who came with specific traditions for locating water, digging wells, and stoning them for stability. Thoreau specifically recalls "an old settler and original proprietor . . . who remembers so well when he first came here with his divining-rod, saw a thin vapor rising from the sward, and the hazel pointed steadily downward, and he concluded to dig a well here." That place would have been the center of Walden's western basin, the epicenter of deepest meltdown, directly above its focus. There lay the "richest vein," in this case a vein of H_2O ice, and "so by the divining-rod and thin vapors I judge; . . . I will begin to mine." "As for the stones . . . that, unfortunately, . . . is no longer a mystery to me. I detect the paver." Here, Thoreau writes like a mystery novelist, letting us know that he has discovered the paver's identity, but refuses to reveal it to his readers at this stage.[45]

So, who was this paver? In context, Jeffrey Cramer concludes it was the "glacier, which deposited boulders and drift." This is strongly supported by Thoreau's more nuanced clues about the paver's identity:

> I have occasional visits in the long winter evenings, when the snow falls fast and the wind howls in the wood, from an old settler and original proprietor, who is reported to have dug Walden Pond, and stoned it, and fringed it with pine woods; who tells me stories of old time and of new eternity . . . a most wise and humorous friend, whom I love much, who keeps himself more secret than ever did Goffe or Whalley; and though he is thought to be dead, none can show where he is buried.

In this passage, Thoreau moves from historic metaphor to glaciological allegory. Born on "long winter evenings" when the "snow falls fast and the wind howls" was the Laurentide Ice Sheet, the continent-sized dome that that "crept down as a glacier," "dug" Walden Pond with meltdown subsidence, and stoned it with a bouldery talus before fringing it with tundra. A small remnant of that great ice sheet remains today, a "secret" buried beneath the modern snows of the Barnes Ice Cap in the northeast Canadian Arctic archipelago. In Thoreauvian lingo, "perchance" the original proprietor "remains alive yet," estivating during the summer, and "keeping to himself" in the bleakness of the north.[46]

With the physical Walden in place, the stage was now set to infuse that cold, rubbly landscape with interglacial life. Enter the "ruddy and lusty old dame" in whose "odorous herb garden," Thoreau passed his Walden pastoral. Through her "genius of unequaled fertility" she transformed that place into a living ecosystem powered by the warmth of sunlight. Using that power, she drew upward through plant roots the "expressed juice" of the hills, and combined it with carbon drawn from the atmosphere through leaves, to create the wood, bark, leaf, humus, pollen, plankton, and invisible dissolved substances on which every living creature at Walden depends. What had been for several millennia a nearly sterile pool of water before about 15,000 years ago became by Thoreau's era a kettle of dilute organic soup simmering with life. "Shall I not have intelligence with the earth?" Thoreau asks. "Am I not partly leaves and vegetable mould myself?"[47]

During that creation, the physicality of the old settler and the fertility of the "lusty old dame" created a child named Walden whose birth was attended to by the demigod "Thaw," who used the "gentle persuasion" of slow meltdown during the delivery. Thoreau was very clear about the symbolic marriage between organic and inorganic nature. "If Nature is our mother, then God is our father." And as a family: "God is my father & my friend—men are my brothers—but nature is my mother & my sister."[48]

Walden was born naked, a child of ice. As an infant, it wore clothes of tundra. In adolescence, those of boreal forest. In maturity, those of pitch pine and shrub oak forest. In old age, the culmination would be cut to rags by wood-cutters seeking energy to drive progress into a new geological epoch.

That ice and snow are prominently featured in *Walden* is consistent with the idea that Thoreau knew, or at least suspected, its glacial origins. The book's symbolic year begins in "Economy" when he framed his house amidst "flurries of snow" and at a time when ice floes edged the pond. It ends when the ice melts, the sun brightens, the lake enlivens, and frost comes of the ground. Between those literary bookends, snow and cold pervade the text as a major theme for five chapters in a row: "House-Warming" through "Spring." Collectively, they narrate the seasonal cycle from freeze-up through melt-down. I find it no accident that he used the word "winter" in the titles for the central three chapters out of five. And that the central chapter of the five-chapter sequence, "Winter Animals," opens with the culmination of winter, arguably a stand-in for the culmination of a glacial age. The land is everywhere gripped by cold. "When the ponds are firmly frozen, they afforded" the "new views" described in previous chapters: "Baffin's Bay . . . snowy plain . . . and the fishermen . . . with their wolfish dogs, passed for sealers or 'Esquimaux.'" After becoming dangerously cold on one winter day, Thoreau realized that creating an actual—rather than imagined—glacial world in

Concord would require little more than an increased frequency and duration of bad winter weather, namely a sequence of historic "Cold Fridays and Great Snows" one after the other.[49]

Could the "intruder from Hudson's Bay" that Thoreau described in *Walden* have been a creeping ice sheet, instead of the allegorical solitary Canadian goose? Later driven northward "out of Concord horizon" by denizens of the interglacial forest, in this case the "hoo-hoo" of the "cat owl?" This highly speculative reading of Concord's "new views" is circumstantially supported by shared elements of three separate texts: Elisha Kane's vision of mid-latitude glaciation from his *Personal Narrative;* Thoreau's vision of Concord's glaciation from his *Journal* entry for February 3, 1852; and Thoreau's surrogate Arctic vision of the snowy expanse of Flint's Pond from *Walden.* All take place in deep snow. All use the name Baffin Bay. All mention the Esquimaux and their wolfish dogs, either directly by name or indirectly by description and allusion. The *Journal*'s glacial passage and *Walden's* icy scenes both feature the hoot owl in the same, prominent symbolic role. This last connection, if valid, would confirm that Thoreau deliberately veiled his knowledge of Walden's glacial origin.[50]

The Fitchburg locomotive plowing its way through the deep snows of winter may be an even more obscure allegorical reference to the power of glaciation. "On this morning of the Great Snow, perchance, which is still raging and chilling men's blood . . . and I behold the plowmen covered with snow and rime, their heads peering, above the mould-board which is turning down other than daisies and the nests of field mice, like bowlders of the Sierra Nevada, that occupy an outside place in the universe." In his *Fully Annotated* edition of *Walden,* Jeffrey Cramer notes that the source of Thoreau's allusion to the "bowlders" remains unidentified. By my reading, they are glacial erratics which, by the nineteenth-century definition, were large boulders lying beyond the place where they properly belong. During the Walden years, those of the Sierra Nevada had become a source of great mystery to the Gold Rush prospectors and the explorers associated with the great western surveys. Some probably saw these colossal and often rounded boulders of Sierra granite in the foothills, and concluded what the explorers of the Jura Mountains had a century earlier. That the "bowlders of the Sierra Nevada" had been turned down from the mountains along with pre-existing alpine vegetation, here symbolized as "daisies and the nests of field mice.[51]

Finally, Thoreau speculates on a repeat performance of the ice age that must inevitably return: "Nor need we trouble ourselves to speculate how the human race may be at last destroyed. It would be easy to cut their threads any time with a little sharper blast from the north. We go on dating from Cold Fridays and Great Snows; but a little colder Friday, or greater snow would put a period to man's existence on the globe." Here, Thoreau's words unambiguously express a common early Victorian fear, not of Armageddon, but of global refrigeration "chilling men's blood" and the certain annihilation that would follow. This fear was a logical corollary of accepting Agassiz's extremist version of ice age catastrophism, based on the theory that if an *Eizeit* can happen once, it can happen again.[52]

Enough speculation. After ten chapters, the time has come to descend toward the end of this book. Thoreau's path to simplicity was downward.

11

SIMPLICITY

Rocky, rugged, stormy, and cirque-bitten. Crowning the Presidential Range of New Hampshire, Mountain Washington is New England's highest peak. Thoreau climbed it with his older brother John Jr. in September 1839 near the end of their famous river voyage, later memorialized as *A Week on the Concord and Merrimack Rivers*. They summited the mountain in what biographer Robert Richardson called "the most laconically reported excursion of . . . [Thoreau's] career." The complete journal entry for that day is this: "Sept. 10. Ascended the mountain and rode to Conway." Alan Hodder also noticed Thoreau's eagerness to "gloss over the four days in the mountains in as few words as possible." Prompted by his suggestion, I counted and inspected all seventy-seven of those words. Not one from five days in the mountain is either positive or negative. All are neutral gray. Far more pleasing to Thoreau were the lower and smoother summit domes of southern New England's monadnocks: Mounts Greylock (Saddleback), Monadnock, and Wachusett, listed here in order of diminishing altitude. More pleasing still were the lowland lakes, swamps, and dead streams where he elected to spend most of his free time.[1]

Maine's Mount Katahdin was another high place Thoreau explored. When walking alone on its high rocky plateau, he did not even bother to summit the peak, rationalizing that his friends might be impatient for his return. Alternatively, he may have held himself back due to an unspoken fear—based on a recurrent nightmare—that he might die if he reached the top. Though Thoreau never completed this solo climb, his later reflections are universally recognized as a major turning point in his life, and certainly the most powerful one having to do with nature. The source of that power was primitive "wildness," stripped of all ro-

Figure 34. One of Walden's "magic circles" shrouded by fog and frozen in time. A hole punctured in the glaze of clear ice allowed an upwelling of liquid water that froze overnight.
January 2013.

manticism. Less well appreciated, however, is the source of that wildness. Beyond the human isolation, most of it had to do with rock, practically the only thing visible on that misty day in the "cloud factory" of Katahdin. Primitive rock. Granite rock. Jagged blocks with interstices resembling "bear's dens." A sea of broken rock not yet claimed by life and not yet readied for human habitation.[2]

Thoreau's close encounter and intimate contact with that raw granite on that elevated *felsenmeer* suggests a motive for his climbing pursuits. Only above tree line could he drench himself in the rock reality he so craved, rather than merely touch it here and there as a ledge or boulder poking through the nearly ubiquitous "alluvion" of New England. In landscapes of the arid American southwest, which Thoreau never experienced, rock is exposed mainly below the tree line, disappearing upward as the moisture increases with coolness. In the humid northeast, however, the situation is reversed. There, inland from the wave-beaten outer coast, one must ascend above the tree line in order to descend down to rock as the dominant surface material.[3]

Katabasis

Going low instead of high was a point of departure between Ralph Waldo Emerson and Henry David Thoreau. After buying his eleven-acre woodlot north of Walden Pond, Emerson made preliminary plans to "build me a cabin or turret there high as the treetops and spend my nights as well as days in the midst of a beauty which never fades for me." Thoreau, of course, built his domicile "low in the woods," beneath not only the local moraine peaks, but the delta plain as well. Later, when Emerson purchased additional land on Walden's southern shore, his impulse was to build there on a bedrock promontory dubbed "Emerson's Cliff" by Thoreau, a name that stuck. There, on "a rocky head, perhaps sixty feet above the water," wrote Emerson, "I think to place a hut, perhaps it will have two stories & be a petty tower, looking out to Monadnoc & other New Hampshire Mountains."[4]

His plans were fairly typical of his day. Long before John Ruskin's lifelong celebration of mountain scenery, leaders of the Romantic movement in both Europe and America, sought what art historian Albert Boime has called the "magisterial . . . gaze of command, or commanding view—as it was so often termed in the nineteenth-century literature . . . the perspective of the American on the heights searching for new worlds to conquer." In Emerson's case, this was a spiritual world. Regardless of motive, the purpose was to reach sublimity by climbing upward, a precedent set by Moses in the Judeo-Christian mythos and by the nineteenth-century poets who rambled the peaks of the English Lake District. At such heights, wrote environmental historian William Conon, "one had more chance than elsewhere to glimpse the face of God."[5]

Thoreau did appreciate mountain summits. But he remained ambivalent about them his entire life. From extensive reading and personal experience, he learned that the native inhabitants preferred to worship them from below. And because high peaks are so inhospitable, Thoreau found it vaguely unsettling to ascend beyond the upper threshold of life, akin to trespassing on the sacred. Climbing for the challenge alone was, in Thoreau's mind, something reserved for "daring and insolent men." Perhaps most importantly, he found mountains more geometrically boring than lowlands: "The higher the *mt.* on which you stand, the less change in the prospect from year to year, from age to age. Above a certain height, there is no change." This observation describes what in mathematics is called the asymptote.[6]

From any summit, the focus dissipates outward because the rays of vision diverge: you get more space because it is less concentrated. Conversely, the focus for every hollow concentrates inward because all the rays converge. When seen from afar, the center of Walden Pond is one such place, especially if one looks through its transparent water. For a visually sensitive person like Thoreau, mountains might have revealed too much at once, making them almost overwhelming. Perhaps this is why, when on the summit of Mount Wachusett, he felt more like a cartographer than a climber: "There lay Massachusetts, spread out before us in its length and breadth, like a map."[7]

Henry preferred an outward and slightly downward view, based on the frequency of his repeat visits to Concord's lower hills, and to the power of his emotions when upon them. This was especially true for the cliff at Fairhaven, which he dubbed his "observatory." At nearby Walden, he also situated his house to provide an outward and slightly downward view, tucking it back against a thicket of trees and building it to face a distant lake. To measure the view from his house site, I set up an engineering transit and quantified the angles from what would have been the height of his eyes when sitting in his doorway. From that exact three dimensional coordinate, he could have seen at least 43° of horizontal arc over an average length of about a third of a mile. The vertical arc between the distant horizon and closest water was 9°, most of which was slightly downward.[8]

Thoreau's instinct to move his body downward and look downward was especially strong when he was young. As a twenty-one-year-old, he felt it "a luxury . . . to cuddle down under a gray stone." In the first draft of *Walden* he went "*down* to the woods" and then "*down* to the pond" to "live deliberately." In summer, he sought the lake's "deepest resort." In winter, he enjoyed lying down on the ice to peer "down into the quiet parlor of the fishes," concluding that "Heaven is under our feet is well as over our heads." Getting his "feet down to the earth" was a precondition for discovering truth. Getting half way down through the bog of confusion was not enough. His head felt like "an organ for burrowing" in search of deeper reality.[9]

When living at Walden, Thoreau went "down to the pond" to get a drink, swim, walk out on the snow, or simply stare. The lower he went, the more reverential—rather than "magisterial"—became the view. From his "pond-side," the view was outward toward the opposite shore, upward to the opposite bank, further upward to the forested kettle rim, and then upward even more to the tops of the clouds. From a rowboat at the lake's epicenter, this reverential gaze would have been radially symmetrical, giving him visual unity. Looking over the gunnels and downward into the ultra-clear water offered a second reverential gaze, one expanding downward into a "lower heaven itself so much the more important." There, in place of the dome of the sky, was a bowl of crystal clear water. Between them was the nexus of the lake surface.[10]

Symmetry

In a private letter to his confidant H. G. O. Blake, Thoreau wrote: "When the mathematician would solve a difficult problem, he first frees the equation of all incumberances, and reduces it to its simplest terms, So simplify the problem of life." Following his suggestion, I hereby simplify Walden to its lowest geometric terms. This requires assuming that the lake basin is the water-filled bowl it only crudely approximates, an ideal hemisphere. Doing so yields a *volume* given by the equation: $V = 2/3\ \pi\ r^3$, where "π" is the famous constant with a value of 3.14159 . . . and "r" is the radius. Inside that volume of water, a loon could fly underwater in any direction, zigzagging approximately a quarter mile and spiraling at whim.[11]

Ratcheting that complexity down one exponential notch, we can envision, not a volume of a hemisphere, but its *surface area* given by $A = 2\ \pi\ r^2$, one that floors the curved bottom of Walden Pond. Anywhere upon it, a milkweed seed drifting in from far away might come to rest, or a dust grain from the nearby Bean Field. Taking this aerial complexity down another notch brings us to the ring of its shore, a *perimeter* defined by $P = 2\ \pi\ r^1$ where wavelets swash the stones, and frogs hide between them. The final downward step toward simplicity brings the exponent to zero, which yields a constant, in this case with a fixed value of $2\ \pi$. Size no longer matters.

We have been doing calculus, a subject Thoreau enjoyed thinking about. At each simplifying step, we have taken what is called the *derivative* of the previous formula, a step toward mathematical simplification. For example, within the volume of Walden's hollow, every point can be thought of as the tip of a vector with three degrees of freedom: compass bearing, downward angle, and length. Vectors drawn to its bottom require only two degrees of freedom because all of the lengths are the same. Those drawn to the circle of its shore are simpler still because there is only one degree of freedom, that of azimuth: all points are of equal distance and have zero depth. In the final step of simplification, there is no vector at all. Nothing remains but a constant, in this case with a value of 6.2832 . . . There is no end to the series of numbers to the right of the decimal point because π is an irrational number, meaning it cannot be expressed as a ratio between integers. This makes it a *transcendental number,* the definition of which we will not bother with here because it involves the roots of polynomials. My point? Walden's magic circle simplifies downward to transcendence in mathematics, rather than in philosophy.[12]

In calculus, the reverse of derivation is a process called integration. Simple things are made more complex by ratcheting up the exponents, rather than stepping them down. Beginning with Walden's transcendental constant, the lake shoreline becomes the integral of its radius, the bottom becomes the integral of that shoreline, and the volume becomes the integral of that bottom.

Moving from math to physics, the integration of Walden begins with a material constant, the "angle of repose" for its cleanly washed delta sand, which governs the steepness of its slopes, especially the straight ones flanking the inner basin. In the ideal case of an hourglass, the steady flow of dry sand through its nexus creates a loss of support immediately above that point. This causes the downward and inward flow of sand toward it. The resulting dimple at the surface is a perfect cone with straight sideslopes governed by the frictional physics of uniform sand, which yields a steady-state angle of repose somewhere near 30°.

Moving from the ideal of physics to the "real world" of geology, meltdown of stagnant ice within Walden's western basin caused a loss of support for the loose sand and gravel of the overlying delta plain. This caused a downward and inward collapse of more complex (heterogeneous) materials via a much more complex agency. From gravitational ground zero at the focus (deepest intact granite), the collapse propagated upward and outward. Integrating the geology, the rays lengthening above Walden's *point d'appui* (its focus) became a crude circle (edge of incipient crater) that became a crude hemisphere (larger hollow) that became an organic machine (the culminating natural system) with a radial symmetry. Thoreau took the additional intellectual step of integrating that machine into a literary masterpiece.[13]

This quasi-mathematical treatment brings me to a mathematical theorem developed in the early twentieth century by Emmy Noether (1882–1935). She was an influential German mathematician and physicist who worked during an era when women were generally excluded from academic positions. A Jew from Erlangen in Bavaria, she fled to the United States in 1933 to take up a position at Bryn Mawr College, where she soon died. The essence of her theorem states that geometric symmetry and the conservation of energy are directly related. Without going into the details, nature's simple shapes are the inevitable consequence of the first law of thermodynamics, the law of conservation of energy. Mathematically, the ideal symmetry of the sphere mirrors the maximum conservation of energy, which is another way of saying the simplest condition.[14]

In the process, symmetry and simplicity become one and the same. Thus, Noether's theorem links the geometric holism of Concord's Walden with Thoreau's stated purpose of *Walden*: the search for "Simplicity, simplicity, simplicity!" Pushing this idea to the limit, the propagation of Walden above its focus gave rise to the holism, purity, and stability of *Walden*. Taking these attributes in sequence: The holism was a straightforward geometric consequence of radial collapse. The purity was a geochemical consequence of that geometry combined with the percolation of pure water through what is effectively crushed granite. The stability is a consequence of both shape and substance. Thus, to understand *Walden* at its deepest level, we must get to the bottom of Walden. To its focus. To its central *point d'appui* on glacially polished granite.[15]

Renewal

Thoreau's upper heaven ran on celestial time. Above Walden Pond, the amount of sunlight reaching the outer edge of the troposphere rose and fell in a perfect sine curve of warming and cooling. Four special thresholds marked each of Thoreau's years: peak warmth, peak cold, the switch from warm to cool, and the switch from cool to warm. These are the summer and winter solstices and the fall and spring equinoxes, respectively. Everything "up there" ran as regular as clockwork. The arrival of spring could be predicted to the nearest second.[16]

Happily this was not the case at Walden, which lies at the bottom of Earth's active atmosphere, its *troposphere*. There, instead of gradual change, Thoreau encountered complex interactions between various climatic and meteorological variables—prevailing winds, air masses, maritime influences, topography, land cover, tropical and extra-tropical storms—that rendered the clockwork of astronomy into the chaos of pond-side weather. These interactions determined when the first frog would croak in the spring, and when that season might be canceled by a frigid blast of Canadian air, re-setting the thermal clock back to winter. Indeed, springtime at Walden was never a moment, but a poorly defined, drawn out, and highly erratic composite transition lagging far behind the steady orbital forcing.[17]

Below this chaos was a second progressively more regular and well-integrated climatic signal. Within the leaf litter, hourly temperatures mirror those on the surface. Deeper into the humus, the ground temperature oscillates only once per day, and the amplitude varies perhaps ten degrees. The temperature of Thoreau's root cellar varied less than that, and the Boiling Spring less than even that, just a few degrees above or below about 45 degrees. Speleologists know that somewhere at even greater depth, the temperature does not change at all, a place where the annualized net loss of heat at the surface is exactly balanced by the net gain from a constant geothermal flux. Geologists call this the "depth of zero annual amplitude." Just one nick above this thermal equilibrium is the most conservative climate signal on earth, a slight annual rise and fall of temperature unaccompanied by obvious chemical or biological responses. Though very stable, and fully integrated, this annual signal would have made a lousy symbol for Thoreau to appropriate for the onset of Spring. First, it cannot be seen, thereby negating his favorite mode of engagement. Second, it came far too late, lagging several months behind the dramatic change at the surface because every calorie of heat exchanged had to be conducted through a thick blanket of insulating soil, sandy meltdown debris, and thermal rock mass. And worst of all, this signal is lifeless.

Because the arrival of spring signified spiritual rebirth for Thoreau, he made three serious attempts to get a firm grasp of it. His first was independent of the astronomical calendar. In search of a pre-biological Spring, he worked his way down into the earth and back in time to the "frost coming out of the ground," trumpeting this success in *Walden*'s climactic passage. Though straightforward and unambiguous at any given site, this bellwether of spring came far too early—sputtering on and off from the December solstice into mid-March—and was far too individualistic, melting in one place while freezing in another. His second attempt to identify the arrival of spring was based entirely on the astronomical calendar. Rather than look for one salient indicator—like ground frost—regardless of the date, he attached to the calendar the statistical noise of all indicators to create what he later called his Kalendar. The premise was simple: if he gathered enough data over enough years, he would be able to predict and celebrate the seasonal transition at its peak moment. The problem with this approach is that it converts the joyful anticipation of spring into a statistical measure, for which one could never get enough data. Thoreau died trying, his phenology unfinished.

Luckily, very late during *Walden*'s composition, Thoreau discovered a signal for spring that was particular enough to be celebrated (like the first robin), but integrated enough to manifest the whole Walden system. I refer to the turnover of Walden's waters when the ice went out. These are his words. The italics are mine:

> Such is the contrast between winter and *spring*. Walden was dead and is *alive* again. . . . The change from storm and winter to serene and mild weather, from dark and sluggish hours to bright and classic ones, is a memorable crisis which *all things* proclaim. It is seemingly *instantaneous* at last. Suddenly an influx of light filled my house . . . I looked out the window, and lo! where yesterday was cold gray ice there lay the transparent pond already calm and full of hope as on a summer evening, reflecting a summer evening sky in its bosom, though none was visible overhead, as if it had *intelligence* with some remote horizon.[18]

Dead to alive. Storm to serene. Winter to warmth. Dark to bright. Sluggish to classic. Cold gray to transparent. A sudden influx of light. These seemingly separate indicators integrate and unify "all things" into a single "memorable crisis" that happens only once per year: the loss of slab ice. Using Thoreau's famous proclamation with only two words changed, "This is the" *ice* "coming out of the" *lake* "this is Spring." This marvelous event took place well after the first flickers of phenological spring, having been delayed by the thermal cold sink of the kettle, the low sun angle, the latent heat associated with thick ice, and the thermal inertia of millions of gallons of water. But to my mind, this delayed but fully integrated "crisis"—in the best sense of this word—is a far better signal for spring than the earlier, but more patchy, loss of frost in the ground.

By the time this annual "crisis" arrived, the zenith of the sun was high enough to rapidly warm the floating layer of cold meltwater to its critical temperature. This sent it plunging downward where, upon reaching the bottom, it pushed the slightly warmer water of Walden's deep hole back up to the surface. When this was mixed with fresh meltwater, it quickly cooled back to its critical temperature, forcing a second descent, which forced a second rise, and so forth. Round and round this process went in a positive feedback loop that kept going until every drop of the lake was homogenized to the uniform temperature of 39°F. At the end of the crisis, Walden was about seven degrees warmer than before. Rising up with the heat was also the chemistry of life. Before the upsurge, the nutrients and exhaust gasses of vigorous benthic life had been slowly dissolving into the stagnant hypolimnion. On their way up from the bottom, however, the pressure dropped, allowing some of the gas to ventilate. Perhaps the upsurge of heat from below, accompanied by the rich smells of rising life was the "intelligence" Thoreau sensed was present, but not "visible overhead"?

Here, in the annual behavior of Walden, we find an analogy for Thoreau's mind. Both were intransitive systems. The lake had two stable equilibrium states: one for summer and one for winter, each with its own characteristic properties. The same was true for his mind, which toggled back and forth between poetic and scientific modes.

Recall that the spring turnover caused by *ice-melt* is the second of two that take place only a few months apart. The first occurs less dramatically and less forcefully during *ice-freeze*. Taken together, these double downward pulses of oxygenated water are analogous to the "lub-dub" sounds of a human heart, followed by a pause of eight or nine months, a time of astonishing stability at depth unmatched by anything at the surface. Ironically, as life shuts down at the surface in mid-December, it surges anew at depth. But to see that life, you have to go down, not only to the pond, but also to its bottom of that pond to find it.[19]

Immediately after Thoreau's description of spring turnover in *Walden* is this line: "I heard a robin in the distance, the first I had heard for many a thousand years, methought, whose note I shall not forget for many a thousand more." Thus, the first birdsong of spring was due not to the sun, nor to the southwest wind, but to the surge of liquid water rising up from below. Thoreau's "lower heaven" had risen up to greet him and the rest of his "Brute Neighbors."[20]

Descendentalism

In American historical culture, the word "transcendentalism" is unambiguously associated with those orbiting Emerson's parlor during the 1830s–1850s. Thoreau, whose orbit had previously been wayward, was powerfully captured by the great man's gravitas. During the days of the *Dial,* he became the elder man's novice, some say his clone, accepting without question Emerson's big idea that nature was a collection of natural facts awaiting correspondence with spiritual facts. Thoreau's early poetry, essays, and journal entries place him firmly in this camp through the 1849 publication of *A Week*. But the closer the reader moves toward the 1854 publication of *Walden,* the less well the label "transcendentalist" applies. By that time, concluded expert Philip Gura, he had become atypical of the movement, "content to be in and of the world rather than to transcend it."[21]

In my trusty, second-hand, and duct-taped 1955 *Oxford English Dictionary,* the root word for "transcendental" is *scendere,* to climb over. To transcend *(trans-scendere)* something is thus to "pass over or go beyond" it. Each hurdle in a race, for example, must be transcended, or each puddle jumped. In contrast, to ascend *(a-scendere)* is to "go or come up" to it. And to descend *(de-cendere)* is to "move or pass from a higher to a lower level." Perhaps it is my geological perspective that is confusing me. But when I read *Walden* against the early *Journal,* I feel like I am on a very different journey, one that is descending inward toward the simplicity of a lower heaven in Walden's waters, rather than ascending to an outward heaven via the stepping-stones of the Peterboro Hills, or flights of fancy to the stars. Whenever I read Thoreau's clear statement of purpose in Walden—"to live deep and suck out all the marrow of life, . . . to drive life into a corner, and reduce it to its lowest terms"—my subconscious thoughts attach to the words "deep" and "lowest," which are physically downward, and to the word "reduce," which is mathematically downward. To my literal and geospatial mind, geo-*Walden* is a downward journey toward ultimate spiritual truth on a bedrock unconformity.[22]

"Simplicity, simplicity, simplicity!" In *Walden* this direction is downward in space, backward in time, and toward the foundation of theory. In *ontogeny* he sought the wisdom of a newborn child, whose whole universe lay within a mother's arms. In *etymology* he sought the "father tongue," of the Greek, Latin, and Norse roots of words, and their Indo-European precedents. In *anthropology* he stripped the dross from modern civilizations to find "the Esquimaux and the Patagonian" as signposts to the point of departure between humans and animals, a time when "men" were "nearer of kin to the rocks and to wild animals than we." This explained his "natural yearning" for the cave, which is documented within both his childhood memories and his adult impulses. In *history* it was down to "an older era than the agricultural" from which he was "convinced" his genius was born. "The Aegean Sea is but Lake Huron still to the Indian." In *literature* it was down to "mythology . . . the most ancient history and biography," containing, "only enduring and essential truth, ethics, history, poetry . . . skeletons of still older and more universal truths." In *philosophy* it was down to the nitty-gritty of "particularity," as "an antidote to the philosophical idealism of the age." In *architecture* it was down from Babelesque towers built of "hammered stone" to the lowly fieldstone wall: "More sensible is a rod of stone wall that bounds an honest man's field than a hundred-gated Thebes." In *religion,* wrote Henry Canby, it was down "again and again into the aqueous regions of the intuition." For Joseph Wood Krutch, this was down from Puritan to Pagan.[23]

In natural *science* it was downward to fundamental laws through induction. In *astronomy* it was down through the planets, comets, and moons to the "Chaos and Old Night," responsible for creating our "morning star." In *geometry* it was inward to the center of all things, and then downward toward the apex of a cone or the pole of a hemisphere. In *psychology* it was downward to the bottom of the pond of consciousness, from which everything radiated. In *acoustics*, it was down from the high frequency of complex noises to the "vibration of the universal lyre." In *paleontology* it was down to the everlasting designs of anatomy, to "some life prowling about" beneath the surface. In *geology* it was downward to the "subterranean fire" beneath the granite basements of our tallest mountains. In river *hydrology* it was downward either to the sluggishness of dead streams (which were most alive) or to the ancestral sea (from which life came). In *physics* it was down from the inherent instability of complexity toward the stability of entropy. In *technology* it was down to the versatility of simple tools like the jackknife, or, better yet, the flint tools of its prototype. Back to "flint and steel," even when he had "a matchbox in his pocket," a point that early critic James Russell Lowell clearly misunderstood.[24]

Thoreau's descent to simplicity made swamps and bogs far more interesting places than mountain summits. "The very sight of this stagnant pond-hole . . . is agreeable and encouraging to behold, as if it contained the seeds of life, the liquor rather, boiled down . . . They speak to our blood, even these stagnant slimy pools." Indeed: "There are sermons in stones, aye and mud turtles at the bottoms of the pools." To hear such sermons Thoreau imagined himself as an amphibian, neck deep in a swamp. On Lake Walden he imagined flying underwater with the loon, or prowling through the muck with the pouts, whose "dull uncertain blundering purpose" he found so intriguing. When in the woods, he asked his readers to "probe the earth to see where your main roots run." On the cliff he asked us to get down to the level of lichens, our "kith and kin," as alive as we, yet elemental. In our houses, he asks that we strip our material needs down to the basics of food, shelter, clothing, and fuel. Given his "peculiarly wide nature, which so yearns toward all wildness," he asks that we follow his example and "channel" our inner fox.[25]

In the 1960s Melvin Lyon recognized that "the word 'well' is the most common synonym for 'pond' in the book." Indeed, during his daily descent to the pond, Thoreau more closely followed the biblical model of Rebecca at the well than of Moses on the Mount. There he experienced *katabasis,* the Greek word for a downward journey that, in Thoreau's case, was toward spiritual discovery. There, in Walden's peaceful waters, he found a cultural antidote to the violence associated with the proverbial "city upon a hill," whether Old World Jerusalem, or the New World Puritan settlements."[26]

"If you have built castles in the air," Thoreau advises us, "your work need not be lost; that is where they should be. Now put the foundations under them." This was his ultimate downward move. Beneath the "castles in the air" of New England transcendentalism he put Acadian bedrock foundations. The steeple of *Walden*'s uniformitarian natural theology rests on the granite unconformity beneath Walden.[27]

Instinct

In his manifesto "Where I Lived and What I Lived For" Thoreau asks us to wedge downward below the "alluvion" of our lives to make contact with the nature of human nature, the bedrock of our biological selves. To instinct, that primitive place where, deep in "Solitude," he became "suddenly sensible of such sweet and beneficent society in Nature." There, he was "distinctly made aware of the presence of something kindred." This "sensed" presence was almost certainly the well-documented human affinity for the ecosystems in which humans co-evolved. An instinctive appreciation for the unity of all living things. Later, in "Brute Neighbors," he probes even deeper by extending his inquiry to all objects, asking: "Why do precisely these objects which we behold make a world?"[28]

The modern answer from astrophysics and cosmology is called the anthropic principle, the recognition that the life we *do* know is the only one we *could* know. As Stephen Hawking put it, when "intelligent beings ask the question: 'Why is the universe the way we see it,' the answer is then simple: If it had been different, we would not be here!" This, of course, is common sense. But there is also plenty of cosmological evidence to back up the anthropic argument, beginning with the grand homogeneity of the universe and the reason why so many of the universe's fundamental constants lie within the narrow range required for life.[29]

There are strong and weak versions of the anthropic principle, both with various spin-offs. The weak version, which states that things are the way they are or we would not notice, obviates the need to invoke a creator. The strong version comes closer to requiring an intentional designer because it seeks to explain why the cosmos was made "just right" for us, delicately fine-tuned for very particular human needs. There is even a participatory subversion of the strong version based on quantum physics. It posits that the universe requires conscious observers in order to exist at all.

To critics, the anthropic principle is an unfalsifiable, and therefore unscientific, tautology. To others, it is chauvinistic to suggest that carbon-based "life as we know it" is the only kind there is. To still others, it is little more than science posing as religion, the closest an atheist can get to theism without invoking God. This, of course, sounds like the Thoreau I've come to know. Though labeled an atheist by his Christian contemporaries—and a heathen to boot—he was intensely spiritual and more religious than most.

The reader may be wondering why I am introducing such a grand conundrum at the end of a long book. My answer is that I do not really know. I am relying on instinct. The philosophical ambiguity (or vacuity) of the semi-scientific anthropic principle strikes me as a fitting end to an extended critique of *Walden,* a semi-scientific book written by a semi-scientific author living in the semi-scientific nineteenth century. Even if correct only as metaphor, this principle still provides a useful philosophical mirror for us to examine ourselves. Thoreau chose Lake Walden to be his anthropic mirror. The reflection he saw was *Walden.*

Apparently, we live in a universe where the gravitational constant and the coefficient of expansion are finely tuned to each other. And a universe in which the electromagnetic forces between atomic particles are neither too small, nor too large, but just right. This allows the sun to burn hydrogen and helium, and allows late-stage stars to explode as supernovae to create heavier elements; without which we could not think, write, and read. Ditto for the age of the universe, which is just right for us. Ditto for the fact that our eyes are tuned to the frequencies of the solar spectrum in its most energetic range. Certainly, this is no coincidence. Having evolved from fruit-eating arboreal primates in a dense forested habitat, our vision during the last forty million years was slowly and naturally selected to discriminate the colors of ripe fruit in dim light. Without this fine-tuning and advantage-taking, our ancestors would have gone extinct, leaving us unimagined and, in the quantum version of the anthropic principle, the universe not in existence.

Perhaps something this primitive accounts for Thoreau's response when he was "unexpectedly struck with the beauty of an apple tree" during the summer solstice of 1852, when sunlight was striking those apples with maximum power. "We pluck and eat in remembrance of Her," he wrote with some ceremony on a day worshipped by all pagans. Doing so was "a sacrament, a communion." It is no wonder that Thoreau fancied *Wild Fruits* more than any other food, especially the taste of hand-picked huckleberries and gnarly wild apples. The act of eating them was ceremonial, an anthropic descent to the place of his ecological home and to the time of his evolutionary origin.[30]

Like most primates, Thoreau loved the chaos of forest foliage with its branching patterns far more than he did the poverty and austerity of the heavens. "I do not get much from the blue sky, these twinkling stars, and bright snow-fields reflecting an almost rosaceous light. But when I enter the woods I am fed by the variety," he wrote during the autumn of 1851. Deepening this thought during a succeeding winter:

> I pine for a new world in the heavens as well as on the earth . . . a wilderness of stars and of systems . . . [the] . . . sky does not make that impression of variety and wildness that even the forest does, as it ought. It makes an impression, rather, of simplicity and unchangeableness . . . It does not affect me as that unhandselled wilderness which the forest is.

Here, Thoreau is harkening back to his arboreal primate origins.[31]

It was in that primitive forested world that Thoreau felt most comfortable, which helps explain why *Walden* bears the subtitle *Life in the Woods*. "As Cowley loved a garden, so I a forest," he concluded, shortly after asking himself: "Why should just these sights and sounds accompany our life? Why should I hear the chattering of blackbirds, why smell the skunk each year?" And at the broader scale of the Concord landscape: "What are these rivers and hills, these hieroglyphics which my eyes behold?" Prompted by these questions, he committed himself to look as deeply into nature as possible: "I would fain explore the mysterious relation between myself and these things. . . . [to] . . . know why just this circle of creatures completes the world." More specifically: "How adapted these forms and colors to my eye! . . . I am made to love the pond and the meadow, as the wind is made to ripple the water?" And finally: "This enchantment is no delusion . . . By some fortunate coincidence of thought or circumstance I am attuned to the universe, I am fitted to hear, my being moves in a sphere of melody." This last sentence contains a clear and compelling statement of the anthropic principle written long before it was defined by modern science. This makes Thoreau a prophet.[32]

Being smitten by the notion that he was "attuned to the universe" brought Thoreau to the God question: "Who placed us with eyes between a microscopic and a telescopic world?" Why is "the universe . . . so aptly fitted to our organization, that the eye wanders and reposes at the same time. On every side there is something to soothe and refresh this sense." These were rhetorical questions because Thoreau had answered them several years earlier, after being inspired by a rainbow hanging above a waterfall: "Plainly the Maker of the universe sets the seal to his covenant with men . . . Designed to impress man . . . Kosmos or beauty.

We live, as it were, within the calyx of a flower." And beneath his calyx were his plant roots seeking sustenance from "inanimate matter,—rocks or earth." Having discovered the anthropic principle, Thoreau wrote: "The mind of the universe" is a "kindred mind with mine."[33]

"I only know myself as a human entity, the scene, so to speak, of thoughts and affections, and am sensible of a certain doubleness by which I can stand as remote from myself as from another." This "doubleness," which he so often felt, was the spectator-critic of his Pleistocene cognition perched on the shoulder of his Pliocene ego. The hypertrophic intelligence of *Homo sapiens* perched on the biological perceptions of "here-and-now" human nature. The latter was unquestionably guided by primate evolution through the geological epochs. Nature was indeed "designed to impress man" with its autovisceral responses to beauty and ugliness, food and famine, sex and abstinence, safety and fear. If such instincts can lie preserved beneath "the thick hides of cattle and horses, like seeds in the bowels of the earth," then why should they not also be preserved beneath our skins as well? Could it be that our purest, most primitive instincts also respond "to the same influence which inspires the birds," to which we are quite closely related in the grand scheme of things? "Shall not men be inspired as much as cockerels?"[34]

Thoreau sensed this evolution at work in his own body. "My feet are much nearer to foreign or inanimate matter or nature than my hands; they are more brute, they are more like the earth they tread on." And "Man's eye is so placed as to look straight forward on a level best, or rather down than up." Any modern specialist in human origins would agree, given our phylogeny as bipedal ground foragers with few natural enemies from above or below. Thoreau put these evolved human traits in the same context as those of other vertebrates, the horse being a "helpless creature . . . out of his element or off his true ground" and the hawk whose "breast and belly pure downy white" suggest an "unfitness . . . in contact with the earth."[35]

Much of Thoreau's life had been spent searching for the "base-alloy" of human beings, from which all humanity arose before their inevitable divergence from animals. The "distinction" of thought that "survives" in every "true man" is "something essentially public," meaning shared by every member of our species as a common evolutionary heritage. Looking back to human origins, Thoreau traced the biology of our species from the "civilized man" to the "savage" and then to the "brute," suspecting comparable spans of time between these stages. He believed that "wild men living along the shores of the Frozen Ocean" were still with us. And after reading Darwin's account of Tierra del Fuego, he knew that ethnographically primitive humans lived in the southern hemisphere as well. These struck him as even wilder than their counterparts in the north: allegedly cannibals who were largely insensitive to cold, and who subsisted by scavenging on the edge of survival. Yet, even under these animal-like circumstances an "old man . . . muttering" before a meal of putrid whale blubber amidst the "profound silence" of his near-naked clan, struck Thoreau as "religious worship." He concluded "that even the animals may have something divine in them and akin to revelation,—some inspirations allying them to man as to God."[36]

There were many forces at work in bringing Thoreau to Walden Pond. Most important was the push-pull of society vs. nature, respectively. Society pushed him out from the village to a place on the margin where he might make a clear philosophical statement regarding self-reliance and material simplicity. This was *Walden*–Part I. Nature pulled him into a place where he could experience the "drenching" immersion of anthropic totality. This was *Walden*–Part II. But why did he choose Lake Walden, as opposed to the Weird Dell or the Hallowell Farm, which he also seriously considered as sites for his experiment? And why did he build his house so far in from the shore of the lake?

My answer involves the downward pull of human prehistory during the australopithecine transition, from an arboreal past to an open-land future. Downward from the tree canopy to the edge of the savannah where, with "brute" feet on the earth, the primal human consciousness looked outward from the refuge of the forest toward an open clearing, preferably with a "field of water" somewhere in the picture. All of Thoreau's potential living places met these criteria. Lake Walden met them best.

From the doorway of his tidy house, the view was outward, over a clearing, through the trees, and slightly downward to open water in the distance. This was an atavistic view from human evolution, and an anthropic one as well. Going "down to the pond" for water, or to pick up a stone, is an instinct that I feel in my own bones; one that I suspect most readers of this book can feel in theirs as well. This instinct is nicely captured by the title of Richard Leakey's co-authored book about human origins. Indeed, we are all *People of the Lake*.[37]

CONCLUSION

Midway through this book project I drove to Concord, Massachusetts, to hear Professor Laura Dassow Walls read from her new book *Passage to Cosmos*. The venue at the Thoreau Society headquarters was intimate, just a dozen chairs pulled together in the parlor of Thoreau's recently restored birthplace home on Virginia Road. In the room were a few of my new colleagues. I felt like I belonged.

On the way to the talk, I detoured into Sleepy Hollow Cemetery to visit Thoreau's grave, which is one of my favorite places. The small slab of marble inscribed "Henry" is no more significant within his family burial plot than those of his unmarried sisters, Sophia and Helen, his more charming brother John Jr., and his nature-loving parents, regal Cynthia and quiet John. All are equally mellowed to a creamy tan by calcite dissolution, and are equally splashed by raindrop dirt. Today, however, Henry's marker always stands out, invariably decorated by pilgrims from around the world bearing gifts of pencils, pine cones, flowers, paper messages, coins, and occasionally a well-worn paperback copy of *Walden*. For reasons I cannot explain, I have never felt the urge to leave a token gift. In fact, my urge is to simplify his grave by cleaning the mess up.

When tourists visit Author's Ridge, it is easy for them to stay focused on the ridge, given its narrow topography and the eye-catching monuments of those associated with the nineteenth-century American Renaissance, and the townsfolk who made their cultural work possible. Notables include the Alcotts, Emersons, Hawthornes, and Ripleys, and individuals like William Ellery Channing, Daniel Chester French, Elizabeth Peabody, and Franklin Sanborn. Of these, the stone for Ralph Waldo Emerson draws the most attention. Unshaped. Massive. An polychrome chunk of vein quartz ranging from pastel pink to rusty yellow. It is a

Figure 35. Thoreau's grave marker at Sleepy Hollow Cemetery. Decorated with simple gifts, and spattered with earth from a recent rainstorm. July 2011.

quarried raw block from New Hampshire, crystallized from the steaming geothermal fluids within the ancient mountain root that became New England. Emerson's gravestone is so geochemically stable that it will likely outlast everything else in the park, especially the marble slabs of the Thoreau family, which may not survive the carbonic acid rains of the next millennium.

During my impromptu visit to Sleepy Hollow that day, I found myself being drawn not to the famous ridge, but to the pair of landforms on either side that create it. To the north is the collapsed margin of the sprawling kame delta that grew southward away from the dying glacier. Its steep, straight slope lies at the angle of repose above the wooded swamp below. To the south, however, the slope is lower, gentler, and gracefully curved, wrapping around on itself to become Sleepy Hollow, a closed depression.

Pausing for a moment, I was struck with great force by a simple idea. Thoreau's grave occupies nearly the same setting as his house foundation at Walden Pond. Both were dug near the top of a pair of sand-gravel deltas that stand as island plateaus above the flat, poorly drained floors of the glacial lakes responsible for the Sudbury meadows, the Musketaquid marshes, and the Bedford Flats where Thoreau was born. At Sleepy Hollow, Thoreau's grave is marked by cut marble. At Walden, his house site is marked by cut granite. Fronting both places is a glacial kettle. The one at Walden collapsed below the water table, giving rise to a namesake lake. The one at Sleepy Hollow bottomed out above the water table, leaving a namesake hollow. Both places had little economic utility during Thoreau's lifetime, given their distance from the village center and their elevated, excessively drained soils. Hence, both of these outbacks remained available as pastoral retreats during their respective eras: Walden for Thoreau's experiment in deliberate living, and Sleepy Hollow for its park-like ambience near a rapidly commercializing nineteenth-century town.

When Thoreau's body was interred at Sleepy Hollow, he was granted one of his most fervent wishes: to be buried within the warm sands of a glacial delta. That wish was boldly declared in early November 1851 at the climax of an extended sojourn with his friend and frequent walking companion William Ellery Channing. Their day had been full of adventures. After "going across lots" in Thoreau's favorite direction, they explored "some moraine-like hills covered with cedars"; played a "game of 'Go away Jack, come against Gill' " with the sparkling water of Dudley Pond; and shared the adrenaline-pumping thrill of being "chased by an ox, whom we escaped over a fence while he gored the trees instead of us."[1]

Later, when walking along eastern side of Long Pond in Cochituate—near where millions of automobiles cross it on the Massachusetts Turnpike today—Henry and Ellery clambered across a shoreline bluff gashed by storm erosion. This may have been the first time Thoreau encountered a deep, natural exposure of one of Glacial Lake Sudbury's kame deltas, the "familiar" earthen materials into which the kettles of Walden Pond and Sleepy Hollow were also sunk. With great delight, he saw with clarity the "naked flesh of New England, her garment being blown aside." And perhaps with his narrow escape from death in mind, he continued: "Dear to me to lie in, this sand; fit to preserve the bones of a race for thousands of years to come. And this is my home, my native soil; and

I am a New Englander. Of thee, O earth, are my bone and sinew made; to thee, O sun, am I brother . . . To this dust my body will gladly return to its origin. Here have I my habitat. I am of thee."

That was a pretty strong rush of feelings, even for Thoreau. The earthy sensuality of lying down naked with Mother Nature stripped of her sod clothing. The more maternal appreciation of her soil as the flesh from which his "bone and sinew were made." And the peaceful acknowledgment that his body must inevitably return "to dust." In that moment, Thoreau came to understand that his "habitat" and his body were one in the same. The golden delta sand that seemed so "familiar" was, in fact, the nexus between his life and his land.

Thoreau's surge of emotion continued to gush into his *Journal* for a second night: "These are my sands not yet run out. Not yet will the fates turn the glass . . . In this clean sand my bones will gladly lie . . . Here ever springing, never dying, with perennial root I stand; for the winter of the land is warm to me . . . I am such a plant, so native to New England, methinks, as springs from the sand cast up from below." He knows—as we all do—that his sands will some day run out. That the hourglass of his life must be turned so that another spirit can take his place in the "clean sand." There his "bones will gladly lie." New life will arise, "ever springing, never dying."[2]

Thoreau's wish to be buried within these "familiar" warm sands was granted eleven years later in 1862, originally in the Dunbar family plot, and then later on Author's Ridge in the 1870s. In those soils, plant roots probed through the dry sand seeking the nutrients of his decomposing body and that of its wooden coffin. As these disappeared, a void was created, causing the overlying soil to sag downward and inward: a dribble of sand here, a clod of earth there. If noticed by the cemetery sextons, the sag would have been filled unceremoniously during routine maintenance: a little earth now and then, followed by some raking. Today, after a century and a half, only Thoreau's hard bones likely remain at depth, entombed beneath a surface that has since become completely stabilized by vegetation.

Something analogous happened at Walden Pond, though on a much larger scale and much earlier. A "living" glacier died. Its final remains—a stagnant block of ice and the crushed rock it contained—were buried in sand. As the ice disappeared, a void formed beneath the surface, creating a sag that was initially filled by sediments of a glacial river, but which later remained unfilled. The sag continued. A crater formed. And then a puddle, a pool, a pond, and finally the coalesced lake of global fame, all within the same hollow. Though no bones have been found within Walden's collapsed residues, those of the ice-age mastodon and their predators are a distinct possibility. In place of bones that might be there are hard granite boulders that surely are there: torn from ancient ledges, milled into shape during transport, and let down gently on the bedrock below.

These boulders now lie entombed by the lake sediment that—ever so slowly—is filling Walden Pond. Like the ink of his original *Walden* manuscript, this geological archive contains tangible evidence of his life at the pond.

ABBREVIATIONS

Thoreau's Works Cited

CC *Cape Cod* (Mineola, NY: Dover Publications, 2004, originally Boston: Ticknor and Fields, 1865).

JR *The Journal of Henry D. Thoreau: In Fourteen Volumes Bound as Two,* ed. Bradford Torrey and Francis H. Allen (New York: Dover Publications, 1962, originally Boston: Houghton Mifflin Company, 1906).

MW *The Maine Woods: A Fully Annotated Edition,* ed. Jeffrey Cramer (New Haven, Yale University Press, 2009).

NH "Natural History of Massachusetts," *Excursions,* ed. Leo Marx (Gloucester, MA: Peter Smith, 1975), 37–72, originally Boston: Ticknor and Fields, 1863, and *The Dial* 3, no. 1 (July 24, 1842).

WA *Walden, Or Life in the Woods,* ed. Jeffrey Cramer (New Haven: Yale University Press, 2004, originally Boston: Ticknor and Fields, 1854).

WG "Walking," *Excursions,* ed. Leo Marx (Gloucester, MA: Peter Smith, 1975, originally Boston: Ticknor and Fields, 1863), 161–214.

WK *A Week on the Concord and Merrimack Rivers,* ed. Thomas Blanding (Orleans, MA: Parnassus Imprints, Inc., 1987, reprinted from Boston: James Monroe & Co., 1849).

WT "A Walk to Wachusett," *Excursions,* ed. Leo Marx (Gloucester, MA: Peter Smith, 1975, originally Boston: Ticknor and Fields, 1863), 73–96.

WW "A Winter Walk," *Excursions,* ed. Leo Marx (Gloucester, MA: Peter Smith, 1975, originally Boston: Ticknor and Fields, 1863), 109–134.

REFERENCES

Acaster, Martin, and M. E. Bickford. "Geochronology and Geochemistry of Putnam-Nashoba Terrane Metavolcanic and Plutonic Rocks, Eastern Massachusetts: Constraints on the Early Paleozoic Evolution of Eastern North America." *Geological Society of America Bulletin* 111 (1999): 240–253.

Ackerly, Spafford C. "Reconstructions of Mountain Glacier Profiles, Northeastern United States." *Geological Society of America Bulletin* 101 (1989): 561–572.

Adams, Frank D. *The Birth and Development of the Geological Sciences.* New York: Dover Publications, 1938.

Adams, Stephen, and Donald Ross Jr. *Revising Mythologies: The Composition of Thoreau's Major Works.* Charlottesville: University of Virginia Press, 1988.

Agassiz, Louis. *Studies on Glaciers: Preceded by the Discourse of Neuchatel* (2 vol. with plates). Translated and edited by Albert V. Carozzi. New York: Hafner Publishing, 1967, originally Newchatel: Louis Agassiz, 1840.

American Association for the Advancement of Science. "About AAAS: History and Archives." Accessed September 17, 2012. http://archives.aaas.org/.

Anderson, Charles R. *The Magic Circle of Walden.* New York: Holt, 1968.

Anderson, Robert S., and Suzanne P. Anderson. *Geomorphology: The Mechanics and Chemistry of Landscapes.* Cambridge, UK: Cambridge University Press, 2010.

Angier, Natalie. "The Mighty Mathematician You've Never Heard Of." *New York Times,* March 27, 2012. D4 (L).

Association of American Geologists and Naturalists. "Abstract of the Proceedings of the Fourth Session." *American Journal of Science and Arts* 45, Article 10 (1843): 310–353.

Bacon, Francis. *Sylva Sylvarum.* London: William Lee, 1639.

Barosh, Patrick J. "Bedrock Geology of the Walden Woods." In *Thoreau's World and Ours: A Natural Legacy,* edited by Edmund A. Schofield and Robert C. Baron, 212–221. Golden, CO: North American Press, 1993.

Baym, Nina. "Thoreau's View of Science (slightly revised)." *Journal of the History of Ideas* 26 (1963): 221–234.

Bedell, Rebecca. *The Anatomy of Nature: Geology & American Landscape Painting, 1825–1875.* Princeton, NJ: Princeton University Press, 2002.

Benn, Douglas I., and David J. A. Evans. *Glaciers & Glaciation.* London: Edward Arnold, 1998.

Benoit, Raymond. "Walden as God's Drop." *American Literature* 43, no. 1 (1971): 122–124.

Black, Adam, and Charles Black. *Black's Picturesque Guide to the English Lakes.* 2nd ed. Edinburgh: Adam and Charles Black, 1844.

Blancke, Shirley. "The Archaeology of Walden Woods." In *Thoreau's World and Ours: a Natural Legacy,* edited by Edmund A. Schofield and Robert C. Baron, 242–253. Golden, CO: North American Press, 1993.

Bickman, Martin. *Walden: Volatile Truths.* New York: Twayne Publishers, 1992.

Boime, Albert. *The Magisterial Gaze.* Washington, DC: Smithsonian, 1991.

Bolles, Edmund Blair. *The Ice Finders: How a Poet, a Professor, and a Politician Discovered the Ice Age.* Washington, DC: Counterpoint Press, 1999.

Borst, Raymond R. *The Thoreau Log: A Documentary Life of Henry David Thoreau 1817–1862.* New York: Macmillan, 1992.

Botkin, Daniel B. *No Man's Garden: Thoreau and a New Vision for Civilization and Nature.* Washington, DC: Island Press, 2001.

Boudillion, Daniel V. "Nashoba Hill: The Hill that Roars." Accessed September 10, 2012. http://www.boudillion.com/nashobahill/nashobahill.htm.

Brain, J. Walter. "The Meaning of Walden." *Lincoln Journal* (1998).

Briggs, Charles Frederic. "A Yankee Diogenes." In *Walden and Resistance to Civil Government: Authoritative Texts, Thoreau's Journal, Reviews and Essays in Criticism, Second Edition,* edited by William Rossi, 314–317. New York: W. W. Norton & Company, 1992, originally *Putnam's Monthly Magazine* 4 (Oct 1854): 443–448.

Briggs, John. *Fractals: the Patterns of Chaos.* New York: Simon and Schuster, 1992.

Brooks, Van Wyck. *The Flowering of New England: 1815–1865.* New York: E. P. Dutton & Co., Inc., 1936.

Buell, Lawrence. *The Environmental Imagination: Thoreau, Nature Writing, and the Formation of American Culture.* Cambridge, MA: Harvard University Press, 1995.

———. "Thoreau and the Natural Environment." In *The Cambridge Companion to Henry David Thoreau,* edited by Joel Meyerson, 171–193. Cambridge: Cambridge University Press, 1995.

———. Forward to *Thoreau's Sense of Place*, edited by Richard J. Schneider, ix–x. Iowa City: University of Iowa Press, 2000.

Calderwood, Tom. "An Astronomer Reads Thoreau." *Thoreau Society Bulletin* 267 (Summer 2009): 7–8.

Cameron, Kenneth W. *Thoreau's Fact Book: In the Harry Elkins Widener Collection the Harvard College Library Annotated and Indexed, Volume II.* Hartford, CT: Transcendental Books, Drawer 1080, 1966.

———. *The Massachusetts Lyceum During the American Renaissance.* Hartford, CT: Transcendental Books, Drawer 1080, 1969.

Cameron, Sharon. *Writing Nature: Henry Thoreau's Journal.* New York: Oxford University Press, 1985.

Canby, Henry Seidel. *Thoreau.* Boston: Houghton Mifflin, 1939.

Carlson, Eric W. "The Transcendentalist Poe: A Brief History of Criticism." *Poe Studies/Dark Romanticism* 32, no. 1–2 (January/December 1999): 47–66.

Carroll, James. *Jerusalem, Jerusalem: How the Ancient City Ignited Our Modern World.* Boston: Houghton Mifflin Harcourt, 2011.

Castle, Robert O., H. R. Dixon, E. S. Grew, Andrew Griscom, and Isidore Zeitz. *Structural Dislocations in Eastern Massachusetts.* Washington, DC: U.S. Geological Survey Bulletin 1410, 1976.

Cavell, Stanley. *The Senses of Walden.* New York: Viking, 1972.

———. "Captivity and Despair in Walden and 'Civil Disobedience.'" In *Walden and Resistance to Civil Government: Authoritative Texts, Thoreau's Journal, Reviews and Essays in Criticism, Second Edition,* edited by William Rossi, 390–405. New York: W. W. Norton & Company, 1992.

———. *Philosophy the Day after Tomorrow.* Cambridge: Belknap Press of Harvard University Press, 2005.

[Chambers, Robert]. *Vestiges of the Natural History of Creation.* London: John Churchill, 1844.

Channing, William Ellery. *Thoreau, the Poet Naturalist.* Edited by F. B. Sanborn. Boston: Charles E. Goodspeed, 1902.

Chura, Patrick. *Thoreau the Land Surveyor.* Gainesville, FL: University Press of Florida, 2010.

Clark, Peter U., Arthur S. Dyke, Jeremy D. Shakun, Anders E. Carlson, Jorie Clark, Barbara Wohlfarth, Jerry Mitrovica, Stephen W. Hostetler, and A. Arshall McCabe. "The Last Glacial Maximum." *Science* 325 (August 7, 2009): 710–714.

Colman, John A., and Paul J. Friesz. "Geohydrology and Limnology of Walden Pond, Concord, Massachusetts." U.S. Geological Survey Water-Resources Investigation Report 01–4137 (2001).

Colman, John A., and Marcus C. Waldron. "Walden Pond, Massachusetts: Environmental Setting and Current Investigations." *U.S. Geological Survey Fact Sheet FS-064–98* (1998).

Cook, Reginald L. *Passage to Walden.* Cambridge, MA: Houghton Mifflin Company, Riverside Press, 1949.

Cramer, Jeffrey C. Introduction to *I to Myself: An Annotated Selection from the Journal of Henry D. Thoreau,* edited by Jeffrey C. Cramer, xv–xxx. New Haven: Yale University Press, 2007.

Cronon, William. "The Trouble with Wilderness; or, Getting Back to the Wrong Nature." In *Uncommon Ground: Rethinking the Human Place in Nature,* edited by William Cronon, 69–70. New York: W. W. Norton & Company, 1996.

Darwin, Charles R. *The Voyage of the Beagle.* Edited by Charles W. Eliot. New York: P. F. Collier & Son Corporation, The Harvard Classics, Volume 29, 1909, 1937, 1963, originally New York: Harper & Brothers, 1845.

———. *Third Part, Geological Observations of South America.* London: Smith & Elder, 1846.

———. *The Origin of Species* and *The Voyage of the Beagle,* with an introduction by Richard Dawkins. New York: Alfred A. Knopf, Everyman's Library, 2003.

———. *Evolutionary Writings.* Edited by James Secord. Oxford, UK: Oxford University Press, 2008.

Davis, W. M. "Structure and Origin of Glacial Sand Plains." *Bulletin of the Geological Society of America* 1 (1890): 195–202 and plate 3.

Dean, Bradley P. "Natural History, Romanticism, and Thoreau." In *American Wilderness: A New History,* edited by Michael Lewis, 73–89. New York: Oxford University Press, 2007.

Deevey, Edward S., Jr. "A Re-examination of Thoreau's Walden." *The Quarterly Review of Biology* 17, no. 1 (1942): 1–11.

Demos, John. *Circles and Lines: The Shape of Life in Early America.* Cambridge, MA: Harvard University Press, 2004.

Dickinson, Emily. *The Complete Poems of Emily Dickinson.* Edited by Thomas H. Johnson. Boston: Little, Brown and Company, 1960.

Donahue, Brian. "Henry David Thoreau and the Environment of Concord." In *Thoreau's World and Ours: A Natural Legacy,* edited by Edmund A. Schofield and Robert C. Baron, 181–189. Golden, CO: North American Press, 1993.

———. *The Great Meadow, Farmers and the Land in Colonial Concord.* New Haven: Yale University Press, 2004.

Dott, Robert H., Jr. "James Hall's Discovery of the Craton." In *Geologists and Ideas: A History of North American Geology,* edited by Ellen T. Drake and William M. Jordan, 157–168. Denver: Geological Society of America, 1985.

Dunaway, Finis. "Gleason's Transparent Eyeball." In *Natural Visions: The Power of Images in American Environmental Reform,* 1–30. Chicago: University of Chicago Press, 2008.

Eaton, Amos. *Geological Textbook*. Albany: Amos Eaton, 1830.
Edel, Leon. *Stuff of Sleep and Dreams: Experiments in Literary Psychology*. New York: Harper & Row, 1982.
Eiseley, Loren. *The Unexpected Universe*. New York: Harcourt, Brace & World, Inc., 1964.
Elder, John. Introduction to *Nature/Walking*, vii–xviii. Boston: Beacon Press, 1991.
Eliade, Mircea. *The Myth of the Eternal Return*. Translated by Willard R. Trask. New York: Pantheon Books, Bollingen Series XVI, 1954.
Emerson, Ralph Waldo. Preliminary Note to "Natural History of Massachusetts" by Henry D. Thoreau, 19. *The Dial* 8, no. 1 (July 1842).
———. "Thoreau." In *Excursions,* by Henry David Thoreau, edited by Leo Marx, 7–33. Boston: Ticknor and Fields, 1863.
———. Entry dated July 13, 1833, from *The Journals of Ralph Waldo Emerson*. In Robert Finch and John Elder, *Nature Writing: The Tradition in English,* 145. New York: W. W. Norton & Company, 2002.
———. "The American Scholar." Accessed September 15, 2012. http://www.emersoncentral.com/amscholar.htm.
Fagan, Brian. *The Little Ice Age: How Climate Made History, 1300–1850*. New York: Basic Books, 2000.
Fail, Rodger T. "Evolving Tectonic Concepts of the Central and Southern Appalachians." In *Geologists and Ideas: A History of North American Geology,* edited by Ellen T. Drake and William M. Jordan, 19–46. Denver: Geological Society of America, 1985.
Faison, E. K., D. R. Foster, W. W. Oswald, B. C. Hansen, and E. Doughty. "Early Holocene Openlands in Southern New England." *Ecology* 87 (2006): 2537–2547.
Flannery, Tim. *Here on Earth: A New Beginning*. London, UK: Penguin Books, Allen Lane, 2010.
Fleeger, G. M., and J. D. Inners. "Henry David Thoreau and his Mentors on the 'Diluvial' Geology of New England." *Geological Society of America Annual Meeting Abstracts with Programs* 38, no. 7 (2006): 38.
Forbes, James D. *Travels Through the Alps of Savoy and Other Parts of the Pennine Chain, with Observations on the Phenomena of Glaciers*. Edinburgh: Adam and Charles Black, 1843.
Foster, David R. *Thoreau's Country: Journey through a Transformed Landscape*. Cambridge, MA: Harvard University Press, 2001).
Friedman, Robert S. "Surveying the Empirical Sublime: Thoreau, Literary and Scientific." In *Restoring the Mystery of the Rainbow: Literature's Refraction of Science,* edited by Valereia Tinkler-Villani and C. C. Barfoot, 743–759. Amsterdam: Rodopi, 2011.
Friedrich, Paul. *The Gita Within Walden*. Albany: SUNY Press, 2008.

Friesen, Victor Carl. *The Spirit of the Huckleberry: Sensuousness in Henry Thoreau.* Edmonton: University of Alberta Press, 1984.

Friesz, Paul J., and John A. Colman. Poster from *Hydrology and Trophic Ecology of Walden Pond, Concord, Massachusetts.* U.S. Geological Survey Water Resources Investigations Report 01–4153, 2001.

Garrard, Greg. "Wordsworth and Thoreau: Two Versions of Pastoral." In *Thoreau's Sense of Place,* edited by Richard J. Schneider, 194–206. Iowa City: University of Iowa Press, 2000.

Gatta, John. *Making Nature Sacred.* New York: Oxford University Press, 2004.

Geological Society of America. *Geologic Map of North America: Scale 1:5,000,000,* compiled by John C. Reed Jr., John O. Wheeler, and Brian Tucholke. Boulder, CO: Geological Society of America, 2005 (updated in digital form in 2009).

Gessner, David. *My Green Manifesto: Down the Charles River in Pursuit of the New Environmentalism.* Minneapolis, MN: Milkweed Editions, 2011.

Gleason, Herbert W. Map of Concord. In *The Journal of Henry D. Thoreau in Fourteen Volumes Bound as Two Vols.,* edited by Bradford Torrey and Francis H. Allen, follows page 1751. New York: Dover Publications, 1962, originally Boston: Houghton Mifflin Company, 1906.

Gleick, James. *Chaos: Making a New Science.* New York: Penguin, 1987.

Godfrey-Smith, Peter. *Theory and Reality: An Introduction to the Philosophy of Science.* Chicago: University of Chicago Press, 2003.

Goldthwait, J. Walter. "The Sand Plains of Glacial Lake Sudbury." *Bulletin of the Museum of Comparative Zoology at Harvard College* 42 (1905).

Gould, Stephen Jay. *Bully for Brontosaurus: Reflections on Natural History.* New York: W. W. Norton & Company, 1992.

———. *Time's Arrow, Time's Cycle: Myth and Metaphor in the Discovery of Geological Time.* Cambridge, MA: Harvard University Press, 1987.

Gribbin, John. *The Scientists: A History of Science Told Through the Lives of Its Greatest Inventors.* New York: Random House, 2002.

Gross, Robert. "That Terrible Thoreau: Concord and Its Hermit." In *A Historical Guide to Henry David Thoreau,* edited by William E. Cain, 181–241. New York: Oxford University Press, 2000.

Gura, Philip F. *American Transcendentalism: A History.* New York: Hill and Wang, 2007.

Gustavson, T., and J. C. Boothroyd. "A Depositional Model for Outwash, Sediment Sources, and Hydrologic Characteristics, Malaspina Glacier Alaska: A Modern Analog of the Southeastern Margin of the Laurentide Ice Sheet." *Geological Society of America Bulletin* 99 (1987): 187–200.

Harding, Walter. *Thoreau: Man of Concord.* New York: Holt, Rinehart and Winston, Inc., 1960.

———. *The Days of Henry Thoreau*. New York: Dover, 1982, originally New York: Alfred A. Knopf, Inc., 1965.

———. *A Thoreau Handbook*. 5th printing. New York: New York University Press, 1976.

Harding, Walter, and Carl Bode. *The Correspondence of Henry David Thoreau*. Westport, CT: Greenwood Press, 1974, reprinted from New York: New York University Press, 1958.

Hawking, Stephen. *A Brief History of Time*. London: Bantam Books, 1988.

Hawking, Stephen, and Leonard Mlodinow. *The Grand Design*. New York: Bantam Books, 2010.

Hawthorne, Nathaniel. *Mosses from an Old Manse*. London: Wiley & Putnam, 1846.

Herbert, Sandra. *Charles Darwin, Geologist*. Ithaca, NY: Cornell University Press, 2005.

Hildebidle, John. *Thoreau: A Naturalist's Liberty*. Cambridge, MA: Harvard University Press, 1983.

Hitchcock, Edward. *Final Report on the Geology of Massachusetts in Four Parts*. Northampton: J. Butler, 1841.

———. "First Anniversary Address before the Association of American Geologists and Naturalists." *American Journal of Science* 41 (1841): 232–275.

———. *Illustrations of Surface Geology*. New York: D. Appleton & Co., Division Phinney & Co. Boston, 1860. Accessed May 2012. http://name.umdl.umich.edu/AJP8876.0001.001.

Hodder, Alan. *Thoreau's Ecstatic Witness*. New Haven: Yale University Press, 2001.

Hovenkamp, Herbert. *Science and Religion in America, 1800–1860*. Philadelphia: University of Pennsylvania Press, 1978.

Howarth, William. *Walking with Thoreau: A Literary Guide to the Mountains of New England*. Boston: Beacon Press, 2001.

Imbrie, John, and Katherine Palmer Imbrie. *Ice Ages: Solving the Mystery*. Hillside, NJ: Enslow Publishers, 1979.

Inners, Jon D. "A Yankee Saunterer: Geographical and Geological Influences on the Life of Henry David Thoreau." *Geological Society of America Annual Meeting Abstracts with Programs* 38, no. 7 (2006): 37.

Irmscher, Christoph. *Louis Agassiz: Creator of American Science*. Boston: Houghton Mifflin & Harcourt, 2013.

Jackson, Charles. T., M. D. *First Report on the Geology of the Public Lands in the State of Maine*. Boston: Dutton and Wentworth, 1837.

———. *Second Annual Report on the Geology of the Public Lands, Belonging to the Two States of Maine and Massachusetts*. Augusta, ME: Luther Severance, Printer, 1838.

———. "Communication on the Subject of Drift (Abstract of the Proceedings of the Fourth Session of the Association of American Geologists and Naturalists." *American Journal of Science and Arts* 45, no. 10 (1843): 320–323.

Jacobson, George L., Jr., Thompson Webb, and Eric C. Grimm. Poster from "Changing Vegetation Patterns of Eastern North America During the Past 18,000 Years: Inference from Overlapping Distribution of Selected Pollen Types." *North America and Adjacent Oceans During the Last Deglaciation,* edited by William Ruddiman. Boulder, CO: Geological Society of America, Decade of North American Geology, vol. K-3, 1987.

James, William. *Varieties of Religious Experience.* New York: Random House, 1929. Accessed August 2012. http://etext.lib.virginia.edu/toc/modeng/public/JamVari.html.

Johnson, Barbara. "A Hound, a Bay Horse, and a Turtle Dove: Obscurity in Walden." *Walden and Resistance to Civil Government: Authoritative Texts, Thoreau's Journal, Reviews and Essays in Criticism, Second Edition,* edited by William Rossi, 444–450. New York: W. W. Norton & Company, 1992, originally 1987.

Johnson, Rochelle L. *Passions for Nature: Nineteenth-Century America's Aesthetics of Alienation.* Athens, GA: The University of Georgia Press, 2009.

Jones, Sir William. *Institutes of Hindu Law; or, the Ordinances of Menu, According to the Gloss of Culluca, Compromising the Indian System of Duties, Religious and Civil.* London: Government of Great Britain, 1796.

Jopling, Alan V. "Early Studies on Stratified Drift." *Glaciofluvial and Glaciolacustrine Sedimentation,* edited by Alan V. Jopling and Barrie C. McDonald, 4–21. Tulsa, OK: Society of Economic Paleontologists and Mineralogists, 1975.

Jorgensen, Sven E. *Evolutionary Essays: A Thermodynamic Interpretation of Evolution.* London: Elsevier, 2008.

Kane, Elisha Kent Kane, M. D. *Adrift in the Arctic Ice Pack: From the History of the First U.S. Grinnell Expedition in Search of Sir John Franklin.* Edited by Horace Kephart. New York: The Macmillan Company, 1926, reprint of Kane, 1853.

———. *The U.S. Grinnell Expedition in Search of Sir John Franklin. A Personal Narrative.* U.S. Navy New Edition. New York: Harper & Brothers, 1857. Adapted from Philadelphia: Childs & Peterson, 1856. Accessed August 2012. http://quod.lib.umich.edu/m/moa/AJA5420.0001.001?rgn=main;view=fulltext.

Keillor, Garrison. "Guy Noir." *A Prairie Home Companion.* Accessed September 2012. http://prairiehome.publicradio.org/programs/2004/04/24/scripts/guy_noir.shtml.

Koster, Drte, Reinhard Pienitz, Brent B. Wolfe, Sylvia Barry, David R. Foster, and Sushil S. Dixit. "Paleolimnological Assessment of Human-Induced

Impacts on Walden Pond (Massachusetts, USA) Using Diatoms and Stable Isotopes." *Aquatic Ecosystem Health & Management* 8 (2005): 117–131.

Koteff, Carl. "Glacial Lakes near Concord, Massachusetts." *U.S. Geological Survey Professional Paper 475-C* (1963): C142–C144.

———. Map GQ-331, 1:24,000. *Surficial Geology of the Concord, Quadrangle, Massachusetts.* Washington, DC: U.S. Geological Survey, 1964.

Koteff, Carl, and Fred Pessl Jr. *Systematic Ice Retreat in New England.* Washington, DC: U.S. Geological Survey Professional Paper 1179, 1981.

Koteff, Carl, Gilpin R. Robinson, Richard Goldsmith, and Woodrow B. Thompson. "Delayed Postglacial Uplift and Synglacial Sea Levels in Coastal Central New England." *Quaternary Research* 40 (1993): 46–54.

Kruckeberg, Arthur R. *Geology and Plant Life: The Effects of Landforms and Rock Types on Plants.* Seattle: University of Washington Press, 2002.

Krutch, Joseph Wood. Introduction to *Walden.* New York: Bantam Books, 1962.

Kuhn, Bernhard. *Autobiography and Natural Science in the Age of Romanticism: Rousseau, Goethe, Thoreau.* Farnham, Surrey, UK: Ashgate, 2009.

Kuhn, Thomas S. *The Structure of Scientific Revolutions.* Enlarged 2nd ed. Chicago: University of Chicago Press, 1970.

Lewis, Ralph S., and Byron D. Stone. "Multiple Glacial Development of Upper Georges Bank." *2012 Abstracts with Programs.* Hartford, CT: Geological Society of America, Northeastern GSA Section Meeting 44, no. 2 (2012): 23.

Lorenz, Edward N. "Climatic Determinism." *Meteorological Monographs* 8, no. 30 (1968): 1–4.

Lowell, James Russell. "Thoreau." *Walden and Resistance to Civil Government: Authoritative Texts, Thoreau's Journal, Reviews and Essays in Criticism, Second Edition,* edited by William Rossi. New York: W. W. Norton & Company, 1992, originally Boston: Literary Essays, 1890.

Luria, Sara. "Thoreau's Geopoetics." *GeoHumanities: Art, History, Text at the Edge of Place,* 126–138. London: Routledge, 2011.

Lurie, Edward. *Louis Agassiz: A Life in Science.* Chicago: University of Chicago Press, 1960.

Lyell, Charles. *Principles of Geology, First Edition, Volume I: 1830.* Edited by Martin J. S. Rudwick. Chicago: University of Chicago Press, 1990.

Lyon, Melvin E. "Walden Pond as a Symbol." *PMLA Transactions and Proceedings of the Modern Language Association* 82, no. 2 (May 1967): 289–300.

Macfarlane, Robert. *Mountains of the Mind.* New York: Pantheon Books, 2003.

Marx, Leo. Introduction to *Excursions,* by Henry D. Thoreau, edited by Leo Marx. Gloucester, MA: Peter Smith, 1975, reprinted from Boston: Ticknor and Fields, 1863, v–xiv.

———. Introduction to *Henry David Thoreau,* by Franklin B. Sanborn, xi–xxiv. New York: Chelsea House, 1980, reprinted from Boston: Houghton Mifflin, 1882.

———. *The Machine in the Garden: Technology and the Pastoral Ideal in America.* New York: Oxford University Press, 1964.

———. "Walden as Transcendental Pastoral Design." *Walden and Resistance to Civil Government: Authoritative Texts, Thoreau's Journal, Reviews and Essays in Criticism, Second Edition,* edited by William Rossi, 377–390. New York: W. W. Norton & Company, 1992.

———. "The Struggle Over Thoreau." *New York Review of Books* 46, no. 11 (June 24, 1999), 60.

Massachusetts Department of Conservation and Recreation, "Geology." Accessed September 15, 2012. http://www.mass.gov/dcr/parks/walden/geology.htm.

Matthiessen, F. O. "Walden: Craftsmanship vs. Technique." *American Renaissance,* 166–175. Oxford: Oxford University Press, 1941.

Maynard, Barksdale W. "Following Drought, Walden's Sandbar Makes Rare Appearance." *The Concord Saunterer,* 242 (2003): 7–8.

———. *Walden Pond, A History.* New York: Oxford University Press, 2004.

———. "Emerson's *Wyman Lot*: Forgotten context for Thoreau's House at Walden." *The Concord Saunterer, New Series* 12/13 (2004/2005): 61–84.

McGrath, James G. "Ten Ways of Seeing Landscapes in *Walden* and Beyond." *Thoreau's Sense of Place,* edited by Richard J. Schneider, 149–164. Iowa City: Iowa University Press, 2000.

McGregor, Robert Kuhn. *A Wider View of the Universe: Henry Thoreau's Study of Nature.* Champaign: University of Illinois Press, 1997.

McKibben, Bill. Introduction to *Walden: With an Introduction and Annotations by Bill McKibben.* Boston: Beacon Press, 2004, xii–xxiii.

"Measuring worth." Accessed August 12, 2012. http://www.measuringworth.com/index.php.

Merrill, George P. "The First One Hundred Years of American Geology, Contributions to the History of American Geology." *Report of the U.S. National Museum Under the Directorate of the Smithsonian Institution for the Year Ending June 30, 1904.* Washington, DC: U.S. Government Printing Office, 1904.

Michaels, Walden Benn. "Walden's False Bottoms." *Walden and Resistance to Civil Government: Authoritative Texts, Thoreau's Journal, Reviews and Essays in Criticism, Second Edition,* edited by William Rossi, 405–421. New York: W. W. Norton & Company, 1992, originally 1977.

Miller, Perry. *Consciousness in Concord: The Text of Thoreau's Hitherto "Lost Journal" (1840–1841) Together with Notes and a Commentary.* Boston: Houghton Mifflin Company, 1958.

Mills, H. H. "Apparent Increasing Rates of Stream Incision in the Eastern United States during the Late Cenozoic." *Geology* 28, no. 10 (2000): 955–957.

Minnaeret, M. *The Nature of Light & Colour in the Open Air.* Translated by H. M. Kremer-Priest, revised by K. E. Brian Jay. New York: Dover Publications, 1954.

Mitchell, John Hanson. *Living at the End of Time.* Boston: Houghton Mifflin Company, 1990.

Mitchell, Stephen. *Bhagavad Gita: A New Translation.* New York: Three Rivers Press, 1988.

Mooney, Edward F. *Lost Intimacy in American Thought: Recovering Personal Philosophy from Thoreau to Cavell.* New York: Continuum International Publishing, 2009.

Morgan, Patrick. "Aesthetic Inflections: Thoreau, Gender, and Geology." *The Concord Saunterer: New Series* 18 (2010): 46–67.

Myerson, Joel, Sandra Harbert Petrulionis, and Laura Dassow Walls, eds. *The Oxford Handbook of Transcendentalism.* Oxford: Oxford University Press, 2010.

Neuendorf, Klaus K. E., James P. Mehl Jr., and Julia A. Jackson, eds. *Glossary of Geology.* 5th ed. Washington, DC: American Geological Institute, 2005.

"North America Glacial Varve project." Accessed September 15, 2012. http://geology.tufts.edu/varves/.

Oates, Joyce Carol. Introduction to *Walden,* by Henry David Thoreau. Princeton: Princeton University Press, 1989, ix–xviii.

Oxford Universal Dictionary. 3rd ed. Oxford: Clarendon Press, 1955.

Papa, James A., Jr. "Water Signs: Place and Metaphor in Dillard and Thoreau." *Thoreau's Sense of Place,* edited by Richard J. Schneider, 70–79. Iowa City: University of Iowa Press, 2000.

Parini, Jay. *Promised Land: Thirteen Books That Changed America.* New York: Doubleday, 2008.

Peck, H. Daniel. *Thoreau's Morning Work: Memory and Perception in "A Week on the Concord and Merrimack Rivers," the "Journal," and "Walden."* New Haven: Yale University Press, 1990.

———. "The Worlding of Walden." *Walden and Resistance to Civil Government, Authoritative Texts, Thoreau's Journal, Reviews and Essays in Criticism, Second Edition,* edited by William Rossi, 451–463. New York: W. W. Norton & Company, 1992.

———. "Lakes of Light: Modes of Representation in Walden." *Nineteenth-Century Prose* 31, no. 2 (Fall 2004): 172–185.

Pergallo, Thomas. "Soils of the Walden Ecosystem." *Thoreau's World and Ours: A Natural Legacy,* edited by Edmund A. Schofield and Robert C. Baron, 254–259. Golden, CO: North American Press, 1993.

Peteet, D. M., M. Beh, C. Orr, D. Kurdyla, J. Nichols, and T. Guilderson. "Delayed Deglaciation or Extreme Arctic Conditions 21–16 Cal. Kyr at Southeastern Laurentide Ice Sheet Margin." *Geophysical Research Letters* 39, L11706 (2012): 1–6.

Pijanowski, B. C., L. J. Villanueva-Rivera, S. L. Dumyahn, Al Farina, B. L. Krause, B. M. Napeoletano, S. H. Gage, and N. Pieretti. "Soundscape Ecology: The Science of Sound in the Landscape." *Bioscience* 61, no. 3 (March 2011): 203–216.

Pipkin, John S. "Hiding Places: Thoreau's Geographies." *Annals of the Association of American Geographers* 91, no. 3 (2001): 527–545.

Playfair, John. *Illustrations of the Huttonian Theory of the Earth.* New York: Dover Publications, 1956, originally Edinburgh: William Creech, 1802.

Poetzsch, Markus. "Sounding Walden Pond: The Depths and 'Double Shadows' of Thoreau's Autobiographical Symbol." *American Transcendental Quarterly* 22, no. 2 (June 2008): 387–401.

Porte, Joel. *Consciousness and Culture: Emerson and Thoreau Reviewed.* New Haven: Yale University Press, 2004.

Prigogine, Ilya. *The End of Certainty.* London: Free Press, 1997.

Repcheck, Jack. *The Man Who Found Time: James Hutton and the Discovery of the Earth's Antiquity.* New York: Perseus Publishing, 2003.

Richardson, Robert D., Jr. *Henry Thoreau: A Life of the Mind.* Berkeley: University of California Press. 1986.

Ridge, John C., Brandy A. Canwell, Meredith A. Kelly, and Sharon Z. Kelley. "Atmospheric ^{14}C Chronology for Late Wisconsin Deglaciation and Sea-Level Change in Eastern New England Using Varve and Paleomagnetic Records." *Deglacial History and Relative Sea-Level Changes, Northern New England and Adjacent Canada,* edited by T. K. Weddle and M. J. Retelle, 171–189. Boulder, CO: Geological Society of America, Special Paper 351, 2001.

Robbins, Roland Wells. *Discovery at Walden.* Concord, MA: The Thoreau Society, 1999, reprinted from 1947.

Rossi, William, ed. *Walden and Resistance to Civil Government: Authoritative Texts, Thoreau's Journal, Reviews and Essays in Criticism, Second Edition.* New York: W. W. Norton & Company, 1992.

Rudwick, Martin. Introduction to *Principles of Geology,* by Charles Lyell. Chicago: University of Chicago Press, 1992, reprinted from London: John Murray, 1830–1833.

Salt, Henry S. *Life of Henry David Thoreau.* Edited by George Hendrick, Willene Hendrick, and Fritz Oehlschlager. Urbana and Chicago: University of Illinois Press, 1993, originally London: Bentley, 1890.

Sanborn, Franklin B. *Henry David Thoreau.* Boston: Houghton Mifflin, 1882.

Sattelmeyer, Robert. *Thoreau's Reading: A Study in Intellectual History, with Bibliographical Catalogue.* Princeton, NJ: Princeton University Press, 1988.

———. "The Remaking of Walden." *Walden and Resistance to Civil Government: Authoritative Texts, Thoreau's Journal, Reviews and Essays in Criticism, Second Edition,* edited by William Rossi, 428–444. New York: W. W. Norton & Company, 1992.

———. "Depopulation, Deforestation, and the Actual Walden Pond." *Thoreau's Sense of Place,* edited by Richard J. Schneider, 235–243. Iowa City: University of Iowa Press, 2000.

Schama, Simon. *Landscape and Memory.* London: HarperCollins, 1995.

Schneider, Richard J. "Reflections in Walden Pond: Thoreau's Optics." *ESQ* 21 (1975): 65–75.

———. "Walden." *The Cambridge Companion to Henry David Thoreau,* edited by Joel Meyerson, 92–106. Cambridge: Cambridge University Press, 1995.

———, ed. *Thoreau's Sense of Place.* Iowa City: University of Iowa Press, 2000.

Schofield, Edmund A. "The Ecology of Walden Woods," *Thoreau's World and Ours: A Natural Legacy,* edited by Edmund A. Schofield and Robert C. Baron, 155–171. Golden, CO: North American Press, 1993.

Schofield, Edmund A., and Robert C. Baron, eds. *Thoreau's World and Ours: a Natural Legacy.* Golden, CO: North American Press, 1993.

Secord, James. "Introduction" in Darwin, Charles, *Evolutionary Writings,* edited by James Secord. Oxford, UK: Oxford University Press, 2008, vii–xxxvii.

Shanley, J. Lyndon. *The Making of "Walden," with the Text of the First Version.* Chicago: University of Chicago Press, 1957.

Shattuck, Lemuel. *A History of the Town of Concord; Middlesex County, Massachusetts, from Its Earliest Settlement to 1832.* Boston: Russell, Odiorne, and Company, 1835.

Skinner, Brian, J., and Barbara Narendra. "Rummaging Through the Attic: or A Brief History of Geological Sciences at Yale." *Geologists and Ideas: A History of North American Geology,* edited by Ellen T. Drake and William M. Jordan, 355–376. Denver: Geological Society of America, 1985.

Stone, Janet R., and Byron D. Stone. "Surficial Geologic Map of the Clinton-Concord-Grafton-Medfield 12-Quadrangle Area in East Central Massachusetts," Scale 1:24,000. *U.S. Geological Survey Open File Report 2006–1260A.* DVD. 2006.

Sullivan, Robert M. *The Thoreau You Don't Know: What the Prophet of Environmentalism Really Meant.* New York: HarperCollins, 2009.

Taton, Rene, ed. *Science in the Nineteenth Century (History of Science).* Translated by A. J. Pomerans. London: Thames and Hudson Limited, 1961, translation published by New York: Basic Books, 1965.

Thoreau, Henry David. "The Service" (unpublished essay). New York: Library of America, July 1840.

[———] (unsigned edition). "The Laws of Menu." *The Dial* 3, no. 3 (January 1843): 331.

[———] (unsigned edition). "Ethical Scriptures: Sayings of Confucius." *The Dial* 3, no. 3 (January 1843): 493.

[———] (unsigned edition). "Ethical Scriptures: Chinese Four Books." *The Dial* 4, no. 2 (October 1843): 205.

———. *A Week on the Concord and Merrimack Rivers.* Edited by Thomas Blanding. Orleans, MA: Parnassus Imprints, Inc., 1987, reprinted from Boston: James Monroe & Co., 1849.

———. *A Yankee in Canada:* A Thoreau Reader. Originally published 1850. Accessed June 2012. http://thoreau.eserver.org/Canada1.html.

———. "Dec. Map 31a, 4½×7½." *[Ralph Waldo Emerson] Lot by Walden.* Concord, MA: Concord Free Public Library Special Collection "Thoreau Surveys," 1857.

———. *Excursions.* Edited by Leo Marx. Gloucester, MA: Peter Smith, 1975, originally Boston: Ticknor and Fields, 1863.

———. "Natural History of Massachusetts." *Excursions,* edited by Leo Marx, 37–72. Gloucester, MA: Peter Smith, 1975, originally Boston: Ticknor and Fields, 1863, and *The Dial* 3, no. 1 (July 24, 1842).

———. "A Walk to Wachusett." *Excursions,* edited by Leo Marx, 73–96. Gloucester, MA: Peter Smith, 1975, originally Boston: Ticknor and Fields, 1863.

———. "A Winter Walk." *Excursions,* edited by Leo Marx, 109–134. Gloucester, MA: Peter Smith, 1975, originally Boston: Ticknor and Fields, 1863.

———. "Walking." *Excursions,* edited by Leo Marx, 161–214. Gloucester, MA: Peter Smith, 1975, originally Boston: Ticknor and Fields, 1863.

———. *Cape Cod.* Mineola, NY: Dover Publications, 2004, originally Boston: Ticknor and Fields, 1865.

———. *The Journal of Henry D. Thoreau: In Fourteen Volumes Bound as Two.* Edited by Bradford Torrey and Francis H. Allen. New York: Dover Publications, 1962, originally Boston: Houghton Mifflin Company, 1906.

———. *The Correspondence of Henry David Thoreau.* Edited by Walter Harding and Carl Bode. Westport, CT: Greenwood Press, 1958.

———. *Wild Fruits: Thoreau's Rediscovered Last Manuscript.* Edited by Bradley P. Dean. New York: W. W. Norton & Company, 1999.

———. *Walden, Or Life in the Woods, A Fully Annotated Edition.* Edited by Jeffrey Cramer. New Haven: Yale University Press, 2004, originally Boston: Ticknor and Fields, 1854.

———. *I to Myself: An Annotated Selection from the Journal of Henry David Thoreau.* Edited by Jeffrey Cramer. New Haven, Yale University Press, 2007.

———. *The Maine Woods: A Fully Annotated Edition.* Edited by Jeffrey Cramer. New Haven: Yale University Press, 2009.

Thorson, Robert M. *Beyond Walden: The Hidden History of America's Kettle Lakes and Ponds.* New York: Walker & Company, 2009.

———. "Geology." *Encyclopedia of New England: The Culture and History of an American Region,* 567–569. New Haven: Yale University Press, 2006.

Thorson, Robert M., and C. A. Schile. "Deglacial Eolian Regimes in New England." *Geological Society of America Bulletin* 107 (1994): 751–761.

Tinkler-Villani, Valeria, and C. C. Barfoot. *Restoring the Mystery of the Rainbow: Literature's Refraction of Science,* vols. 1–2. Amsterdam: Rodopi, 2011.

Torrey, Bradford. Editor's Introduction to *The Journal of Henry D. Thoreau: In Fourteen Volumes Bound as Two,* 9–17. New York: Dover Publications, 1962, originally Boston: Houghton Mifflin Company, 1906.

Totten, Stanley M., and George W. White. "Glacial Geology and the North American Craton." *Geologists and Ideas: A History of North American Geology,* edited by Ellen T. Drake and William M. Jordan, 125–141. Denver: Geological Society of America, 1985.

Van Doren, Philip Stern. *The Annotated Walden. Walden; or, Life in the Woods, by Henry David Thoreau.* New York: Clarkson, Crown Publishers, 1970.

Walker, Eugene H. "Minerals of Concord." *The Concord Saunterer* 9, no. 3 (1974): 1–6.

Walls, Laura Dassow. "Walden as a Feminist Manifesto," *Interdisciplinary Science, Literature, and Environment (ISLE)* 1, no. 1 (1993): 137–144.

———. *Seeing New Worlds: Henry David Thoreau and Nineteenth-Century Natural Science.* Madison, WI: University of Wisconsin Press, 1995.

———, ed. *Material Faith: Thoreau on Science.* Boston: Houghton Mifflin, 1999.

———. "Believing in Nature: Wilderness and Wildness in Thoreauvian Science." *Thoreau's Sense of Place,* edited by Richard J. Schneider, 15–27. Iowa City: University of Iowa Press, 2000.

———. "Science and Technology." *The Oxford Handbook of Transcendentalism,* edited by Joel Myerson, Sandra Harbert Petrulionis, and Laura Dassow Walls, 572–582. Oxford: Oxford University Press, 2010.

———. *The Passage to Cosmos.* Chicago: University of Chicago Press, 2011.

———. "Henry David Thoreau: Writing the Cosmos." *The Concord Saunterer* New Series 19/20 (2011–2012): 1–21.

Werner, A. G. *Kurze Klassification und Beschreibung der verschiedenen Gebirgsarten*. Dresden, 1787.

Wetzel, Robert G. *Limnology, Lake and River Ecosystems*. 3rd edition. New York: Academic Press, 2001.

White, E. B. "Walden-1954." *Walden and Resistance to Civil Government: Authoritative Texts, Thoreau's Journal, Reviews and Essays in Criticism, Second Edition,* edited by William Rossi, 359–366. New York: W. W. Norton & Company, 1992.

White, George W. Introduction and Biographical Notes to *Illustrations of the Huttonian Theory of the Earth*. New York: Dover, 1956, originally Edinburgh: William Creech, 1802.

White, Gilbert. *The Natural History of Selborne and the Naturalist's Calendar, A New Edition*. Edited by G. Christopher Davies. London: Frederick Warne and Co., 1879.

White, Richard. *The Organic Machine*. New York: Hill and Wang, 1995.

Whitney, Gordon G., and William C. Davis. "Thoreau and the Forest History of Concord, Massachusetts." *Journal of Forest History* 30 (April 1986): 70–81.

Whittlesey, Charles. "On the Fresh-Water Glacial Drift of the Northwestern States." *Smithsonian Contributions to Knowledge 197*. 1886. Originally *Smithsonian Contributions to Knowledge* 15. 1867. Contains *Map Illustrating Limits of the Glacier Drift of North America*. 1864.

Willis, Charles G., Brad R. Ruhfel, Richard B. Primack, Abraham J. Miller-Rushing, Jonathan B. Losos, and Charles C. Davis. "Favorable Climate Change Response Explains Non-Native Species Success in Thoreau's Woods." *PLoS ONE* 5, no. 1 (2010): e8878. Accessed November 1, 2012. doi:10.1371/journal.pone.0008878.

Wilson, Edward O. *Consilience: The Unity of Knowledge*. New York: Vintage Press, 1998.

Winchester, Simon. *The Map That Changed the World: William Smith and the Birth of Modern Geology*. New York: Harper Collins, 2001.

"Windfinder." Data for Logan Airport from 1/2007 to 5/2012 for winds at or above 4 on the Beaufort scale. Accessed April 2012. http://www.windfinder.com/windstats/windstatistic.htm.

Winkler, Marjorie G. "Changes at Walden Pond during the Last 600 Years." *Thoreau's World and Ours: A Natural Legacy,* edited by Edmund A. Schofield and Robert C. Baron, 199–211. Golden, CO: North American Press, 1993.

Wordsworth, William. *Guide to the Lakes*. 5th ed., reprint. Oxford: Oxford University Press, 1970.

Worster, Donald. *Nature's Economy: A History of Ecological Ideas*. 2nd ed. Cambridge: Cambridge University Press, 1994.

———. "Nature, Liberty, and Equality." *American Wilderness: A New History,* edited by Michael Lewis, 263–272. New York: Oxford University Press, 2007.

Zebuhr, Laura. "Sounding Walden." *Mosaic: A Journal for the Interdisciplinary Study of Literature* 43, no. 3 (Sept. 2010): 35–50.

Zen, E-an, Richard Goldsmith, N. M. Ratcliffe, Peter Robinson, R. S. Stanley, N. L. Hatch, A. F. Shride, E. G. A. Weed, and D. R. Wones. *Bedrock Geologic Map of Massachusetts. Scale 1:250,000.* Washington, DC: U.S. Geological Survey, 1983.

NOTES

Most notes provide citations for quoted phrases. Each bears a one-word or one-phrase "tag" linking the quote in my text to its original source, and including all quotes since the previous citation. Tags for non-quotes consist of a distinctive word or phrase near the end of the idea being cited. Tags are followed by a colon. The first citation of a work is given in full. All others are by short titles, as listed. Two sources for the same tag are linked by a semicolon. Discrete notes separated by periods.

Introduction

1. Church records: Walter Harding, *A Thoreau Handbook* (New York: New York University Press), 2; Walter Harding, *The Days of Henry Thoreau* (New York: Dover Publications, 1982), 467. Genius: Ralph Waldo Emerson, "Thoreau," *Atlantic Monthly* (August 1862), cited from *Excursions,* ed. Leo Marx (Gloucester: Peter Smith, 1975, originally Boston: Ticknor and Fields, 1863), 24–25.

2. Pythagoras: Stephen Hawking and Leonard Mlodinow, *The Grand Design* (New York: Bantam Publications, 2010), 18–20; Alan Hodder, *Thoreau's Ecstatic Witness* (New Haven: Yale University Press, 2001), 86.

3. Thaw: WA, 298. Ant lion: Henry D. Thoreau, *Cape Cod* (Mineola, NY: Dover Publications, 2004, originally Boston: Ticknor and Fields, 1865), 92. Run out: Ibid.

4. Ice broke up: WA, 276. Something: Patrick Chura, *Thoreau the Land Surveyor* (Gainesville, FL: University Press of Florida, 2010), 24. Claims Thoreau used an axe to chop the holes, though this tool is far less efficient than an ice chisel, which Thoreau had used to cut fishing holes (WA, 173). A Gunter's chain, named after the early seventeenth-century English clergyman Edmund Gunter, is specially designed for surveying in units of rods and feet. I assume he would not borrow or buy a pencil, since his family manufactured them.

5. Bizarre: Lawrence Buell, *The Environmental Imagination: Thoreau, Nature Writing, and the Formation of American Culture* (Cambridge, MA: Harvard University Press, 1995), 276. Removed: Philip Van Doren, *The Annotated Walden; or, Life in the Woods, by Henry David Thoreau* (New York: Clarkson, Crown Publishers, 1970), 38. Satire: Barksdale W. Maynard, *Walden Pond, A History* (New York: Oxford University Press, 2004), 79. Conceit: Bernhard Kuhn, *Autobiography and Natural Science in the Age of Romanticism: Rousseau, Goethe, Thoreau* (Farnham, Surrey, UK: Ashgate, 2009), 136.

6. Mystic: Leo Marx, *The Machine in the Garden: Technology and the Pastoral Ideal in America* (New York: Oxford University Press, 1964), 380. Cauldron: JR, Jul 16, 1851. Environs: WA, 130. Coherence: Leo Marx, introduction to *Henry David Thoreau*, by Franklin B. Sanborn (New York: Chelsea House, 1980, originally Boston: Houghton Mifflin, 1882), xxi.

7. The technical term for the focus of an earthquake is "hypocenter," counterpoint to "epicenter." Deepest Resort: WA, 187. Thoreau reached the deepest point he could have, barely into the muck below the deepest water. Directly below this are older sedimentary layers, and then bedrock. Bullheads: Thoreau's frequent mention of "pouts" or "horned pouts" refers to a species of catfish *(Ameiurus nebulosus)* known today as the brown bullhead, and which are bottom-feeders (and likely introduced).

8. Roar: JR, Jul 6, 1845. Proprietor: WA, 113. As it expanded downward and outward, the cone of collapse reached the bottom of the basin, which truncated the original talus slopes. Fans of finer-grained sediment then created a more central, less steeply inclined cone inside that complex cone. With the drape of suspended sediment, the cone began to assume a more rounded, hemisphere form.

9. Monadnock: WA, 280. "Is not this the rule also for the height of mountains?" (Though he didn't mention it explicitly, Monadnock, was his favorite mountain, and perhaps motivated this thought.) Plateau: Ibid., 85.

10. Famous: Harding, *Handbook*, 45. Continental: Edward Lurie, *Louis Agassiz: A Life in Science* (Chicago: University of Chicago Press, 1960), 124. Accident: Thomas S. Kuhn, *The Structure of Scientific Revolutions*, enlarged 2nd ed., (Chicago: University of Chicago Press, 1970), 4. Not all Americans were against the glacial theory, though institutional science clearly was.

11. Historians: Stanley M. Totten and George W. White, 1985, "Glacial Geology and the North American Craton," *Geologists and Ideas: A History of North American Geology*, ed. Ellen T. Drake and William M. Jordan (Denver: Geological Society of America, 1985), 125–141. They were the first to draw my attention to one of the most important lines of inquiry in this book. They cite the published transactions of the Association of American Geologists and Naturalists, 1843, p. 324–325, which I have reread. Refer also to John S. Pipkin, "Hiding Places: Thoreau's Geographies," *Annals of the Association of American Geographers* 91, no. 3 (2001): 527–545; Nina Baym, "Thoreau's View of Science (slightly revised)," *Journal of the History of Ideas* 26 (1963): 26; Robert

Sattelmeyer, *Thoreau's Reading: A Study in Intellectual History, with Bibliographical Catalogue* (Princeton, NJ: Princeton University Press, 1988), 79.

12. Till: Modern word for "hardpan," "boulder-clay," or unstratified "drift," introduced by Charles Lyell, "On the Boulder Formation or Drift, and Associated Freshwater Deposits Composing the Mud Cliffs of Eastern Norfolk," *Proceedings of the Geological Society of London* 3 (November 1838–June 1842): 171–179.

13. All of these short quotes are passages from *Walden* that will be more thoroughly cited and explicated in later sections. *"Eizeit"* translates to ice age from German, where the term was first used by Karl Schimper, a colleague of Louis Agassiz's.

14. Sense: Richard J. Schneider, ed., *Thoreau's Sense of Place* (Iowa City: University of Iowa Press, 2000), ix. Biocentrism: Lawrence Buell, *The Environmental Imagination: Thoreau, Nature Writing, and the Formation of American Culture* (Cambridge, MA: Harvard University Press, 1995), 138.

15. Reality (and preceding quotes): WA, 187, 37, and 95–96. Laura Dassow Walls, "Henry David Thoreau: Writing the Cosmos," *The Concord Saunterer* New Series 19/20 (2011–2012): 1–21. Argues that Thoreau's *"point d'appui"* is not rock, but the whole of Walden Pond as a place. This seems too abstract and general to qualify.

16. Hard Science: Primary references to Walden Pond are two U.S. Geological Survey reports published as a monograph (John A. Colman and Paul J. Friesz, "Geohydrology and Limnology of Walden Pond, Concord, Massachusetts," U.S. Geological Survey Water-Resources Investigation Report 01–4137, 2001, 68 p.), and as a poster (Paul J. Friesz and John A. Colman, *Hydrology and Trophic Ecology of Walden Pond, Concord, Massachusetts,* U.S. Geological Survey Water Resources Investigations Report 1–4153, 2001). Many studies explore the link between Thoreau's physical science and his literature, but not the science itself, for example: Sara Luria, "Thoreau's Geopoetics," *GeoHumanities: Art, History, Text at the Edge of Place* (London: Routledge, 2011), 126–138; Richard J. Schneider, "Reflections in Walden Pond: Thoreau's Optics," *ESQ* 21 (1975): 65–75. My geological terminology throughout this book follows the discipline "standard": Klaus K. E. Neuendorf, James P. Mehl Jr., and Julia A. Jackson, eds., *Glossary of Geology, Fifth Edition* (Washington, DC: American Geological Institute, 2005, updated online), 800.

17. New Thoreau: Henry Seidel Canby, *Thoreau* (Boston: Houghton Mifflin, 1939), 227.

18. Boston: Edward S. Deevey Jr., "A Re-examination of Thoreau's Walden," *The Quarterly Review of Biology* 17, no. 1 (1942): 1–11.

19. New Thoreau: JR, Mar 11, 1852. Sounding of White Pond dates the end of the data-gathering stage, begun five years earlier in February (?) 1847. Pathetic: Canby, *Thoreau,* 326. Definition: U.S. Supreme Court, *Daubert v. Merrell Dow Pharmaceuticals, Inc.*, 509 U.S. 579, 590, 1993. Frogpondian: Eric W. Carlson, "The Transcendentalist Poe: A Brief History of Criticism," *Poe*

Studies/Dark Romanticism 32, no. 1–2 (January/December 1999): 47; Laura Dassow Walls, *The Passage to Cosmos* (Chicago: University of Chicago Press, 2011), 256. Physicality: Rochelle L. Johnson, *Passions for Nature: Nineteenth-Century America's Aesthetics of Alienation* (Athens, GA: The University of Georgia Press, 2009), 21; Crystalline: Philip F. Gura, *American Transcendentalism: A History* (New York: Hill and Wang, 2007), 269.

20. Mineral: William Howarth, *Walking with Thoreau: A Literary Guide to the Mountains of New England* (Boston: Beacon Press, 2001), 13. Sermons: JR, n.d.,1850). Thoreau's mineral collection is curated at the Fruitlands Museum, Harvard, MA. Rocks: Thoreau's use of the term "limestone" is a utilitarian label for any rock from which lime can be extracted. The high grade of metamorphic rocks in Concord precludes "limestone" as geologists use it today. Rather, it is a poor grade of marble. Selenite: JR, Feb 3, 1852. Emerald: WA, 286. Azure: Ibid. Flatulency: Ibid., 264. See also Eugene H. Walker, "Minerals of Concord," *The Concord Saunterer* 9, no. 3 (1974): 1–6, for a review of Thoreau's interest in minerals.

21. Renaissance: F. O. Matthiessen, "Walden: Craftsmanship vs. Technique," *American Renaissance* (Oxford: Oxford University Press, 1941), 166–175. Hero-worship: WA, 55. Compare with Roland Wells Robbins, *Discovery at Walden* (Concord, MA: The Thoreau Society, 1999), 43–46. Symbol: Martin Bickman, *Walden: Volatile Truths* (New York: Twayne Publishers, 1992), 26. Green: David Gessner, *My Green Manifesto: Down the Charles River in Pursuit of the New Environmentalism* (Minneapolis, MN: Milkweed Editions, 2011). Urban: Robert M. Sullivan, *The Thoreau You Don't Know: What the Prophet of Environmentalism Really Meant* (New York: HarperCollins, 2009), 354.

22. Hairshirt: E. B. White, "Walden—1954," *Walden and Resistance to Civil Government, Second Edition,* ed. William Rossi (New York: Norton, 1992), 366. Trail: Edmund A. Schofield, "The Ecology of Walden Woods," *Thoreau's World and Ours: A Natural Legacy* (Golden, CO: North American Press, 1993), 161.

23. Refracts: Robert S. Friedman, "Surveying the Empirical Sublime: Thoreau, Literary and Scientific," *Restoring the Mystery of the Rainbow: Literature's Refraction of Science,* ed. Valereia Tinkler-Villani and C. C. Barfoot (Amsterdam, Rodopi, B. V., 2011), 743–759; Luria, *Geopoetics,* 126–138. Ecocriticism: Walls, *Passage,* 243. Appeals for more science. Astronomy: Tom Calderwood, "An Astronomer Reads Thoreau," *Thoreau Society Bulletin* 267 (Summer 2009): 7–8. Parasitic: WA, 298.

24. I spent much of my early career on geo-archaeology and taught dinosaur paleontology for more than a decade. Drunk: White, *Walden—54,* 363.

25. Remote: Robert Sattelmeyer, "Depopulation, Deforestation, and the Actual Walden Pond," *Thoreau's Sense of Place,* ed. Richard J. Schneider (Iowa City: University of Iowa Press, 2000), 236.

26. Limestone: WA, 35, 47. The local bedrock, dominated by synorogenic granite, gneiss, and sillimanite-grade metamorphism, preclude the survival of limestone. Bloom: Harding, *Days*, 264 (citing Sophia Thoreau and Thoreau's *Journal* IX, 83).

27. Rebirth: Melvin E. Lyon, "Walden Pond as a Symbol," *PMLA Transactions and Proceedings of the Modern Language Association* 82, no. 2 (May 1967): 289–300. Flowery: WA, 298.

28. Impelled: WA, 185; H. Daniel Peck, "The Worlding of Walden," *Walden and Resistance to Civil Government, Second Edition,* ed. William Rossi (New York: Norton, 1992), 457.

29. Rain-splash: JR, Nov 3, 1862.

30. Massachusetts Department of Conservation and Recreation: *Geology.*

31. Literary wing, Masterpiece: Harding, *Days,* 333. Vitamin: White, *Walden—54,* 360. Monumental: Reginald L. Cook, *Passage to Walden* (Cambridge: Riverside Press, 1949), 235. Summit: Robert Sattelmeyer, *Thoreau's Reading: A Study in Intellectual History, with Bibliographical Catalogue* (Princeton, NJ: Princeton University Press, 1988), xi; Johnson, *Passions,* 217. For his literary importance: Maynard, *History,* 121; Jay Parini, *Promised Land: Thirteen Books That Changed America* (New York: Doubleday, 2008), 108; Stephen Adams and Donald Ross Jr., *Revising Mythologies: The Composition of Thoreau's Major Works* (Charlottesville: University of Virginia Press, 1988), 9. Stylist, Joyce Carol Oates, introduction to *Walden* (Princeton: Princeton University Press, 1989), xii. Environmental wing, Prophet: Bill McKibben, introduction to *Walden* (Boston: Beacon Press, 2004), x. General: Laura Dassow Walls (citing Buell, *Imagination*), "Believing in Nature: Wilderness and Wildness in Thoreauvian Science," *Thoreau's Sense of Place,* ed. Richard J. Schneider (Iowa City: University of Iowa Press, 2000), 17; Simon Schama, *Landscape and Memory* (London: HarperCollins, 1995), 7. Saint: WA, 574; Buell, *Imagination,* 115. For an example of recent research on Thoreau's ecological observations: Charles G. Willis et al., "Favorable Climate Change Response Explains Non-native Species Success in Thoreau's Woods," *PLoS ONE* 5, no. 1 (2010): e8878, doi:10.1371/journal.pone.0008878.

32. Plumbing: Robert M. Thorson, "Geology," *Encyclopedia of New England: The Culture and History of an American Region* (New Haven: Yale University Press, 2006): 567–569. Premises: Buell, *Imagination,* 2.

33. Hermit: Robert Gross, "That Terrible Thoreau: Concord and Its Hermit," in *A Historical Guide to Henry David Thoreau,* ed. William E. Cain (New York: Oxford University Press, 2000), 181–241. Afterbirth: Perry Miller, *Consciousness in Concord: The Text of Thoreau's Hitherto "Lost Journal" (1840–1841) Together with Notes and a Commentary* (Boston: Houghton Mifflin Company, 1958), 126–127. Phallic: Paul Friedrich, *The Gita within Walden* (Albany, NY: SUNY Press, 2008), 41. Feminism: Patrick Morgan, "Aesthetic Inflections: Thoreau, Gender, and Geology," *The Concord Saunterer: New Series* 18 (2010): 48. Voyeurism: Sharon Cameron, *Writing Nature: Henry Thoreau's Journal* (New York: Oxford University Press, 1985), 103, 106. Magnetic: Chura, *Surveyor,* 41.

34. Letter: Van Doren, *Annotated,* 313. Annotation 29 documents that "two sentences on ms 956–957 are written on the back of a letter dated February 26, 1854."

35. Stanley Cavell, *The Senses of Walden* (New York: Viking, 1972), 117. I use these phrases as titles of Part I and Part II of this book.

36. Lake vs. Pond: For background on definitions, see Thorson, *Beyond*, 53–54. Thoreau's use: WA, 180. Observation: JR, Jul 2, 1852.

1. Rock Reality

1. Brickbatful: JR, Jan 2, 1854. Target: Edward Hitchcock, *Final Report on the Geology of Massachusetts in Four Parts* (Northampton: J. Butler, 1841), 188. Hitchcock (and no other of Thoreau's sources) classifies the cinnamon stone as "essonite," linking him to Thoreau's word choice. The context of the *Journal* paragraph supports this interpretation by connecting it indirectly to Edward Hitchcock's "iceberg" drift theory and Louis Agassiz's catastrophic glacial theory (via the latter's trip to the "Pictured Rocks" of Lake Superior). Both "men of science" were major sources of frustration for Thoreau.

2. Copies: Hitchcock, *Final Report,* 5. Geologist: Ibid., title page. Alignment: Herbert Hovenkamp, *Science and Religion in America, 1800–1860* (Philadelphia: University of Pennsylvania Press, 1978), 47.

3. Quartz: JR, Jun 24, 1853. Hornstone: Howarth, *Walking,* 158. Mica: Ibid., Apr 11, 1852. Limonite: This is a general term for a group of hydrated iron oxides including goethite (named for Johann Wolfgang von Goethe, Thoreau's intellectual hero), lepidocrocite (which is a bright golden yellow in color), and jarosite (which has been found on Mars).

4. Magnetite: JR, Feb 27, 1851.

5. Persistent: Garrison Keillor, "Guy Noir," *A Prairie Home Companion,* August 24, 2004, http://prairiehome.publicradio.org/programs/2004/04/24/scripts /guy_noir.shtml. Creation: Jack Repcheck, *The Man Who Found Time: James Hutton and the Discovery of the Earth's Antiquity* (New York: Perseus Publishing, 2003), 42. Alternate times for creation assigned to the Ussher chronology are 9:00 AM and noon on Monday, October 23: Simon Winchester, *The Map That Changed the World* (New York: Harper Collins, 2001), 13.

6. Dates: Chronometric dates for the universe and earth are based on astrophysical measurements and radiometric dating of earth/moon/asteroid rock, respectively. Vestige-prospect: Thoreau would have read Hutton's quote in Charles Lyell, *Principles of Geology*, 1st ed., vol. 1, ed. Martin J. S. Rudwick (Chicago: University of Chicago Press, 1990). Lyell quotes John Playfair, *Illustrations of the Huttonian Theory of the Earth* (New York: Dover Publications, 1956, reprinting Edinburgh: William Creech, 1802). Bhagavad: Stephen Mitchell, *Bhagavad Gita: A New Translation* (New York: Three Rivers Press, 1988). Luther: Repcheck, *Hutton,* 41.

7. Rock Reality: WA, 95, 96. *Point d'appui:* Walter Benn Michaels, "Walden's False Bottoms," *Walden and Resistance to Civil Government, 2nd*

ed., ed. William Rossi (New York: Norton, 1992), 409. Michaels expands the discussion by confirming that "the issue is solidity, not depth." Alluvion: A curious word choice, the archaic version of "alluvium," which, in Thoreau's era, included all unconsolidated material demonstrably related to modern geologic agencies.

8. Lichens: JR, Mar 5, 1852. Columbine: Ibid., May 16, 1852. Concord River alluvium: JR, Aug 30, 1853. For a review of the link between geology and plants, consult Arthur R. Kruckeberg, *Geology and Plant Life: The Effects of Landforms and Rock Types on Plants* (Seattle: University of Washington Press, 2002). Alluvion and alluvium were different spellings for the same thing. Darwin used the latter term for soils and sediments similar to those of Concord: Charles Darwin, *The Voyage of the Beagle,* ed. Charles W. Eliot with Introduction and Notes (New York: P. F. Collier & Son Corporation, The Harvard Classics, Volume 29, 1909, 1937, 1963, originally New York: Harper & Brothers, 1845).

9. Regional Geology: Geological Society of America, *Geologic Map of North America: Scale 1:5,000,000* (Boulder, CO: Geological Society of America, 2005). State geology: E-an Zen et al., *Bedrock Geologic Map of Massachusetts: Scale 1:250,000* (Washington, DC: U.S. Geological Survey, 1983). Local geology: Patrick J. Barosh, "Bedrock Geology of the Walden Woods," *Thoreau's World and Ours: A Natural Legacy,* eds. Edmund A. Schofield and Robert C. Baron (Golden, CO: North American Press, 1993), 212–221.

10. Granite: Neuendorf, *Glossary.* History: Frank D. Adams, *The Birth and Development of the Geological Sciences* (New York: Dover Publications, 1938).

11. General: Barosh, *Bedrock,* 216. Xenolith compositions vary. Marble, gneiss, quartzite, and basalt, then called "trap." The photo of the Andover granite used for this book shows xenoliths in the lower right corner.

12. Poor Granite: Walker, *Minerals,* 3; Lemuel Shattuck, *A History of the Town of Concord; Middlesex County, Massachusetts, from Its Earliest Settlement to 1832* (Boston: Russell, Odiorne, and Company, 1835), 198. Massachusetts had many major nineteenth-century quarries for granite dimension stone, including those in Westford, Milford, Northbridge, Dedham, and Acton (personal communication, Joseph Kopera, February 2013).

13. Clast size/shape: During basal glacial transport, average clast size diminishes with distance from source due to crushing and attrition. There is also horizontal diffusion, meaning the concentration of particular lithologies diminishes.

14. Aeromagnetic: Robert O. Castle et al., *Structural Dislocations in Eastern Massachusetts* (Washington, DC: U.S. Geological Survey Bulletin 1410, 1976). Bedrock: Colman and Friesz, *Geohydrology,* 5. Figure 2 shows seismic imaging and lithological gamma log.

15. Werner: Abraham G. Werner, *Kurze Klassification und Beschreibung der verschiedenen Gebirgsarten* (Dresden, 1787). History: Adams, *Geohistory,* 221.

Saussure: Ibid., 390. Younger sedimentary rocks, especially granular rocks like sandstone, were also considered marine, though not precipitates.

16. Burnet: Adams, *Geohistory,* 209. Stephen Jay Gould, *Time's Arrow, Time's Cycle: Myth and Metaphor in the Discovery of Geological Time* (Cambridge, MA: Harvard University Press, 1987), 21–59. Whiston: Repcheck, *Hutton,* 98.

17. Fairyland: Adams, *Geohistory,* 210. Normal science: Kuhn, *Structure,* 15.

18. Linnaeus: Adams, *Geohistory,* 128. Exact descriptions: Henry David Thoreau, *A Week on the Concord and Merrimack Rivers* (Orleans, MA, Parnassus Imprints, Inc., 1987, originally Boston: James Monroe & Co., 1849), 407. Hostile: Adams, *Geohistory,* 248. Faust: Ibid. Professor: Walls, *Cosmos,* 28.

19. Continental: Winchester, *Map,* 226. Geognosy: Repcheck, *Hutton,* 173. Herbert, *Darwin,* 32–33. Absurdity: Charles Darwin, "Recollections of the Development of My Mind and Character," in *Evolutionary Writings,* ed. James Secord (Oxford: Oxford University Press, 2008), 373.

20. Maclure: George W. White, introduction and biographical notes to *Illustrations of the Huttonian Theory of the Earth* (New York: Dover, 1956, originally Edinburgh: William Creech, 1802), vii. Silliman: Ibid., viii. Standard text: Hovencamp, *Theology,* 125.

21. Eaton: White, *Illustrations,* vii. Concord: Shattuck, *History.* Rained rocks: MW, 56.

22. Illustrations: White, *Illustrations,* v. Summarizes Hutton's earlier work, *Theory of the Earth; or an Investigation of the Laws Observable in the Composition, Dissolution, and Restoration of Land upon the Globe* (1788), and a series of published lectures he had given at the University of Edinburgh beginning in 1785. Subservient: Playfair, *Illustrations,* 129. Thoreau's "revolutions" were at the contact between the Andover Granite and the overlying glacial sediments, and between the glacial sediments and the overlying "alluvion."

23. Maclurin: Repcheck, *Hutton,* 60.

24. Morality: Playfair, *Illustrations,* 125. Impiety: Ibid., 119, 126. Unpopular: Hovencamp, *Theology,* 94. Bigotry: Winchester, *Map,* 283.

25. General sources: Martin Rudwick, ed., introduction to *Principles of Geology* (Chicago: University of Chicago, 1992, originally London: John Murray, 1830–1833); John Gribbin, *The Scientists: A History of Science Told Through the Lives of Its Greatest Inventors* (New York: Random House, 2002); Gould, *Times Arrow.* Geology was a recognized field by 1810, but lacked the understandable paradigm Lyell provided.

26. Geologica: Baym, *View,* 28. The actual title was printed in English as *Principles of Geology.* Same: Playfair, *Illustrations,* epigraph.

27. Audience: Geology was well established in England by 1810, having been included in the *Encyclopedia Brittanica*, a standing solidified by the 1815 publication of William Smith's important map of England (Winchester, *Map,* 25).

Great works: Kuhn, *Structure,* 10. Pages: Sandra Herbert, *Charles Darwin, Geologist* (Ithaca: Cornell University Press, 2005), 80. Recommended: Robert D. Richardson Jr., *Henry Thoreau: A Life of the Mind* (Berkeley: University of California Press, 1986), 82.

28. Menu: Sir William Jones, *Institutes of Hindu Law; or, the Ordinances of Menu, According to the Gloss of Culluca, Compromising the Indian System of Duties, Religious and Civil* (London: Government of Great Britain, 1796). Consciousness: Richardson, *Life,* 107. Thoreau relied on Lyell throughout *A Week,* for example, the theory of coral atolls (WK, 451–452). Pretensions: JR, n.d. June 1841. See also Henry David Thoreau (unsigned ed.), "The Laws of Menu," *The Dial* 3, no. 3 (Jan. 1843): p. 331; Henry David Thoreau (unsigned ed.), "Ethical Scriptures: Sayings of Confucius," *The Dial* 3, no. 3 (Jan. 1843): 493; Henry David Thoreau (unsigned ed.), "Ethical Scriptures: Chinese Four Books," *The Dial* 4, no. 2 (Oct. 1843): 205.

29. Subterranean: WK, 154. Frost-work: NH, 69. Revolution: Playfair, *Hutton,* 125.

30. Phosphorescence: JR, May 5, 1852. Truth: JR, Jan 26, 1853.

31. Granite: MW, 64. Facilitates comparison of granite at Walden and Katahdin. Star surface: MW, 64. Solitary: JR, Sep 7, 1851.

32. Celestial: JR, May 5, 1852.

33. Esquimaux: JR, Mar 29, 1852; CC, 89. Quoting Pierre Jean Edouard Desor, unsourced. Covered: Lyell, *Principles,* 88. Crust: Tim Flannery, *Here on Earth: A New Beginning* (London: Penguin Books, Allen Lane, 2010), 43. Summarizes the argument for a biochemical origin of Earth's crust.

34. Hindu: Jones, *Manu,* 2. Climate: JR, Sep 7, 1851.

35. Penokian: Geosociety, *Map.* An excellent source for all the tectonic events.

36. Contact!: MW, 64.

37. Alma: JR, Sep 15, 1838. Letter: Walter Harding and Carl Bode, *The Correspondence of Henry David Thoreau* (New York: New York University Press, 1958, reprinted Westport, CT: Greenwood Press, 1974), 319. Dated January 21, 1854. Pyrrha: WA, 4–5. Raleigh: Ibid. See also annotations 23–25.

38. Burrowing: WA, 96. Bedrock: Zen, *MassMap.*

39. Chain of ponds: WA, Apr 19, 1852. Geology: Barosh, *Bedrock,* 216. Opposite: JR, Aug 3, 1852.

40. Revolutions: Playfair, *Hutton,* 2. Cycle: Ibid., 128.

41. Agricola: Adams, *Geohistory,* 342, 343.

42. Steno: Adams, *Geohistory,* 358. Moro: Ibid., 365.

43. Contraction: Rodger T. Fail, "Evolving Tectonic Concepts of the Central and Southern Appalachians," *Geologists and Ideas: A History of North American Geology,* ed. Ellen T. Drake and William M. Jordan (Boulder, CO: Geological Society of America, 1985), 24. For a primer on plate tectonics, consult any college-level introductory geology text.

44. Acadian Age: Lead ages are based on 207^{Pb}–204^{Pb} and tectonics from petrology sessions at Geological Society of America Northeastern Meeting, 18–20 March, 2012. The Avalon Terrane is thought to have been sliced away from the West African craton at some much earlier time.

45. Nashoba: Martin Acaster and M. E. Bickford, "Geochronology and geochemistry of Putnam-Nashoba terrane metavolcanic and plutonic rocks, eastern Massachusetts: Constraints on the early Paleozoic evolution of eastern North America," *Geological Society of America Bulletin* 111 (1999): 240–253. Nashoba was created and docked about 430–470 and 400 million years ago, respectively.

46. Spine: JR, May 10, 1853. Foundations: WK, 213.

47. Lithology and map units: Zen, *MassMap;* Barosh, *Bedrock.* Acaster and Bickford, *Geochronology,* 240. The authors report that petrologically, the original lithology of the Nashoba block near Concord is calc-alkaline, which is consistent with the tectonic framework (ocean-continent subduction) and syntectonic, peraluminous origin of the Andover Granite, which indicates partial melting of crustal rocks. In places it qualifies as a migmatite, meaning a rock on the genetic boundary between metamorphic and igneous. Some of Nashoba remained soft until the end of the Paleozoic.

48. Map Units: Hitchcock, *Final Report,* Plate 52 (fourth edition of hand-tinted foldout map).

49. Barefoot: Harding, *Days,* 11–12. Cows: Ibid., 20. Traits: William Ellery Channing, *Thoreau, the Poet-Naturalist,* ed. F. B. Sanborn (Boston: Charles E. Goodspeed, 1902); Walter Harding, *Thoreau: Man of Concord* (New York: Holt, Rinehart and Winston, Inc., 1960); Franklin B. Sanborn, *Henry David Thoreau* (Boston: Houghton Mifflin, 1882).

50. Academy: Harding, *Days,* 27.

51. Curriculum: Canby, *Thoreau,* 42. Sattelmeyer, *Reading,* 9. Diploma: Harding and Bode, *Correspondence,* 190 (Nov 14, 1847, to Emerson).

52. Books: JR, Mar 23, 1853. Intelligence: Ibid., Jan 30, 1854. Lichen: Ibid., Jan 15, 1853. Lichens: Ibid., Jun 18, 1852. Hard terms: JL, Mar 1, 1852.

53. Clematis: JR, Jan 26, 1853.

54. Moose: Van Wyck Brooks, *The Flowering of New England: 1815–1865* (New York: E. P. Dutton & Co., Inc., 1936), 363.

55. Subtitle of "Natural History" is "Reports: On the Fishes, Reptiles, and Birds; the Herbaceous Plants and Quadrapeds; Insects Injurious to Vegetation; Invertebrate Animals of Massachusetts." Contrivance: Robert Kuhn McGregor, *A Wider View of the Universe: Henry Thoreau's Study of Nature* (Champaign: University of Illinois Press, 1997), 53–54.

56. Minerals: Shattuck, *History,* 198. Main Street: Harding, *Days, 265.*

57. Information provided by Michael Volmar, Chief Curator at Fruitlands Museum.

58. According to written communication (Feb 4, 2013) from Joseph P. Kopera (Senior Research Fellow, Massachusetts Geological Survey), all of Thoreau's mineral and rock specimens "could be found in the northeast, and in Massachusetts," and that the Nashoba Terrane (local bedrock) is "positively drowning in magnetite." Though one of Thoreau's specimens is labeled as being from Martha's Vineyard, a place Thoreau may not have visited in the Walden years, he could have collected it elsewhere on Cape Cod or received it from someone else.

59. Osgood: Harding, *Days,* 103. A more detailed study of Thoreau's collection is underway.

60. Michigan: Harding, *Days, 197.* One of Thoreau's contemporaries, Henry Rowe Schoolcraft, was hired to do a similar job

61. Basalt: JR, Sep 19, 1850. Sandstone: JR, Feb 24, 1854. Laminae: JR, Feb 26, 1852. Temper: JR, Apr 21, 1852. Indian hoe: JR, Jun 24, 1853. White quartz: JR, Jun 24, 1853. Flint's stones: JR, Oct 10, 1851. Imported stone: Ibid., Jun 24, 1853.

62. Geologist: JR, Apr 27, 1852. Merges his local observations with those of Lyell.

63. Cliffs: JR, Apr 27, 1852. Batemans: JR, Oct 5, 1851. His question mark. He later named this outcrop "Curly Pate Hill" after small-scale contortions of the rock "bald" highlighted by thin marble beds.

64. Roads: Barosh, *Bedrock,* Figures 2 (p. 214) and 3 (p. 216). Of the two block faults at Walden, one is mapped, and the other inferred.

65. Nashoba: Daniel V. Boudillion, "Nashoba Hill: The Hill that Roars." Accessed September 10, 2012. http://www.boudillion.com/nashobahill/nashobahill.htm.

66. Anon: MW, 56. Disjointed: JR, Apr 21, 1852.

2. Landscape of Loss

1. Meadows: WK, 1. Regardless of basement, New England topography dates to Triassic rifting, whereas those in the others date to late Cretaceous closure of Tethys. The date of 17,000 years approximates failure of the ice dam creating the lake.

2. River: JR, Aug 30, 1853. Describing lower Sudbury and Concord. Tract: JR, Aug 3, 1852.

3. Needle: Henry David Thoreau, "Walking," *Excursions,* ed. Leo Marx (Gloucester, MA: Peter Smith, 1975, originally Boston: Ticknor and Fields, 1863), 175. Settle: Ibid., 176. Magnetism: Ibid. Needle: Ibid. West: Ibid.

4. Range: Henry David Thoreau, "A Walk to Wachusett," *Excursions,* 80. Alignment: Ibid., 92.

5. Wonder: JR, Sep 5, 1838.

6. River: JR, Dec 15, 1841. Stealthy-paced: WK, 154. Perseverance: JR, Sep 11, 1851.

7. Today, the New England crust is about thirty kilometers thick below New England, whereas it may have once exceeded forty-five kilometers.

8. The geological content of the Nashoba block is fixed, whereas its plate-embedded location has drifted about the surface of the earth and changed orientations. Thus, terms like "east of" must have a specified time frame and reference frame.

9. Antiquity: WK, 188.

10. Beagle: Darwin, *Voyage*. Fitness: JR, Mar 21, 1853. Twisted: Ibid., Jun 29, 1852. Remarkable: Ibid., Aug 21, 1851. Fitted: Ibid., Feb 19, 1854. Nature has produced: Ibid., Aug 21, 1851.

11. Fish: JR, Apr 18, 1852. The streamlined design includes a body shaped to minimize resistance through viscous fluids, a detection-and-feeding system up front, and a propulsion system behind. Others: WK, 196.

12. Fossils: JR, Aug 31, 1851. Bulls: Ibid., Feb 27, 1851; Tortoise: WG, 196.

13. Trilobite: Harding, *Days,* 278. Horse: JR, Jun 11, 1851. Megatheriums: JR, Sep 19, 1850, and Nov 8, 1850. Griffins: WG, 196.

14. Creation: JR, Apr 27, 1852. Lichens: Ibid., Jun 6, 1853. Alder: Ibid., Feb 11, 1854. Cape Cod: CC, 131. Pools: JR, Jun 15, 1852. Inorganic: Ibid., Jan 23, 1852.

15. Antiquity: WK, 188. Thoreau's quote was also noted by Richardson, *A Life,* 82. Richardson paraphrases Lyell, *Principles,* 29.

16. Rockwork: JR, May 16, 1852.

17. Farmer: JR, Mar 15, 1852. Next comes: Ibid., May 16, 1852. Lawns: WW, 117.

18. Muskrat: JR, Mar 29, 1853. Boulders: Ibid., Jun 13, 1852. Rock stain: Ibid., Jun 23, 1852. Indian hue: Ibid., Jan 29, 1852. Virgin mould: WK, 313.

19. Bog ore: Walker, *Minerals,* 2. Assabet: JR, Jul 10, 1852. His question mark. Explanation: Chemically, the process is $H_2O + CO_2 \rightarrow H_2CO_3. \rightarrow H^+ + HCO_3^-$.

20. Mussels: JR, Dec 3, 1853. Dust again: Ibid., Jan 15, 1853. Dead Sea: Ibid., Jun 9, 1850.

21. Snowmelt: JR, Dec 31, 1850. Prior to recent glaciation, all of this residue was produced by weathering within soils developed on bedrock. Since then, mechanical crushing has become the dominant mechanism.

22. Ice crystals: JR, Nov 25, 1853. Ant hill: Ibid., Mar 27, 1853. Cowpaths: Ibid., Aug 3, 1852. House: Ibid., Nov 30, 1851; this description refers both to the original location of the house foundation above Walden Pond and the later one near Lincoln Road. Setting: Ibid., no date, 1850.

23. Raindrops: JR, Aug 23, 185. Rills: Ibid., Sep 19, 1850. Haverhill: Ibid., Apr 27, 1853.

24. Clad: JR, Feb 20, 1852. Curves: Ibid., Aug 3, 1852.

25. Schoolmaster: List from Harding and Bode, *Correspondence,* 186. Letter of September 30, 1847, to Henry Williams Jr. from Concord responding to an

inquiry from his alma mater about his profession; Harding, *Days,* 274. Broadside: Chura, *Surveyor,* 85.

26. Profession: JR, Sep 7, 1851. Vocation: Ibid., Sep 20, 1851. Trivial life: Ibid., Dec 12, 1851. Drudgery: Ibid., Feb 11, 1853. How trivial: Ibid., Aug 6, 1853.

27. Unsold Copies: JR, Oct 27, 1853. Grumbling: Ibid., Dec 22, 1853, and Dec 17, 1853. Sympathy: Ibid., Jan 17, 1854. Prince: WG, 170.

28. Ravines: WA, 17.

29. Great drops: JR, Nov 9, 1853. Drought: Ibid., Jul 7, 1853. Streets: Ibid., Jul 24, 1851.

30. Thin sheet: JR, Apr 4, 1854. Shelter: Ibid., Jul 30, 1852. Evaporation: Ibid., Jan 27, 1852. Thirstiness: Ibid., Mar 10, 1854.

31. Bowels: JR, Aug 19, 1852. River contrast: Ibid., Jul 12, 1852. Flood response: Ibid., Apr 19, 1852. Variable gradient: Ibid., Mar 30, 1853. Pools: Ibid., Mar 29, 1852. Antidune: Ibid., Mar 30, 1853. Shear zones: Ibid., Jul 24, 1853. Floes shearing: Ibid., Feb 12, 1854. Stream power: Ibid., Mar 28, 1852.

32. Ice dams: JR, Feb 12, 1854. Ice breakup: Ibid., Dec 25, 1853. Flood infiltration: Ibid., Mar 10, 1854. High water: Ibid., Apr 4, 1852. Historic flood: Ibid., Jul 18, 1852. Flood range: Ibid., Aug 22, 1852.

33. Bed material: JR, Jul 9, 1852. Golden comb: Ibid., Sep 4, 1851. Fluvial walk: Ibid., Jul 10, 1852. Genial mud: Ibid., Jul 12, 1852. Iron pan: Ibid., Jul 10, 1852. Nut Meadow: Ibid., Mar 10, 1853. Lagoons: Ibid., Apr 7, 1853. Oxbows: Ibid., Jun 20, 1853.

34. Colored streams: JR, Apr 16, 1852. Merrimack: Ibid., Apr 17, 1853. Social: WA, 139.

35. Idol: JR, Mar 17, 1853. Water-worn: Ibid., Mar 29, 1853.

36. Bowels: WK, 413.

37. Robber: WK, 305.

38. Meandering: JR, Oct 15, 1851.

39. Conservation: JR, Jun 15, 1852. Sexualization: Ibid. Straight line: Ibid., Jul 8, 1853. Summary: Ibid., Aug 26, 1853. Compare with WK, 170. Compare to Hitchcock, *Final Report,* 343.

40. Eye at last: JR, Mar 28, 1853.

41. Leech: JR, Mar 28, 1852.

42. Corkscrew: Ibid., Apr 11, 1852. Shelving: JR, Mar 10, 1853.

43. Resistance: JR, Apr 12, 1852. Shipwreck: CC, 3. Bridge: JR, Apr 12, 1852; Hitchcock, *Final Report,* 628. Sulfuret: pyrite is only one of several metal sulfide minerals that quickly disintegrate in the presence of oxygenated water.

44. Monadnock: Howath, *Walking,* 248. Fifty miles: JR, Sep 27, 1852; Hitchcock, *Final Report,* 628. Buttresses: JR, Sep 7, 1852. Spelling was "Monadnoc" in Thoreau's day.

45. Iron-bound: JR, Jul 18, 1852. Conantum: Ibid., Apr 27 and May 3, 1852, respectively. Rock list: Ibid., Jun 9, 1850.

46. River as key: WK, 98. Peneplane: I use the term broadly as the grand sum of all the historically named "peneplanes," of which there are a dozen or so, extending all the way across the Berkshires in Massachusetts. Collectively, they reveal that all of the hard-rock landscape of New England was deeply and thoroughly eroded prior to more recent passive uplift (rejuvenation).

47. Uplift: H. H. Mills, "Apparent increasing rates of stream incision in the eastern United States during the late Cenozoic," *Geology* 28, no. 10 (2000): 955–957.

48. Sermon: JR, May 10, 1853. High blue: JR, May 10, 1853. Sharon Cameron, *Writing,* 11, remarked that in this scene "Thoreau's attention to the landscape is scrupulously chaste, in the sense of being vigilant to the precision of detail." This is consistent with his penetrating geological attention to the scene.

49. Superstructure: JR, May 10, 1853.

50. Table-land: MW, 31 (see also ibid., 58). Bounded: JR, Apr 11, 1852.

51. Ridges: JR, May 7, 1852. Western: Ibid., May 10, 1853. Mountains: Ibid., Jul 27, 1852. Snowiest: Ibid., Mar 7, 1854.

52. Favored: JR, Jul 5, 1852.

53. Map of Concord: Appendix to Volume 2 of *The Journal of Henry D. Thoreau: In Fourteen Volumes Bound as Two* (Boston: Houghton Mifflin Company, 1906), vol. I–VII (1837–October 1855), ed. Bradford Torrey and Francis H. Allen (New York: Dover Publications, 1962).

54. Gentleness: WK, 6. Dead stream: Ibid. Skating: WA, 131. Bridge: WK, 6. Descent: Ibid., 7. The name was "Musketaquid" to local inhabitants (Shattuck, *History,* 2), "Musketicook" to the Maine Penobscots (MW, 131), "Grass-ground River" or "Meadow River" in *A Week* (WK, 1), and "Prairie River" in *Natural History of Massachusetts* (NH, 51).

55. Tilt: J. Walter Goldthwait, "The Sand Plains of Glacial Lake Sudbury," *Bulletin of the Museum of Comparative Zoology at Harvard College* 42 (1905); Koteff, *Glacial Lakes,* C142; Carl Koteff et al., "Delayed Postglacial Uplift and Synglacial Sea Levels in Coastal Central New England," *Quaternary Research* 40 (1993): 46–54. The latter reference reports a slope of 0.852 meters per kilometer at an azimuth of North 28.5W for Glacial Lake Merrimack, which equals 4.516 feet/mile. Goldthwait was the first to recognize and calculate the tilt.

56. Back-tilting: There is no proof that other causes are not involved. The isostatic tilt is simply the most parsimonious explanation at this time. More research is needed to test this hypothesis.

57. Billerica: WK, 70. Lake: JR, Apr 16, 52.

58. Defect: Shattuck, *History,* 15–16. Sabbath: WK, 98.

59. Falling: WK, 104. References to rivers occur throughout the pages of Nathaniel Hawthorne's *Mosses from an Old Manse* (London: Wiley & Putnam, 1846).

3. Thoreau's Arctic Vision

1. Emerson, *Eulogy,* 20.
2. Glacier: JR, Feb 3, 1852.
3. General: Kane, *Narrative,* 1856.
4. Peterboro: JR, Apr 4, 1852. Patagonia: Darwin, *Beagle,* 247. Snowfall: JR, Apr 14, 1952. Snowshoes: Ibid., Feb 10, 1852, and Jan 27, 1854. Orbit: Ibid., Feb 16, 1854. Trigger: Peter U. Clark et al., "The Last Glacial Maximum," *Science* 325 (7 August 2009): 710–714. Thoreau's era occurred during the close of the Little Ice Age. See Brian Fagan, *The Little Ice Age* (New York: Basic Books, 2000), frontispiece illustrations.
5. New Views: WA, 262. Thoreau had written about mastodon bones (JR, Nov 8, 1850) and may have seen them. The earliest evidence for humans in New England is about 11,000 years before present.
6. Globally, the "last glaciation" is Marine Isotope State 2 (MIS2) also known as the "Last Glacial Maximum" (LGM), dating to 19,000–36,000 years before present, peaking about 23,000 years before present. (Clark et al., *Maximum,* Figure 3, 711). For general reconstruction: Robert M. Thorson, *Beyond Walden: The Hidden History of America's Kettle Lakes and Ponds* (New York: Walker & Company, 2009), 11–62. For textbook coverage of ice sheet mechanisms: Douglas I. Benn and David J. A. Evans, *Glaciers & Glaciation* (London: Edward Arnold, 1998). For specific local chronology: John C. Ridge et al., "Atmospheric ^{14}C chronology for late Wisconsin deglaciation and sea-level change in eastern New England using varve and paleomagnetic records," *Deglacial History and Relative Sea-Level Changes, Northern New England and Adjacent Canada,* ed. T. K. Weddle and M. J. Retelle (Boulder, CO: Geological Society of America, Special Paper 351, 2001), 171–189, and updated at "North America Glacial Varve Project," http://geology.tufts.edu/varves/ (accessed 9/15/2012). For specific meltwater sedimentology: Thomas Gustavson and J. C. Boothroyd, "A Depositional Model for Outwash, Sediment Sources, and Hydrologic Characteristics, Malaspina Glacier Alaska: A Modern Analog of the Southeastern Margin of the Laurentide Ice Sheet," *Geological Society of America Bulletin* 99 (1987): 187–200.
7. There is no way to exactly pinpoint the terminus of the flowline over Concord. I use western Martha's Vineyard as the closest familiar place reference based on inferred surface contours inland from lobe boundaries. The thickness is unknown, estimated using modern analogs and two-dimensional numerical models of ice sheet behavior.

8. Chura, *Surveyor,* 15, quotes Thoreau from *The Maine Woods*, 1857.

9. Two book length treatments of the history of the ice age concept: Edmund Blair Bolles, *The Ice Finders: How a Poet, a Professor, and a Politician Discovered the Ice Age* (Washington, DC: Counterpoint Press, 1999); and John Imbrie and Katherine Palmer Imbrie, *Ice Ages: Solving the Mystery* (Hillside, NJ: Enslow Publishers, 1979). See also editor's introduction to Louis Agassiz, *Studies on Glaciers: Preceded by the Discourse of Neuchatel* (2 vol. with plates), trans. and ed. by Albert V. Carozzi (New York: Hafner Publishing, 1967, originally Newchatel: Louis Agassiz, 1840). Good reviews for U.S. and Canada: Totten and White, *Craton;* and Alan V. Jopling, "Early Studies on Stratified Drift," *Glaciofluvial and Glaciolacustrine Sedimentation,* ed. Alan V. Jopling and Barrie C. McDonald (Tulsa, OK: Society of Economic Paleontologists and Mineralogists, 1975), 4–21. Saussure: Playfair, *Hutton,* 328 (note 348). Goethe: Bolles, *Finders,* 72.

10. Early History: Lurie, *Agassiz,* 94–105. Charpentier's paper: Jean de Carpentier, "Notice sur la cause probable du transport des blocs erratiques de la Suisse," *Annales des Mines,* ser. 3, vol. 8 (1835), 219–236. London: Bolles, *Finders,* 49.

11. For more detailed story: Bolles, *Finders;* Gribbin, *Scientists,* 465–467; and Imbrie and Imbrie, *Ice Ages.*

12. Lurie, *Agassiz,* viii. See also Christoph Irmscher, *Louis Agassiz: Creator of American Science* (Boston: Houghton Mifflin & Harcourt, 2013). Here, in contrast to more general claims, I list Thoreau as a "scientist," a point more fully developed later in this book.

13. In Thoreau's era a "debacle" referred to a violent catastrophic flood, often containing icebergs.

14. Dewey: Agassiz, *Etudes,* 171. In footnote 31, the author cites *American Journal of Science* 37 (1839), 240–242. New York: Totten and White, *Craton,* 127. Dobson: Lurie, *Agassiz,* 102. Cites p. 619 of the 1924 Yale University Press reprint of George P. Merrill, "The First One Hundred Years of American Geology, Contributions to the History of American Geology," *Report of the U.S. National Museum under the Directorate of the Smithsonian Institution for the Year Ending June 30, 1904* (Washington, DC: U.S. Government Printing Office, 1904).

15. Alps: Agassiz, *Etudes,* 174. Glaciation: Ibid., 169.

16. Great Time: Mircea Eliade, *The Myth of the Eternal Return,* trans. Willard R. Trask (New York: Pantheon Books, Bollingen Series XVI, 1954), ix, 12.

17. Lyell, *Principles,* 29. Ironically, Lyell used these same cliffs more than a century later to subdivide recent geological history. Adams, *Geohistory,* 262.

18. Lyell, *Principles,* 47–48.

19. Cuvier quotes are from Adams, *Geohistory,* 266. See also Repcheck, *Hutton,* 175.

20. Buckland: Repcheck, *Hutton,* 181. Historian of science Sandra Herbert (Herbert, *Darwin,* 184–185) concludes that Buckland intentionally selected the term for its biblical association.

21. America: Hitchcock, *Final Report,* 405–406. For a precedent on the connection between diluvium and transcendental geology, consult G. M. Fleeger and J. D. Inners, "Henry David Thoreau and his mentors on the 'diluvial' geology of New England," *Geological Society of America Annual Meeting Abstracts with Programs* 38, no. 7 (2006): 38. Boulder was spelled "bowlder" during Thoreau's era.

22. Herbert, *Darwin,* 267 (citing Murchison's 1839 *The Silurian System*). Catastrophy: Jopling, *Drift,* 7.

23. Proving Lyell's conversion is a letter sent to Agassiz from Murchison as quoted by Gribbin, *Scientists,* 469. "Lyell has adopted your theory in toto!!"

24. Lyell's method: Gould, *Time's Arrow,* 104. Gordian: Lyell, *Principles,* 6. Writer: Gribbin, *Scientists,* 326.

25. Minority: Rudwick, *Lyell,* x–xi. Evolution: Lyell, *Principles,* 153. Crusade: Rudwick, *Lyell,* xxxviii.

26. The slow uplift is due to continuing isostatic adjustment following ice-sheet deglaciation. Long Island: Bolles, *Finders,* 146.

27. Cosmic: Lurie, *Agassiz,* 97. Dubious: Sattelmeyer, *Reading,* 83. Secondary: Bolles, *Finders,* 184. Creations: Ibid., 71; Hovencamp, *Theology,* 111.

28. Prussia: Lurie, *Agassiz,* 114. Letters: Harding and Bode, *Correspondence,* 243–244.

29. With a dearth of CO^2 in the atmosphere, the planet sought the thermal equilibrium temperature predicted by its astronomical distance from the sun, which is about six degrees Fahrenheit, well below freezing. Luckily for us, volcanic de-gassing restored the carbon dioxide that had previously been sequestered, saving the planet from a permanent grip of ice, and returning us to blue planet conditions.

30. Global dates: Benn and Evans, *Glaciation.* Local glaciations: Ralph S. Lewis and Byron D. Stone, "Multiple Glacial Development of Upper Georges Bank," *Abstracts with Programs* (Hartford, CT: Geological Society of America, Northeastern GSA Section Meeting) 44, no. 2 (2012): 23.

31. River ice: JR, Dec 1, 1853. Details in next chapter. Aerodynamics: Ibid., Jan 1, 1854. Tesselation: Ibid., Feb 12, 1854. Fractures: Ibid., Mar 4, 1854.

32. Kit: Channing, *Poet-Naturalist,* 42. Umbrella: MW, 105. Lacking a standard measure, he marked lengths using knots on the rope or bateau "painter," converted these distances to integer and fractional lengths of his umbrella, and finally converted them using a two-foot rule. Geese: JR, Nov 23, 1853. Frozen ground: Ibid., Feb 24, 1854.

33. Arithmetic: Emerson, *Eulogy,* 8. Mensuration: Ibid., 9. Literary selections: Ibid., 31–32.

34. Dimensions: Chura, *Surveyor,* 18. Hydrometer: JR, Aug 25, 1852. Bronchitis: Harding, *Handbook,* 14. Pathetic: Canby, *Thoreau,* 327.

35. Spy Glass: Canby, *Thoreau,* 326–327.

36. Mastery: Richard J. Schneider, "Walden," *The Cambridge Companion to Henry David Thoreau,* ed. Joel Meyerson (Cambridge: Cambridge University Press, 1995): 92–106.

37. Curvature: JR, Aug 23, 1851. Horizon: Ibid., Jun 11, 1851. Quoting Darwin, *Voyage.*

38. Haze: JR, Mar 30, 1853. Frost: Ibid., Dec 2, 1853. Heat: Ibid., Nov 20, 1853. Mechanics: Ibid., Dec 11, 1853. Hydraulics: Ibid., Nov 15, 1853. Thermodynamics: Ibid., Feb 5, 1853.

39. The onset and count of "mensurations" was based on my first, and therefore least-biased, reading. A second "read-through" would likely yield a different count and perhaps a different onset date, but it would not change the basic conclusion that mensuration showed up dramatically and permanently.

40. Examples from JR, 1850: Micrometeorology: Nov 23. Freeze-up: Dec 17. Seepages: Dec. 17.

41. Examples from JR, 1851: Magnets: Feb 27. White oak: Jun 3. Icebergs: Jun 11. Hydraulic gradient: Jun 22. Swamp deposits: Jul 6. Topography: Sep 7. Plunging folds: Oct 5. Moonlight: Oct 6. Serpentine: Oct 15. Trees: Nov 22. Snowdrifts: Dec 25.

42. Examples from JR, 1852: Snowprints: Jan 8. Snow budget: Jan 27. Streamflow: Mar 28. Ice-out: Mar 26. Cloud velocity: Apr 13. Strike/Dip: Apr 27. Groundwater temperature: Jun 15. Clams: Jul 3. Skull measurements: Jul 25. Haze density: Jun 26. Round Hill: Jul 18. Lake Superior: Jul 20. Canopy storage: Jul 30. Distance perspective: Sep 22. High water: Dec 5.

43. Examples from JR, 1853: Ice bubbles: Jan 8. Pitch pine: Mar 27. Woodchuck: Mar 30. Sine wave: Apr 7. Fractal serrations: May 14. Indian hoe: Jun 24. Weather front: Jun 26. Bangor: Dec 2. Pyramids: Dec 10.

44. Examples from JR, 1854: Snow depth: Jan 2. Ice organ pipes: Jan 11. Snowflakes: Jan 14. Acoustic barometer: Feb 13. Frozen ground: Feb 22. Buoyancy: Feb 24.

45. Breakup: Harding, *Days,* 300. Onset: Bradley P. Dean, "Natural History, Romanticism, and Thoreau," *American Wilderness: A New History,* ed. Michael Lewis (New York: Oxford University Press, 2007), 73–89. Surveying: Harding, *Days,* 274. Broadside: Reprinted and discussed in Chura, *Surveyor,* 84–85.

46. Harding, *Days,* 94.

47. Cirques: Spafford C. Ackerly, "Reconstructions of mountain glacier profiles, northeastern United States," *Geological Society of America Bulletin* 101 (1989): 561–572.

48. Spring: WA, 298.

49. Polished: Thoreau's Monadnock descriptions are from June 3, 1858 (Howarth, *Walking,* 270), which post-date the Walden period. I cite them because he was equally impressed on his final trip on August 2, 1860 (Ibid., 309), as he likely would have been on his pre-*Walden* visits (elaborated later in text). Potent fires followed tree blow-downs of the famous Gale of 1815.

50. Bosses: JR, Jul 9, 1852. Seen in: Hitchcock, *Final Report,* 8a, Figs. 281 and 282. Wachusett: WT, 85. Concord, elements: JR, Oct 5, 1851.

51. Lodgment usually happens when the paste stiffens as water is squeezed from pores by rising normal stress, which is perpendicular to the surface. Hard pan: JR, Oct 2, 1851. Swells: Ibid., Jul 18, 1852. Round Hill: Ibid. Thoreau wrote "grateful" instead of graceful. Green sward: Ibid., Apr 30, 1852.

52. Water table: Technically, the glacier doesn't have a water table. Rather, it has an equipotential surface based on the sum of elevation and pressure. For every ray, the potential rises northward as the ice thickens with a ratio approximating 1/11.

53. Pothole: WK, 308. Canaan: Ibid., 310, cited Hitchcock, *Final Report.* I can't find a town of this name on the named divide between the Merrimack and Connecticut Rivers.

4. After the Deluge

1. Phi Beta Kappa Society at Cambridge, "The American Scholar," last modified August 31, 1837, Accessed September 15, 2012. http://www.emersoncentral.com/amscholar.htm.

2. Geological independence: Robert H. Dott Jr., "James Hall's Discovery of the Craton," *Geologists and Ideas: A History of North American Geology,* ed. Ellen T. Drake and William M. Jordan (Denver: Geological Society of America, 1985), 158. James Hall: Ibid., 157.

3. Charles T. Jackson, *First Report on the Geology of the Public Lands in the State of Maine* (Boston: Dutton and Wentworth, 1837), 17.

4. Committee: Charles T. Jackson, "Communication on the Subject of Drift (Abstract of the Proceedings of the Fourth Session of the Association of American Geologists and Naturalists)," *American Journal of Science and Arts* 45, no. 10 (1843): 320–323.

5. New cause: "Report read by Mr. John Hayes" published with Jackson, *Drift.* Difficulty: Merrill, *First Hundred,* 463.

6. Hills and hollows: CC, 92. Erratics: Ibid., 158. Diluvial: Ibid., 92. Chopped sea: Ibid., 93. Troubled: Hitchcock, *Final Report,* 405. Operandi: Ibid., 368.

7. Thoreau's quip: CC, 93. Mineralogy: Ibid., 158. Hitchcock's map shows the flow lines, but misinterprets them as the flow of diluvial currents.

8. Deluge: Hitchcock, *Final Report,* 350. Noah: Ibid., 403. Hitchcock's logic regarding the absence of human remains is a perfect example of the thesis

of Christian natural theology: not to test ideas, but to find concordance between infallible ideas.

9. Geological Sense: Hitchcock, *Final Report,* Postscript, A3-A11. Retaining: Ibid., 403. Comment: Ibid., 350–406.

10. Government: Hitchcock, *Final Report,* A11. Glaciers exist: Ibid., 403. Demiurgic: Ibid., 749.

11. Set back: Totten and White, *Craton,* 127. Emerson: Maynard, *History,* 20.

12. Jon D. Inners, "A Yankee saunterer: geographical and geological influences on the life of Henry David Thoreau," *Annual Meeting Abstracts with Programs, 2006, 22–25 October, Philadelphia* 38, no. 7 (Boulder, CO: Geological Society of America, 2006): 37.

13. Downmelting based on mass balance from Gustavson, *Meltwater.* Backmelting rates from more well dated ice-margin positions from Ridge, *Varves.*

14. This form of glacial recession is called stagnation-zone retreat.

15. Actual calendar dates for the passage of ice past Concord are 17.2–15.8 calendar years before present (BP) and radiocarbon dates of 13.3–14.3 ^{14}C years before present from Ridge, *Varves* (website last updated 3/30/2008).

16. Moors: JR, Apr 24, 1852. Erratics: Ibid., Jun 11, 1851. Boulders: Ibid., Jun 23, 1852. Brute life: Ibid., Apr 21, 1852. Natick: Ibid., Nov 7, 1851 (see frontispiece).

17. Mountain: JR, May 1, 1851. Ideally carved by rotational flow of cirque glaciers. Tarn: Greg Garrard, "Wordsworth and Thoreau: Two Versions of Pastoral," *Thoreau's Sense of Place,* ed. Richard J. Schneider (Iowa City: University of Iowa Press, 2000), 194. Thoreau owned (Sattelmeyer, *Reading,* 294) a copy of Wordsworth's *The Prelude, Or Growth of a Poet's Mind* (London: E. Moxon, 1850). Katahdin: MW, 58.

18. Favorite: Buell, *Imagination,* 417: "No contemporary travel book interested Thoreau more." Biography: Herbert, *Darwin,* xv.

19. Darwin, *Voyage,* dedication page.

20. Terminology: Herbert, *Darwin,* 60–62. Spectacles: Ibid., 245. Beryl-blue: Darwin, *Voyage,* 226. Icebergs: Ibid., 229. Ice cap: Ibid., 250. Vast piles: Ibid., 246.

21. Alluvium: Darwin, *Voyage,* 191. Tortuous: Herbert, *Darwin,* 245–246.

22. Copious Notes: Richardson, *A Life,* 243. Very close: Ibid., 246. Agree: John Hildebidle, *Thoreau: A Nautralist's Liberty* (Cambridge, MA: Harvard University Press, 1983) also argues for Darwin's influence. Beheld: Herbert, *Darwin,* 134 (the cited phrase is Herbert's). Sympaticos: Loren Eiseley, *The Unexpected Universe* (New York: Harcourt, Brace & World, Inc., 1964), 138–139. Eisely compared Thoreau to the "Scarecrow of Oz," if only "because the city he sought was more elusive and he did not have even the Cowardly Lion for company."

23. Walk: Herbert, *Darwin,* 130. In his conclusion to the *Voyage* [Charles Darwin, *Voyage of the Beagle* (New York: Knopf, Everyman Library, 2003),

516], Darwin described the relative advantages and disadvantages of breadth and depth in scientific studies. Thoreau may have taken this as a choice he could make. Dissipate: JR, Jan 30, 1852. Nailed down: Ibid., Nov 12, 1853. Thoreau and Darwin: Parini, *Thirteen,* 130: "Buell rightly positions Thoreau and Darwin on the same plane."

24. Slow clock: Darwin, *Voyage,* 178. Endless cycle: Darwin, *Voyage,* (Knopf), 506. Herschel: Herbert, *Darwin,* 31. Equator: Christoph Irmsher, *Louis Agassiz: Creator of American Science* (Boston: Houghton Mifflin & Harcourt, 2013), 77.

25. Unknown bird: Torrey and Allen, *Journal of HDT,* 238, Footnote 1. Phenology: JR, Jun 11, 1851. Four days later: Ibid., Jun 15, 1851. Cameron, *Writing,* 5.

26. Near their edge, ice sheets have a roughly parabolic surface profile governed by the physics of ice.

27. Cherry Brook: This was the spillway for the main stage of Glacial Lake Sudbury, named and described by James W. Goldthwait, "The Sand Plains of Glacial Lake Sudbury," *Bulletin of the Museum of Comparative Zoology at Harvard College* 42 (1905). A re-examination of the spillway was done by Carl Koteff, "Glacial Lakes near Concord, Massachusetts," *U.S. Geological Survey Professional Paper 475-C* (1963): C142–C144. The specific situations they describe are general models only because every specific situation was further complicated by moraine and drift dams. There were actually two Cherry Brook spillways very close in elevation, but they are lumped here due to isostatic tilting.

28. Brawling: JR, Nov 4, 1851.

29. Yesterday: JR, Apr 27, 1852. Perfect sea: Ibid., Jul 25, 1852. Summer sea: Ibid., Sep 8, 1851. Chile: Darwin, *Beagle* (Knopf), 267.

30. Primeval: JR, Jul 25, 1852. Goldthwait: Goldthwait, *Sand Plains*, Plate 5. His basic interpretation remains intact (with minor corrections) after major updates by the U.S. Geological Survey, first in the 1960s by Carl Koteff (Koteff, *Glacial Lakes,* C142-C144) and currently by Janet Radway Stone and Byron Stone, also of the U.S. Geological Survey. The lower level mapped by Goldthwait was renamed Glacial Lake Concord by Koteff. See also local surficial geology quadrangle mapping by Carl Koteff, *Surficial Geology of the Concord, Quadrangle, Massachusetts,* Map GQ-331, 1:24,000 (Washington, DC: U.S. Geological Survey, 1964), updated by Janet R. Stone and Byron D. Stone, *Surficial Geologic Map of the Clinton-Concord-Grafton-Medfield 12-quadrangle area in East Central Massachusetts,* Scale 1:24,000, U.S. Geological Survey Open File Report 2006–1260A (DVD-ROM), 2006. Thorson, *Beyond,* 113, describes geological work done between Thoreau and Goldthwait, notably by Warren Upham. Though Upham published work in 1877 and 1879 on the glacial drift of the region, it did not focus on Walden Pond.

31. Period: JR, Dec 5, 1852. Ghost: Ibid., Jul 25, 1852. Delicate: Ibid., Oct 12, 1852.

32. Failures: Harding, *Handbook,* 11. Shanley, *Making,* 18.

33. Salvo: Harding, *Days,* 187, Shanley, *Making,* 27. Not unified: Adams and Ross, *Mythologies,* 51. Adams and Ross cite Ronald Earl Clapper, "*Walden": A Genetic Text* (Los Angeles: UCLA Ph.D. Dissertation, 1967), adopting his designations A–C for Versions I–III, and D–G for Versions IV–VII. Diptych: Cameron, *Writing,* 24.

34. Marx: Leo Marx, "The Struggle Over Thoreau," *New York Times Review of Books* 46, no. 11 (June 24, 1999), C1–C11. Spirit: JR, Oct 12, 1851. Field Notes: Ibid., Mar 2, 1853.

35. Divide: Hodder, *Ecstatic,* 28. Beginnings: Ibid., 32. Formulations: Ibid., 254. Reorientation: Dean, *Romanticism,* 84. Literalness: Johnson, *Passion,* 191. Romantic: Adams and Ross, *Mythologies,* 6. Habits: Robert Sattelmeyer, "The Remaking of Walden," *Walden and Resistance to Civil Government: Authoritative Texts, Thoreau's Journal, Reviews and Essays in Criticism, Second Edition,* ed. William Rossi (New York: W. W. Norton, 1992), 433. Long view: Donald Worster, *Nature's Economy: A History of Ecological Ideas,* 2nd ed. (Cambridge: Cambridge University Press, 1994), 60.

36. Peck: H. Daniel Peck, *Thoreau's Morning Work: Memory and Perception in* A Week on the Concord and Merrimack Rivers, *the* Journal, *and* Walden (New Haven: Yale University Press, 1990), 46. Richardson: Richardson, *A Life,* 265. Walls: Walls, *Cosmos,* 262–263. Dean, *Romanticism,* 84.

37. Divine: JR, Aug 15, 1845. Transmission: Ibid., Jun 11, 1851.

38. Farthest heard: JR, Aug 15, 1851.

39. Transition: Harding, *Days,* 300. Field Evidence: Walls, *Cosmos,* 579. Historian: Hildebidle, *Liberty*, 148.

40. Sandy drop: JR, Mar 10, 1852. Science: Dean, *Romanticism,* 85.

41. Erratic: JR, Jun 11, 1851. Advantage: JR, Jun 12, 1851.

42. Deep Cut: Ibid., Jun 13, 1851.

43. Sinking: JR, Jun 15, 1851.

44. Stay-at-home: JR, Aug 6 and Aug 19, 1851. Discoveries: Ibid., Aug 21, 1851. Profession: Ibid., Sep 7, 1851.

45. Inundation: JR, Sep 8, 1851. Inquisitive: WA, 279. Hard-pan: JR, Oct 2, 1851. Worn smooth: Ibid., Oct 5, 1851. Torrent: Ibid., Nov 4, 1851. Moraine: Ibid., Nov 7, 1851 (Agassiz used this term and Hitchcock used it in his review of Agassiz). Elaborated: Ibid., Nov 7, 1851. Chain of Ponds: Darwin, *Voyage* (Knopf), 455. Foliage: Ibid., Dec 30, 1851.

46. Quotes: JR, Jan 5, 1852. Great slip: Darwin, *Beagle* (Knopf), 221.

47. Shrub oak: JR, Jan 14, 1852. Sunk down: Ibid. Level: Ibid. Flowed: Ibid., Jul 25. Primeval: Ibid. Period: Ibid., Dec 5.

48. Table Lands: CC, 42. Clay: Ibid., 92. Circular: Ibid. Tumble: Ibid. Ant lion: Darwin presents an amusing sketch of ant lion behavior in his *Journal* (*Voyage* [Knopf], 454), which Thoreau read before submitting his Cape Cod essays for publication in November 1852 (Harding, *Days,* 359). Run out: Ibid. Chopped sea: Ibid., 92; Hitchcock, *Final Report,* 367.

49. Glacier: JR, Feb 5, 1852. Icebergs: Ibid., Mar 4, 1852.

50. Impetus: Richardson, *A Life,* 253.

51. There were actually two spillways operating to control the level of Glacial Lake Sudbury, but the elevations are so close (when adjusted for isostatic tilting) that they are lumped here. The "younger lake" draining through Tophet Swamp was named Glacial Lake Concord by Koteff, *Glacial Lakes.* Level: JR, Apr 24, 1852. Meadows: Ibid., Apr 16, 1852.

52. Near the present site of Walden Pond, fine sand and mud deposited from suspension early in the process gave way upward to broadly sweeping lobes of rippled sand, which gave way farther upward to steeply inclined beds of sandy gravel marking the place where the incoming stream dropped its load over the delta edge and gravity took over. The tops of these inclined "foreset" beds mark the height of the lake at the time. Above these inclined beds are nearly horizontal and heavily channeled strata of the "topset," which is essentially an outwash plain built in a sector that had previously been open water. The sediment sequence is documented by Goldthwait, *Sand Plains;* Koteff, *Glacial Lakes;* and lithologic logs in Colman and Friesz, *Geohydrology,* 5 (Figure 2).

53. Delta plain: Goldthwait, *Sand Plains,* 279. Goldthwait describes sand plain elevations hand-surveyed to "200'-180" and at Walden, "195 feet." This is very accurate (Friesz and Colman, *Poster,* contour lines).

54. Irishman: Walker, *Minerals,* 5. Providence: Quoted from Shattuck, *History,* who, in his footnote 2, credits "2 Mass Hist. Coll. Vol. iii, pp. 156–159. Written about 1650."

55. Brister's Hill: Koteff, *Glacial Lakes.* Koteff maps an ice margin near there. Frosty hollow: Maynard, *Wyman,* 73. Pond Hole: WA, 239. Boiling Spring: The bubbling/swirling is due to toggled fluidization, which requires excess porewater pressure.

56. Husbandry: Brian Donahue, *The Great Meadow: Farmers and the Land in Colonial Concord* (New Haven: Yale University Press, 2004), 1. Bricks: Shattuck, *History,* 218. Shattuck reports 300,000 manufactured in 1831. Bog Ore: Ibid., 198.

57. Night Season: Shattuck, *History,* 156–159.

58. Ashamed: Herbert, *Darwin,* 284. Cwm: Ibid., 280. The phrase "Paulinian" refers to the Apostle Paul's religious conversion en route to Damascus.

59. Investing Sea: Bowles, *Finders,* 97–98 (quoting Kane).

60. Rodgers: Bolles, *Finders,* 213.

61. Edward Hitchcock, *Illustrations of Surface Geology* (New York: D. Appleton & Co, Division Phinney & Co. Boston, 1860), E-book "MOA" http://quod.lib.umich.edu/cgi/t/text/text-idx?c=moa;idno=AJP8876.0001.001 (accessed 9/15/2012), 11. Take stand: Hitchcock, *Illustrations,* 71.

62. Large Arrow: Charles Whittlesey, "On the fresh-water glacial drift of the northwestern states," *Smithsonian Contributions to Knowledge* 197 (1886), previously *Smithsonian Contributions to Knowledge* 15 (1867), contains *Map Illustrating Limits of the Glacier Drift of North America* (1864).

63. R. W. Emerson brought Bellew there in July 1855 (Maynard, *History,* 124).

5. Meltdown to Beauty

1. Valhalla: WA, 286.

2. The following geological narration, based on updated references (Stone and Stone, *Surficial;* Clark, *Maximum;* Ridge, *Varves;* Benn and Evans, *Glaciation;* and references cited in Thorson, *Beyond*) differs from that of more widely used sources (Maynard, *History,* 19; and the official web version Massachusetts Department of Conservation and Recreation, "Geology," http://www.mass.gov/dcr/parks/walden/geology.htm [Accessed September 15, 2012]. The main issues involve the mechanism of block detachment, the chronology of deglaciation, the elevation of the water, and the duration of ice-block melting.

3. Fact and myth: JR, Jan 14, 1852, and WA, 176–177, respectively. Navigators: Thorson, *Beyond,* 30. Ohio: Ibid., 116 (kettle cross-sections are exposed in sea-cliffs). Crater: CC, 105.

4. The United States Environmental Protection Agency, which is responsible for managing the quality of lakes, uses ten acres at a cutoff between lake and pond, making Walden a lake six times over. I use the count of four kettles as a reasonable minimum number. Actually, each kettle had minor accessory kettles. This is particularly clear for Wyman Meadow which is the sum of several.

5. Monotonous: WA, 180. Source passage: JR, Sep 1, 1852. Amphitheatre: WA, 185 (original spelling).

6. Eastern rim: JR, Jan 25, 1852. Rule: WA, 281.

7. Indented: JR, Sep 1, 1852; WA, 180. Triangular refers to the bulk shape defined by the bathymetry (the actual tips are rounded, as expected). Identification of coves is based on morphological contrast with central deep basin. Names follow Maynard, *History,* 22–23, except for my use of Railroad Cove, which incorporates his "Little Cove" and replaces his term "Bay." The western tip of Railroad Cove was artificially filled, likely with sediment from the Deep Cut, and was likely much more pointed.

8. Shape: Friesz and Colman, *Poster* (bathymetry). Each of its facets is a straight slope standing at an angle between about ten and fifteen degrees, which is near the angle of repose for saturated fine sand that's never been drained. This

was calculated from the steep gradient on the western submerged slope, which drops approximately 100 feet in 500 feet for a 20 percent slope, the arctangent of which is approximately 12 degrees, which is consistent with the angle of repose for never-drained saturated fine sand. Darker water: WA, 181. Haze: Water clarity due to the dearth of dissolved oxygen, which restricts phytoplankton.

9. Flat bottom: WA, 276–277. Reflection: Colman and Friesz, *Geohydrology,* Figure 2, shows a unit labeled as "till," which (given the general setting, its description as "sand and gravel," and the absence of a sharp discontinuity in the natural gamma radiation) is likely coarse melt-out till from the base of the ice block, rather than hardpan lodgment. This indicates a geothermal source for the melting.

10. Single leap: WA, 173. Specifically, the bisectors of the four coves are independent of the morphology of the inner basin. Also, each cove bottoms out at a different depth between –15 and –35 feet, which rules out an origin based on some fixed water level. Though I have definite ideas about how the coves formed, I do not address them here because they have not yet been subjected to peer review. The presence of deltaic sand and gravel between the basins suggests that melting was first concentrated there, causing sedimentation between the residual blocks as they formed. The lack of this sediment on the floors of the basins themselves indicate that little sediment remained above or within the residual blocks during their melting.

11. Old Man: WA, 185. Caught fish: Ibid., 175. Late 1853: Oct 23, 1853. Not belong: WA, 173.

12. Lowest stage: WA, 185.

13. Colman and Friesz, *Geohydrology,* 5–6 (Figures 2, 3). Elevations are based on U.S.G.S. Topographic map of the Concord Quadrangle: 1:24,000. Tectonic: Barosh, *Bedrock,* 216 (Fig 3). As one kettle, Walden's western basin manifests a radial symmetry. But as one of many kettles, it is a link in a linear chain occupying the Walden paleo-valley.

14. Opening: WA, 1. Public Road: Canby, *Thoreau,* 216 (citing Prudence Ward's letter of Jan 20, 1846). Savage: MW, 72. Hickories: WA, 40. Clearing: Ibid. Johnswort: Ibid., 110. Most scholars, however, see the line of text for what it is, as little more than the prelude to the myth Thoreau tells us he is writing. Disingenuous: Leon Edel, "The Mystery of Walden Pond," *Stuff of Sleep and Dreams: Experiments in Literary Psychology* (New York: Harper & Row, 1982).

15. Rods: WA, 110. Roland Wells Robbins, *Discovery at Walden* (Concord, MA: The Thoreau Society Reprint, 1999, originally, 1947), 10.

16. Road: WA, 126. Pond-side: Ibid., 313. This rounded bench appears to be the wave-beaten eastern rim of a subsidiary small kettle containing a swamp to the west.

17. Sketch: WA, frontispiece. Across: Ibid., 162. Trail: Canby, *Thoreau,* facing page 218 (courtesy Wayside Museums). Squatter: Maynard, *Wyman,* 62;

Henry David Thoreau, *[Ralph Waldo] Emerson Lot by Walden* (Concord: Concord Free Public Library, Special Collection "Thoreau Surveys," 1857), Dec. Map 31a, 4½" × 7½".

18. Thoreau called this terrace a "shrub oak plateau": WA, 85.

19. Trumpet: WA, 50.

20. Imagination: WA, 85.

21. Lapse of time: WA, 108.

22. Foundation: Robbins, *Discovery,* 61. With a Brunton compass, I measured the alignment of the granite posts for the house monument to be 141°, which is four degrees east of Robbin's reported alignment. With a contractor's transit set up in the doorway midway between a sitting and standing height, the modern horizontal cone of vision is 43°. The vertical range of vision from the lowest visible water beyond the foreground (–6°) to the horizon (+3°) is about nine degrees. On this azimuth, the sun at the winter solstice would be about 16 degrees above the horizon at 9:15 a.m.

23. Unburied ice blocks usually exhibit rims of debris on the adjacent terrace moved into place by mass movement from the higher block, and asymmetry in scour channeling or deposition. The central position of the block in the paleovalley and of the deep hole suggests isolated burial. The narrowness and limited mass of collapse fill between the blocks suggests it was let-down from above as one block separated into four. The tops of blocks were likely irregular.

24. Strewn: Maynard, *Wyman,* 62.

25. Koteff, *Glacial Lakes.*

26. The geometry of this meltdown is hard to visualize. Though the end result was a broadly flaring cone truncated on the bottom (tip), this resulted mostly from the meltdown of a disk-shaped mass, the edges of which slanted inward as talus slopes.

27. Thaw: WA, 298. Varves: Ridge, *Varves.* Bog-bottom: D. Peteet et al., "Delayed deglaciation or extreme Arctic conditions 21–16 cal. kyr at southeastern Laurentide Ice Sheet Margin," *Geophysical Research Letters* 39, L11706 (2012): 1–6. Vegetation adds stability by enhancing infiltration, reducing peak flood runoff, retaining loose sediment, and armoring stream banks.

28. Anticyclone: Wind speed estimates from Robert M. Thorson and C. A. Schile, "Deglacial Eolian Regimes in New England," *Geological Society of America Bulletin* 107 (1994): 751–761. Our main study site was the eastern shore of ice-dammed Glacial Lake Hitchcock in the Connecticut River Valley, then comparable to the eastern shore of Glacial Lake Sudbury. Congealed: WW, 113. Frozen dry: JR, Mar 6, 1854.

29. Nauset: JR, Apr 3, 1853.

30. Sand import: WA, 236.

31. Clay Pits: CC, 110.

32. God's Drop: WA, 187, is attributed to Emerson. Refer to Raymond Benoit, "Walden as God's Drop," *American Literature* 43, no. 1 (1971): 122–124.

33. Inequalities: WA, 279.

34. As if water: JR, Jul 6, 1853. Irishman: Ibid., Feb 17, 1854. Spruce: Ibid., May 31, 1853, and Oct 24, 1837 (one of his first).

35. Deevey, *Re-examination,* 6–7.

36. Fifty-six: WA, 278. Fire: Ibid., 169.

37. Disappeared: WA 184. Referring back: JR, May 27, 1841.

38. Geologists look for diagnostic features of the bulk sediment, for example the relative proportion of certain mineral and organic sediments, laminae of clay that might indicate shoreline erosion, or bands of silt from bottom-hugging currents. Chemists use mass spectrometers to report on the stable isotopes of elements within plant tissue, and chromatographs to investigate stable plant pigments. Taxonomists identify the remains of living things that escaped being recycled by bottom feeders. Charcoal is important too because it resists decomposition, making it a good proxy for the frequency and intensity of past fires, whether natural or anthropogenic. For limnological details, consult Robert G. Wetzel, *Limnology, Lake and River Ecosystems,* 3rd ed. (New York: Academic Press, 2001). Paleoecology: One favorite indicator is the hard mouth parts of the bottom-feeding larvae of midges or lake flies, aquatic insects known as *Chironomids.* Another favorite is a type of zooplankton called *ostracods,* crustaceans that superficially resemble tiny clams. More common are *diatoms,* a single-celled photosynthetic algae that leaves behind telltale skeletons of opaline silica called "tests."

39. Cores are from Houghton Pond in the Blue Hills south of Boston, Winneconnet Pond near Taunton, and Mirror Lake in New Hampshire. See Marjorie G. Winkler, "Changes at Walden Pond During the Last 600 Years: Microfossil Analyses of Walden Pond Sediments," *Thoreau's World and Ours: A Natural Legacy,* ed. Edmund A. Schofield and Robert C. Baron (Golden, CO: North American Press, 1993), 199–211 (Note: I have converted Winkler's radiocarbon dates to calendar ages). For a more data-rich, regional perspective, consult George L. Jacobson, Thompson Webb, and Eric C. Grimm, "Changing Vegetation Patterns," *North America and Adjacent Oceans During the Last Deglaciation* (Boulder, CO: Geological Society of America, 1987), poster. The record and sites are updated frequently for the North American Pollen Database.

40. For more detail, see Donahue, *Great Meadow,* 32–35.

41. Collection: Harding, 427. Most of the specimens were on long-term loan from Harvard, who recently reclaimed them. Archaeology: Shirley Blancke, "The Archaeology of Walden Woods," *Thoreau's World and Ours: A Natural Legacy,* ed. Edmund A. Schofield and Robert C. Baron (Golden, CO: North American Press, 1993), 242–253. Skeletons: JR, Mar 24, 1853.

42. Persia: JR, Dec 29, 1853. Implement: Ibid., Oct, 21, 1842. Arrowhead chips: Ibid. Vicinity: Ibid. Indian hoe: Ibid., Jun 24, 1853. Rain: Ibid., Apr 19, 1852. Plow: Ibid., May 28, 1853, and Jun 2, 1853. Prophesy: Ibid., Feb 13, 1851. Earthworks: Ibid., Jun 26, 1852.

43. Blancke, *Archaeology.*

44. Shattuck, *History,* 3. Thoreau: Blancke, *Archaeology,* 5. In addition to the freshwater mussels that provided much of the sustenance, the leftovers included turtle shells, and the bones of upland mammals, birds, and fish. There were also non-food artifacts such as the fire-cracked rocks and carved stone bowls that say something about food preparation. Also present were axes and adzes for working with wood, scrapers and perforators for working skin, plummets for weighting down nets and fishing lines and projectile points—arrows and spears—for hunting. We can only speculate on what is missing, the shelters they lived in, perhaps made of bark, reed, or skins. The wild plants they collected, nuts, berries, tubers, grains, and herbs.

45. Barrenness: MW, 143. Burrow: WA, 37.

46. Subsoiling: Ibid. Plants a town: WK, 58. Virgin soil: Ibid.

47. Dust flies: Ibid., May 6, 1852. Years ago: Ibid., Nov. 9, 1850. Overgrazing: WK, 245. Stone wall: Ibid., Apr 10, 1853.

48. Garden: JR, Jan 8, 1853. Deforestation: Gordon G. Whitney, and William C. Davis, "Thoreau and the Forest History of Concord, Massachusetts," *Journal of Forest History* 30 (April 1986): 70–81. As early: JR, Jun 4, 1853. No natural: WA, 255. Pebbly sand: Ibid., 189; JR, Nov 12, 1853. Worthless: Donahue, *Great Meadow,* 30. Forests: WA, 184.

49. Great Fields: JR, Jun 10, 1853. Deer: Ibid., Aug 19, 1852. Impenetrable: Ibid., Jul 16. 1852. Primitive: Ibid., Jun 26 1852.

50. Extensive: Ibid., Mar 23, 1854. Solitude: Ibid., no date, 1850. Woods: Ibid., Jan 26, 1853. Path: Ibid., Jan 26, 1853. Saunter: Ibid., Apr 26, 1841.

51. Pollen update: Drte Koster et al., "Paleolimnological assessment of human-induced impacts on Walden Pond (Massachusetts, USA) using diatoms and stable isotopes," *Aquatic Ecosystem Health & Management,* 8(2005): 117–131. Most of the analysis postdates *Walden* and is therefore not included here. See also E. K. Faison et al., "Early Holocene Openlands in Southern New England," *Ecology* 87 (2006): 2537–2547.

52. Grain: JR, May 6, 1852. Paddled: JR, Jan 25, 1852. Scene: Harding, *Days,* 4; JR, Aug 6, 1845. There is a rise in silicate content about 7–8 inches down likely due to increased dustiness caused by tillage. This is consistent with Thoreau's report (WA, 35) that "no dust gathers on the grass, unless where man has broken ground."

53. Know: Cavell, *Senses,* 63. Book: Ibid., 117.

6. The Walden System

1. Best book: Buell, *Imagination,* 11; Marx, *Machine.*

2. Rivet: WA, 321. Book title: Richard White, *The Organic Machine* (New York: Hill and Wang, 1995).

3. Engagement: Kuhn, *Autobiography,* 144. Though the scholarly literature contains data and more focused reports on Walden geohydrology and limnology, I restrict my analysis to Colman and Friesz, *Geohydrology* (2001) because it is more than sufficient for my purposes and is widely available and frequently used by Thoreau scholars. New data is not necessary because my period of interest ends in 1854. Thought: Francis Bacon, *Sylva Sylvarum* (London: William Lee, 1639).

4. Illumines: WA, 10.

5. Nauset: JR, Apr 3, 1852.

6. Ripples: JR, Sep 4, 1851. Fingers: Ibid.

7. Title: Harding, *Days.* Bosom: JR, May 5, 1852.

8. Anabolism/catabolism: Sven E. Jorgensen, *Evolutionary Essays: A Thermodynamic Interpretation of the Evolution* (London: Elsevier, 2008), 6.

9. Dimple: WA, 181.

10. Image: WA, 278. Resort: Ibid., 187. The longer lines could have been drawn either inward or outward. Shorter lengths must have been drawn from the center because they were measured from that origin.

11. On the headlands, rays rise and steepen from below to reach a maximum slope midway up the exposed bank before diminishing toward the top. In the coves, the rays rising out of the flat bottom steepen until they reach the floors of the coves before flattening and developing a second concavity higher up. Fluviatile: WA, 180. Midwinter: Ibid., 173. Nitella: Colman and Friesz, *Geohydrology,* 39 (Fig. 16). Newburyport: JR, Dec 6, 1850.

12. Word list is from various portions of *Walden.*

13. Here *delta plain* is shorthand for that portion of the delta plain above the groundwater divide between the kettle and adjacent drainages.

14. Cauldron: JR, Jul 16, 1851. Geomorphology: Anderson and Anderson, *Geomorphology.*

15. Warmest: WA, 218–219. For thermal behavior and stratification, compare Colman and Friesz, *Geohydrology*, and Deevey, *Re-examination*, against the range of temperature measurements in *Walden.*

16. So deep: WA, 23. Chilling: Masses chilled by conduction sink downward to mix and cool the entire epilimnion. General freezing: Ibid., 236–237. Compare with Deevey, *Re-examination,* 5: "Because of its unusual depth and sheltered situation, Walden Pond has an exceptionally cold hypolimnion," in spite of the fact that it "acquires a surprising amount of heat in the summer," estimated to be "17,300 calories per square centimeter of lake surface."

17. Shutter: WA, 184. Eyelids: Ibid., 274. Best in town: Ibid., 176.

18. Breaks up: WA, 289.

19. The waves are sinosoidal and symmetrical with orbits diminishing with depth.

20. Chemistry: slightly acidic, soft water, and of limited hardness. Ascetic: WA, 172. Not fertile: Ibid., 178. Frogs: WA, 172.

21. Shiners: WA, 170. Ascetic: Ibid., 172. Handsomer: Ibid., 179. Boast: Ibid., 178. Thoreau's fish: WA, 178. No fish: Shattuck, *History,* 200; Koster, *Paleolimnology.* Animals: Ibid., 178.

22. Barrel-ful: WA, 307. Deevey, *Re-examination,* 7, reports a bottom fauna dominated by Chironomid larva. The phosphorous budget is discussed by Colman and Friesz, *Limnology.*

23. Stagnant: At that point, oxygen-using (aerobic) bacteria yield to those that operate without oxygen (anaerobic). Samples taken at this time show an absence of oxygen and an abundance of iron, manganese, and ammonia nitrogen.

24. Refrigerators: JR, Jun 18, 1853, and Jun 11, 1851. Vitality: Ibid., Jun 28, 1852. Dog: Ibid. Distinct: Ibid., Aug 31, 1851. Twice as far: Ibid., Feb 5, 1853.

25. Anvil: JR, Dec 12, 1841.

26. Depth: WA, 176.

27. Undulated: WA, 283.

28. Birds: WA, 178.

29. Milkweed: WA, 224. Sinks: JR, Sep 1, 1852.

30. Weald: Maynard, *History,* 17.

31. Tap root: JR, Jan 8, 1853.

32. Humus: Thoreau wrote about humus in his first journal entry about nature. It takes a century or more to build up a good layer in settings like Walden, after which it assumes a steady-state thickness and composition. Fattened: WA 235.

33. Angleworms: WA, 215.

34. Amount: WA, 170. West: WA, 85. Dry-shod: JR, Nov 14, 1853.

35. Aquifer: defined as any mass of earth, porous rock, or fractured rock capable of holding and yielding a significant amount of water. Two simple models help us understand the Walden aquifer. (1) The first involves a clear straight drainpipe filled with saturated fine sand, and with a drain at the bottom. Tipping the pipe more steeply causes the water to run out more quickly. This is the control of gradient. Replacing the fine sand with coarse sand causes it to drain more quickly. This is the control of permeability. Enlarging the diameter of the pipe increases the discharge. This is the control of aquifer cross-section. Combining these terms, the strongest groundwater flows toward Lake Walden occur where the gradient is steep, the material is coarse, and the cross-sectional area of the flow is high. (2) A second model is a clear flexible U-tube with the "U" filled two-thirds of the way up with saturated, very coarse granular sand. Water poured in one end will immediately force a rise in the water level on the other end, causing them to be equal on both sides. This sediment is too coarse to maintain a difference in pressure. If the same tube is filled with fine-to-medium sand like that below Walden, however, it

will pool up on one side, which raises the differential pressure, which then forces the water down one side of the U-tube and up the other side until the pressure gradient is equalized. This process might take several minutes, and will decline in rate as the difference in pressure gets smaller. This delayed and diminishing control of limiting permeability, when operating at the scale of Walden's aquifer, ensures continuous groundwater flow between more rapid infusions of rain or snowmelt. Effectively, the aquifer drainage is never complete.

36. The fining-downward increases the lag times between recharge and discharge and extends the duration for each recharge event. Level: WW, 130.

37. During drenching rains, the lake surface rises directly like a rain gauge, creating a transient layer over the whole system that drains outward in all directions, even to the east, because that same infusion on land has not yet had time to work its way downward to the water table and then sideways to the pond. By the time the aquifer-delayed infusion arrives, however, its height will have been magnified by a factor of three or four because the average porosity of the aquifer seldom exceeds 20 percent. Thus, an inch of rain stored on the lake takes five inches to store in the land. This provides yet another mechanism that ensures continuous flow.

38. Residence time: Colman and Friesz, *Geohydrology,* 17. Balance: Ibid., 17 (Figure 9). The figure shows the measured balance as 55 and 45 percent for groundwater inflow and precipitation, and 74 and 26 percent for outflow and evaporation, respectively. Volume conversion is 3,211,450 cubic meters.

39. Pretend: WA, 175.

40. Sky water: WA, 182. Spring: Ibid., 170. Many water molecules take a shortcut that moves from air to lake as direct precipitation and from lake to air as direct evaporation.

41. Epicycles: Miller, *Consciousness,* 198 (Journal of Dec 27, 1840). Pulse: WA, 173. Seasonal: Ibid., 175. Twenty-five: Ibid.

42. Average elevations from Colman and Friesz, *Geohydrology,* 6 (Figure 3). From the same report, the highest stage of 49.8 meters occurred in 1956, and the lowest of 46.34 meters in 1967.

43. Even Lake Superior, he noted, rose and fell on Walden's long-term schedule. Sympathize: WA, 175. Maynard, *Wyman,* 8, reports that 1965 had the "least rain since 1822. The following year, 1966, was the fourth parched one in a row, and Walden shrank to six feet below 'normal.' The sandbar stood a foot high and a hundred feet long; two trucks could drive abreast all around the shoreline."

44. Other negative feedbacks involved: various attributes of the woodland and its soil; the storage of water within the canopy, snowpack, and unsaturated earth; and the rate-limiting permeability of the aquifer. Groundwater arrivals,

even from rainfalls of uniform distribution, were desynchronized due to differences in flow path length, aquifer permeability, and water table gradients.

45. Fiction: Walls, *Seeing,* 157.

7. Sensing *Walden*

1. Delicious: WA, 125.

2. Trumpet: Emerson, *Eulogy,* 22. Cook: Cook, *Passages,* 144. Friesen: Victor Carl Friesen, *The Spirit of the Huckleberry: Sensuousness in Henry Thoreau* (Edmonton: University of Alberta Press, 1984), xiv and 106, respectively. Epistemology: Walls, *Seeing,* 126. Canonical: Hodder, *Ecstatic,* 71.

3. Woods: WA, 1. Prelude: J. Lyndon Shanley, *The Making of "Walden," with the Text of the First Version* (Chicago: University of Chicago Press, 1957). Excruciating: Adams and Ross, *Mythology,* 58 (Figure 2) and 59. Organicism: Ibid., 166.

4. Blue: JR, Jan 9, 1852.

5. Observation: JR, Jul 2, 1852. Archaeology: JR, Feb 13, 1851. Bottom: Ibid., Feb 22, 1845–47. Cliffs: Ibid., May 10, 1853. Look experiment: Ibid., Apr 10, 1852. Pitch Pine: Ibid., June 23, 1852. Bubbly ice: Ibid., Jan 8, 1853. Walden bubbles: WA, 238. Three ices: JR, Jan 27, 1854, WA, 286.

6. Strains: JR, Jul 16, 1851. Looking: Hodder, *Ecstatic,* 291 (PJ5.343–344). Sympathy: NH, 72 (cited in Worster, *Ecology,* 96). Schneider, *Optics,* provides a good example of this non-thinking sensuality because it deemphasizes the physics to the point of omitting phenomena such as polarization and scattering.

7. Light and color: M. Minnaeret, *The Nature of Light & Colour in the Open Air,* trans. H. M. Kremer-Priest, rev. K. E. Brian Jay (New York: Dover, 1954). Whole book: Friesen, *Huckleberry.*

8. Integral: Harding and Bode, *Correspondence,* 222 (May 2, 1848). Mock sun: WA, 132. For light physics, refer to Wetzel, *Limnology,* and Minnaeret, *Light & Colour.*

9. Flashing: JR, Sep 26, 1852. Nebulous: JR, Sep 7, 1851. Borealis: Ibid., May 19, 1852.

10. Grandest: JR, Jul 25, 1852. Artist: Ibid., Jun 13, 1851. Ethereal: JR, Jan 2, 1854. Vitreous: Ibid., Feb 14, 1852. Blue oases: Ibid., Jun 12, 1853.

11. Storm: JR, Aug 31, 1852.

12. Water color: WA, 170–172.

13. Vanished: JR, Nov 28, 1837. Angry flood: Ibid., Apr 16, 1852. Ink: Ibid., Mar 22, 1853. Shallows: Ibid., Apr 28, 1852. Vivid green: Ibid., Sep 11, 1852.

14. Blue/green: WA, 171. Bluer: Ibid., Oct 26, 1852. Dark green: WA, 171; JR, Aug 27, 1852. Walden green: JR, Jun 17, 1853. White green: Ibid., Jan 24, 1854. Serene: Ibid., Jan 3, 1853. Hollows: JR, Feb 9, 1854.

15. Storm: JR, Feb 9, 1854. Turbid: Ibid., Jun 25, 1852. Better: Ibid., Sep 1, 1852.

16. Luminescent: Lakes and cat's eyes are not luminous, though the combination of reflection and reflection can make them appear to be. Angelo: WA, 172. Refractive bending is governed by Snell's Law.

17. Lakes and flakes: WA, 192 and 183, respectively.

18. Heavens: JR, May 5, 1852. Blackness: Ibid., Aug 31, 1852; Glassy: Ibid., Sep 2, 1852. Leaves: Ibid., Jun 23, 1852.

19. Woke me: JR, Sep 2, 1851. Halo: Ibid., Aug 29, 1851.

20. Invert: JR, Jun 26, 1853. Schooner: Ibid., Aug 31, 1852. Polarization: When light arrives perpendicular to the surface of pure water, only 2 percent is reflected. At lower angles, an increasing proportion of photons are reflected, especially in the plane parallel to the surface of a lake, creating glare. Polarization begins at an angle of incidence of about 40° and diminishes below 60° when virtually all the light is being reflected.

21. Rainbow: JR, Jun 22, 1852. Halo: WA, 195. Colors: JR, Dec 3, 1853. Flower: Ibid., Dec 2, 1853. Invisible flame: Ibid., Mar 4, 1854.

22. Confusedly: JR, Feb 7, 1854.

23. Shimmering: JR, Jun 13, 1854. Medium: Ibid., Jun 18, 1853. Haze: Ibid., May 3, 1852. Gold: Ibid., Jul 27, 1852. Shadow: Ibid., Jun 13, 1851. Moon: Ibid., Mar 28. 1852.

24. Glory: JR, Jun 13; 1854. Dusky: Ibid., Dec 8, 1853. Heavenly: Ibid., Aug 3, 1852.

25. Coldest: JR, Jan 2, 1854. Beheld: Ibid., Feb 4, 1854. Empurpled: Ibid., Oct 5, 1852.

26. Unvaried: JR, Feb 19, 1852. Emerald: Ibid., Jan 30. Cobwebs: Ibid., Feb 12, 1854. Crevices: Ibid., Feb 18, 1852. Winters: Ibid., Feb 21, 1854.

27. Questions: JR, Feb 25, 1851.

28. Reductionism: Physicists use the term differently, to reduce the complexity of nature down to general laws, which is synonymous with induction. Honeybees: JR, Feb 13, 1852. Sketching: Ibid., Mar 28, 1852.

29. Ice: JR, Dec 31, 1853. Slush: Ibid., Feb 28, 1854. Dark ripples: Ibid., Nov 11, 1853. Bird's tracks: Ibid., Nov 11, 1853. Sheets: Ibid., Dec 8, 1853.

30. Creature: JR, Jan 1, 1853. Cannon: Ibid., Dec 12, 1853. Frog-like: Ibid., Dec 25, 1853. Voices: Ibid., Feb 12, 1854. Thunder: Ibid., Feb 12, 1854.

31. Cracks: JR, Jan 4, 1854. Flake-like: Ibid., Jan 9, 1853. Conchoidal: Ibid., Feb 7, 1854.

32. Bubbles: JR, Jan 7, 1853. Holes: Ibid., Nov 19, 1851. New layers: Ibid., Jan 7, 1853. Glaze: Ibid., Dec 11, 1853. Floes: Ibid., Dec 28, 1853. Marbling: Ibid., Dec 11, 1853.

33. Curves: JR, Mar 20, 1853. Thrusts: Ibid., Feb 5, 1853.

34. Icy coat: JR, Jan 1, 1853. Waterfall: Ibid., Jan 26, 1853. Bedposts: Ibid., Jan 11, 1854. Splash: Ibid., Mar 22, 1854.

35. Consilience: Wilson, *Consilience,* 8.

36. Green stone: JR, Aug 18, 1852. Oceans: Ibid., Mar 11, 1852. Rainbows: Ibid., Jun 24, 1852. Flowers: Ibid., Dec 2, 1853. Footprints: Ibid., Jan 8, 1852. Beggar tick: Ibid., Jul 25, 1853. Rain-splashed: Ibid., Sep 19, 1850. Contracted: Ibid., Feb 9, 1854.

37. Boat: JR, Feb 18, 1852.

38. Clouds: JR, Jul 10, 1851. Too cold: Ibid., Apr 11, 1852.

39. Proportion: Harding and Bode, *Correspondence,* 247 (August 10, 1849, letter to Mr. Black). Intelligence: JR, Jan 1, 1854.

40. Higher Laws: WA, 20. God: Jesse Bering, *The God Instinct: The Psychology of Souls, Destiny and the Meaning of Life* (London: Nicholas Brealy Publishing, 2010). Paleolithic: S. Boyd Eaton, Marjorie Shostak, and Melvin Konner, *The Paleolithic Prescription: A Program of Diet & Exercise and a Design for Living* (New York: HarperCollins, 1989).

41. Discovered: JR, Mar 19, 1851. Workman: WA, 130. Obey: JR, Feb 27, 1851. Government: Ibid., Jan 1, 1854. Lawless: Ibid., Feb 27, 1851.

42. Moonlit: JR, Dec 3, 1853. Waves: Ibid., Apr 3, 1853.

43. Snow waves: JR, Jan 22, 1852. In fields: Ibid., Jan 26, 1853 (the bracketed [to] was Thoreau's error corrected by the editors of the 1906 edition). Track: Ibid., Feb 7, 1854. Angle strikes: Ibid., Jan 26, 1853. Good study: Ibid., Jan 13, 1852 (and next three questions). Imbricate: JR, Jul 10, 1851.

44. Hog Island: JR, Jul 25, 1851. Ripple: Ibid. Flints: Ibid., May 20, 1853. Dunes: CC, 145. Water and sand: JR, Feb 7, 1854.

45. Workshop: Miller, *Consciousness,* 192 (Journal of Dec 15, 1840).

46. Shake: JR, Sep 19, 1850. Acoustics: WK, 488. Belfries: JR, Jan 2, 1842. Bedford: Ibid., Jan 21, 1853. Celebration: Ibid., Nov. 15, 1853. Loudest: Ibid., Jun 18, 1853. General: Refer to Bryan C. Pijanowski et al., "Soundscape Ecology: The Science of Sound in the Landscape," *Bioscience* 62, no. 3 (2011): 203–216.

47. Din: JR, May 20, 1853. Roar: Ibid., Aug 9, 1851, and Nov 9, 1853. Tempest: Ibid., Aug 9, 1851. Falling: Ibid., Mar 9, 1852.

48. Rotundo: JR, n.d. 1845.

49. Ring: Ibid., n.d. 1845. Echoes: Ibid., after Sep 19, 1850; WA, 169. Fiction: Hodder, *Ecstatic,* 78.

50. Musical: JR, Apr 12, 1852. Previous word choices were Thoreau's, though not in quotes. Wolf: Ibid., Oct 8, 1852; WA, 226.

51. Stem: JR, Jan 13, 1854. Harp: JR, Mar 12, 1852. Resonates: Ibid., Sep 23, 1851. Insects: Ibid., Jun 11, 1852.

52. Crystallization: JR, Jul 7, 1853. Crystals: Ibid., Feb 5, 1840. Coarsest: Ibid., Nov 20, 1851. Grinder: Ibid., Aug 5, 1851.

53. Basking: JR, Mar 22, 1840. Fireside: WA, 231. Southerly: Feb 12, 1854. November: Ibid., Nov 8, 1850. Leaves: Ibid., Feb 12, 1854.

54. Cellar: JR, Jun 19, 1852. Really cold: Ibid., Jun 15, 1852. Warm springs: Ibid., Mar 20, 1853 (general discussion of thawing differences).

55. Dark: JR, May 16, 1851. Hand: Ibid., Aug 12, 1851. Elevation: Ibid., Feb 22, 1854.

56. Twenty-one: JR, Feb 27, 1854. Temperature: Ibid., Apr 13, 1852.

57. Hound: Henry S. Salt, *Life of Henry David Thoreau,* edited by George Hendrick, Willene Hendrick, and Frit Oehlschlager (Urbana and Chicago, University of Illinois Press, 1993, originally London: Bentley, 1890), 61. Quoting Moncure Conway, *Fraser,* 1866. Musty: JR, Sep 2, 1853. Grapes: Ibid., Oct 9, 1853. Pipe: Ibid., Nov 2, 1853. Assurance: WA, 307. Carrion: Ibid., Jul 21, 1852. Stratification: Ibid., Jun 28, 1853.

58. Meadows: JR, May 16, 1852. Walden: Ibid., Jun 11, 1852.

59. Palate: JR, Dec 31, 1953. Huckleberries: WA, 168.

8. Writing *Walden*

1. Cavell, *Senses,* 120; Ibid., 20.

2. Gem: WA, 190–191.

3. Railroad: WA, 190. Forty: Ibid., 190. Monsters: Ibid., 192.

4. Forty Acres: WA, 190. Light: Ibid., 192. Chopper: Ibid., Sep 29, 1851. Twin: WA, 191.

5. Cliffs: The corridors are the railroad and the river. Refer to the surficial geology map of Concord in this manuscript (Figure 15). Rambles: n.d., 1850. Essay: Harding, *Days,* 69. Staten: Sanborn, *Henry,* 122. Vision: JR, Feb 3, 1852.

6. Description: WA, 168. Flow of time: Ibid., 96.

7. Fundamental: Darwin, *Voyage* (Knopf), 295.

8. Crack: WA, 182. Unsoiled: JR, Sep 1, 1853.

9. Daily: WA, 186 (full span from 1822 to 1845, almost daily span 1845–1854). Live there: Sanborn, *Henry,* 11. Extended world: JR, Aug 6, 1845; Harding, *Days,* 13–14; Richardson, *A Life,* 88. Forest: JR, May 27, 1841.

10. Mistress: Miller, *Consciousness,* 198 (footnote to Journal entry of Dec 27, 1840). Phrases: Harding, *Handbook,* 46–47. Bare in places: WW, 123.

11. Selvage: WA, 180. Ago: Ibid., Jun 14, 1853. Shore: Ibid., Feb 5, 1853. Autonomous: Cameron, *Writing,* 22.

12. Pickerel: WA, 275. Tripe: JR, Jan 13, 1854. Bank: WA, 176. Licked: Ibid. Feb 5, 1853.

13. Resistless: WA, 128. Thunderbolt: JR, Dec 26, 1853. Crashing: JR, Jun 27, 1852.

14. Acute personal losses, social isolation, estrangement with R. W. Emerson, and nightmares of dying were very much a part of Thoreau's psyche. Harbor: WA, 164. Port: Ibid., 20. Embosomed: JR, Jun 20, 1853 (word refers to

a different lake). Skin-flint: Ibid., 189. Waves: Ibid., 188. Seashore: Ibid. Them: Ibid. Freshets: Ibid., 306.

15. Encircling: JR, Nov 4, 1852. Archipelago: MW, 31.

16. Winnowed: JR, n.d., 1845–47, Torrey and Allen, *The Journal of HDT,* 121 (V I, 413).

17. So forth: JR, Aug 3, 1852. Erase: Ibid., Dec 27, 1853.

18. Poem 1494: Emily Dickinson, *The Complete Poems of Emily Dickinson,* ed. Thomas H. Johnson (Boston: Little, Brown and Company, 1960, written 1880?), 629. Sentence: Ibid., Jul 10, 1852.

19. Holist: Walls, *Seeing,* 13.

20. Deep: WA, 170.

21. Theory: Jeffrey C. Cramer, "Introduction" to Henry David Thoreau, *I to Myself: An Annotated Selection from the Journal of Henry David Thoreau* (New Haven: Yale University Press, 2007), xviii.

22. Radiated: WA, 78. Freshets: Ibid., 324.

23. Pure: WA, 188.

24. Rumbling: WA, 123. Source entry: JR, Aug 15, 1845. Acoustics: Ibid., Jun 11, 1851.

25. Fish: WK, 25–34. Potholes: Ibid., 307–311. Observations: David R. Foster, *Thoreau's Country: Journey through a Transformed Landscape* (Cambridge, MA: Harvard University Press, 2001), xii. Philosophizing: Johnson, *Passions,* 200.

26. Source: JR, Apr 27, 1854.

27. Ice: WA, 336–238; JR, Dec 25, 1851. Angleworms: WA, 215.

28. Technically, Walden does "flow," but too slowly to see. The residency time for an average molecule is nearly five years. Mountains: WA, 84.

29. Zephyr: WA, 185. Radiation contrasts: Thermal contrasts are muted by the absence of deciduous leaves and because radial flow is overwhelmed by both greater prevailing storminess and the stronger asymmetry in heating in south-facing vs. north-facing banks, an effect strengthened when the solar zenith is lower.

30. Source Passage: JR, Jan 25, 1852. This recapitulates earlier passages.

31. Cool it: JR, Jul 10, 1852. Evening: Ibid., Jul 21, 1853. Blurred: Ibid., Sep 20, 1852. Season: Ibid., Aug 19, 1853.

32. Leap: WA, 173. Wafted: Ibid.

33. Wind data: "Windfinder," http://www.windfinder.com/windstats/windstatistic.htm (data for Logan Airport from 1/2007 to 5/2012 using winds at or above 4 on the Beaufort scale. Accessed April 2012).

34. Neva: WA, 20. Potomogeton: Ibid., 173.

35. Unobstructed: WA, 176.

36. Flatulency: WA, 263–264. Sensitive: Ibid., 291. Exactly: JR, Feb 12, 1854. The coefficient of thermal expansion for ice is about 5×10^{-5} deg^{-1}, which would yield about two feet of expansion at Walden for a drop of 20 degrees Celsius.

37. Whoop: JR, Dec 25, 1853.

38. Encroachments: WA, 176. Source: JR, Dec 5, 1852. Heaved up: Ibid., Feb 5, 1853.

39. Forced: JR, Feb 5, 1853. Raised: Ibid. Crumbling: Ibid., Mar 21, 1853. Waves: Ibid., Dec 5, 1852.

40. Paving: WA, 173. Paved: Ibid., 176. Hitchcock, *Final Report,* headings on page 303. My written sketch of the process here is greatly oversimplified.

41. What you see: Ibid., Aug 5, 1851; Marx, *Machine,* 390. Model: Hawking and Mlodinow, *Grand Design,* 7. See also Kuhn, *Structure,* 113: "What a man sees depends both upon what he looks at and also upon what his previous visual-conceptual experience has taught him to see." See also B. Kuhn, *Autobiography,* 15.

42. Inaccessible: Cameron, *Writing,* 104. Predicated: Ibid, 6. Distillations: Hodder, *Ecstatic,* 254. Characteristic: Peck, *Morning.* Splintered: Cameron, *Writing,* 22.

43. Abstraction: JR, Mar 29, 1853.

44. On the unstable early Earth, life may have had a sputtering start. It may have begun and gone extinct, only to be rebooted again. Thus the plural "origins of life." Thoreau and Walden were microcosmic of New England life as summarized by John Demos, *Circles and Lines* (Cambridge, MA: Harvard University Press, 2004).

45. Page: Friedrich, *Gita,* 112. Fossil: WA, 298. History: Schama, *Landscape,* 574.

46. Fresh: Schama, *Landscape,* 182. Eternity is an important subject in Walden, present as early as "Economy," i.e., WA, 7.

47. Similar: WA, 308. Left woods: Ibid., 313. Deleted: Van Doren, *Annotated,* 141.

48. Pulse: WA, 172–173. Gun: Ibid., 291. Influence: Ibid., 263.

49. Young: WA, 186.

50. Zigzag: JR, Aug 26, 1853. Thoreau discovered this fractal pattern while surveying. It may have involved meltdown processes associated with glacial crevasses.

51. Imperialism: Ibid., 270. Sesame: Charles R. Anderson, *The Magic Circle of Walden* (New York: Holt, 1968), 213. Radiate: Ibid., 222. Praise: Ibid., 233. One centre: WA, 11. Holism: Ibid., 187. Two diameters: Ibid., 281. Eightieth: WA, 237. Trembling: Ibid., 180.

52. Orbital: Earth's late stage collision with a Mars-sized proto-planet gave earth its significant wobble and tilt, changed its rate of spin, tweaked its orbit, and gave birth to the moon. Without these events, there would be no cycle of the day, no cycle of the moon, no cycle of the seasons, and no cycle of the glaciations.

53. Sailing: WA, 309. Across-lot: Ibid., 17. Obfuscated: Buell, *Imagination,* 185. Shore: WA, 170. World: Ibid., 126. Lonely: Ibid., 132.

54. Squatted: WA, 86. Conflation: Buell, *Imagination,* 304.

55. Description: WA, 170. Feet: Ibid., 172. Bottomless: Ibid., 183. Welled up: Ibid., 183. Spring: Ibid., 170. Crystal: James A. Papa Jr., "Water Signs: Place and Metaphor in Dillard and Thoreau," *Thoreau's Sense of Place,* ed. Richard J. Schneider (Iowa City: University of Iowa Press, 2000), 72. Spirit: WA, 182. Buoyancy: Ibid., 85. God's Drop: Ibid., 187. Heaven: Ibid., 84. Scarcely: Ibid., 169–170. Friendship: Ibid., 187.

56. Drenching: Papa, 95. Religious: Ibid., 85. Azure: Ibid., 132. Luxuriant: JR, Aug 31, 1852.

57. Ground: WA, 35.

58. Furrows: WA, 237. Quartz: Ibid.

59. Conspiracy: JR, Mar 7, 1841. Lasts: Ibid., Sep 13, 1852. Economy: Ibid., Aug 26.

60. Fish: WA, 300. Ruffled: Ibid., 125. Chinks: Ibid., 275.

9. Interpreting *Walden*

1. Limestone: WA, 35. Sister: Harding, *Days*, *264*.

2. Thousands: Walker, *Minerals,* 3. Marble is also reported to outcrop in Conantum, much nearer to Thoreau's house site.

3. Harvesting mussel shells for lime burning was illegal in colonial Concord. It is unclear what he made his plaster with. He could have used lime, but he also had gypsum in his rock collection.

4. Disgust: WA, 35. High: Ibid., 47. Materials: Ibid., 236. Boulders: Robbins, *Discovery,* 30. Dollars: "Measuring worth," http://www.measuringworth.com/uscompare/relativevalue.php (accessed August 12, 2012). Burned: WA, 236. Housewife: Laura Dassow Walls, "Walden as a Feminist Manifesto," *Interdisciplinary Science, Literature, and Environment (ISLE)* 1, no. 1 (1993): 137–144.

5. Institute: WA, 50. Chelmsford: Walker, *Minerals,* 2.

6. Effete: WA, 150. Yellow: Ibid., 151. Glaciofluvial: Thomas Pergallo, "Soils of the Walden Ecosystem," *Thoreau's World and Ours: A Natural Legacy,* ed. Edmund A. Schofield and Robert C. Baron (Golden, CO: North American Press, 1993), 256, 258. Online mapping updates by the U.S. Natural Resources Conservation Service (the successor to the Soil Conservation Service) show Windsor soil is restricted to that part of the Bean Field nearer the road.

7. Divined: JR, Jun 15, 1852. Mingled: WA, 177.

8. Cellar: WA, 43.

9. Cavity: WA, 283.

10. Vale: WA, 119. Drummer: WA, 317.

11. Survey: WA, 276. Wolmer: Gilbert White, *The Natural History of Selborne and the Naturalist's Calendar, A New Edition,* ed. G. Christopher Davies (London: Frederick Warne and Co., 1879). Selborne: Chura, *Surveyor,* 111.

Anchor: Laura Zebuhr, "Sounding Walden," *Mosaic: A Journal for the Interdisciplinary Study of Literature* 43, no. 3 (Sept. 2010): 36.

12. Fathomed: WA, 276. Chisel: WA, 173. Henry used both an ax and an "ice chisel," with which he likely cut more than 100 holes for the pond survey. Chura, *Surveyor,* 24 claims the holes were cut with an ax, though this is far less effective. Map: Chura, *Surveyor,* 23. Sounding: Elaborate discussion of soundings were not included until the fifth draft according to Shanley, *Making,* 65.

13. Mapped: WA, 279.

14. Apart: Chura, *Surveyor,* 42 (baseline between points B and D). Process: Ibid., 22–44, provides a good summary of the survey strategy.

15. Interpretations: WA, 276–282. My text departs from the actual chronology: the survey in February of 1846; the general conclusions of *Walden* in 1849; the pre-planned survey of White Pond in February 17, 1851; and reading Gilpin in July 26, 1852. Method: see "Material to Mythology" chapter in Chura, *Surveyor.* Mythology: Ibid., 22–44. See also James G. McGrath, "Ten Ways of Seeing Landscapes in *Walden* and Beyond," *Thoreau's Sense of Place,* ed. Richard J. Schneider (Iowa City: Iowa University Press, 2000), 149–164 ("a truly scientific project of formulating hypotheses and laws"), and Daniel B. Botkin, *No Man's Garden: Thoreau and a New Vision for Civilization and Nature* (Washington, DC: Island Press, 2001), 73.

16. It is unclear whether Thoreau made his map before laying out the transects. Arguing for this is the fact that Thoreau liked to work from a map. Arguing against it is the fact that his survey transects do not represent maximum length and width.

17. Depth: WA, 276–277. Angle: Ibid., 277.

18. Plate: WA, 277. Chasm: Ibid. Shower: Ibid., 279. Ocean: Ibid.

19. Regularity: Shanley, *Making,* 199, documents the word "astonished" in Version I, relative to "surprised" in Version VII. Deepest Part: Shanley, *Making,* 199, uses "near the middle." Plough: WA, 279. Channel: Ibid.

20. Coves: WA, 280. Diameters: WA, 281 (technically, the lines are one diameter and a line through three basins that aligns with a diameter).

21. Mountains: Ibid., 280. Buttresses: JR, Sep 7, 1852. Pond: WA, 280 Except for Thoreau's Core, these are not really bars, but straight drop-offs between the coves and the main basin.

22. Sea-coast: WA, 280. Formula: Ibid.

23. White Pond: WA, 280. Sixty feet: Ibid., 281.

24. Point: WA, 281.

25. Actual: Sattelmeyer, *Reading,* 439.

26. Ethics: WA, 281. Harding, *Handbook,* 44, describes this as Thoreau writing in "his most pompous and aphoristic vein." Revolve: Henry David Thoreau, "The Service" (unpublished essay, July 1840). Parcel: JR, Apr 2, 1852.

27. Calculation: WA, 281. Wonderful: Ibid.

28. Channing: Channing, *Poet,* title. Salt, *Life,* 63. At her: Ibid., 111.

29. Neither: Deevey, *Re-examination,* 9. Intermixing: Peck, *Morning,* 63. Imagination: Ibid., 81.

30. Intransitive: Edward N. Lorenz, "Climatic Determinism," *Meteorological Monographs* 8, no. 30 (1968), 1. Verge: JR, Jul 18, 1852. Vision: Walls, *Cosmos,* 226. Tropes: JR, May 10, 1853.

31. Moral-aesthetic: Marx, *Excursions,* ix-x.

32. Generalities: Richardson, *A Life,* 272. Description: JR, Jun 27, 1852.

33. Titanic: JR, Jun 27, 1852.

34. Fancy free: JR, Feb 18, 1852. Contrary: Worster, *Economy,* 108.

35. Complexity: Wilson, *Consilience,* 54. Actual life: Walls, *Seeing,* 148.

36. Warfare: NH, 41. Eye: Ibid., 42. Intelligence: JR, Dec 2, 1853.

37. Confidence: JR, Jul 24, 1853.

38. Engineers: JR, Jun 15, 1852.

39. America: Kuhn, *Autobiography,* 1.

40. Prose: JR, Nov 23, 1853. Imaginations: Ibid., Jun 4, 1853. Humanity: Ibid., Feb 13, 1852. Telescope: Ibid., Oct 20, 1852. Rainbow: Ibid., Aug 7, 1852. Lightning: Ibid., Jun 26, 1852. Mythology: Ibid., Nov 8, 1851. Astronomized: Ibid., Apr 2, 1852. Savage: Ibid., Mar 31, 1852. Botanist: Ibid., Feb 5, 1852. Peter-Paul: Ibid., Dec 25, 1851. Mythology: Ibid., Nov 8, 1851. Sawdust: Ibid., Aug 5, 1851.

41. Refusal: JR, Mar 5, 1851 (confirms receipt of invitation). Odd-fellow: WA, 166.

42. Letter: written December 19, 1853. Response: Harding and Bode, *Correspondence,* 309–310.

43. Snarky: JR, Mar 5, 1853. Furious: Miller, *Consciousness,* 183. Defensive: Worster, *Economy,* 89.

44. Significant: JR, May 10, 1853. Only difference: Dean, *Romanticism,* 84.

45. Bonds: JR, Feb 5, 1854. Sailing: WA, 309. Other direction: Baym, *View,* 3. Calculus: Harding and Bode, *Correspondence,* 94 (April 2, 1843, letter to Richard Fuller) and 189 (Nov 14, 1847, letter to R. W. Emerson). Spin: Richardson, *A Life,* 272. Delight: Stephen Jay Gould, *Bully for Brontosaurus: Reflections on Natural History* (New York: W. W. Norton & Company, 1992), 12 (refers to Saint Francis of Assisi, 1181–1226 A.D. and Galilei Galileo, 1564–1642 A.D.).

46. Insight: Richardson, *A Life,* 256. Anticipation: Miller, *Consciousness,* 104. Biblical: Buell, *Imagination,* 170.

47. Sandstone: JR, Feb 24, 1854.

48. Phenomena: WA, 294. Vines: Ibid. Italian: JR, Dec 15, 1837. Inanimate: NH, 66.

49. Beautiful: Shanley, *Making,* 204.

50. Passages: JR, Jun 13, 1851. Note that he summarized his findings only two days after Darwin enters his *Journal.*

51. Cape Cod departure: Harding and Bode, *Correspondence,* 158 (October 9). Organic relic: Hitchcock, *Final Report,* 348 (Figure 59). Copies: Ibid., 5. There is no record in Sattelmeyer, *Reading,* 233, of a library loan or *Journal* entry.

52. Deltas: Goldthwait, *Sand Plains,* Plate 5. Forts: JR, Nov 7, 1851. Methinks: Ibid., Nov 8, 1981.

53. December: JR, Dec 29, 30, and 31, 1851. Parasites: Ibid., Dec 31, 1851. Important: Richardson, *A Life,* 311.

54. Return: JR, Feb 5, 1854 (enters observations made Feb 2, 1854).

55. Origin: Robert Chambers, *Vestiges of the Natural History of Creation,* published anonymously (London: John Churchill, 1844). Slag: WA, 298.

56. Poetry: JR, Feb 8, 1854. Full year earlier: Ibid., Feb 5, 1853. Prime: Ibid., Mar 1, 1854. Inwardly: Ibid., Mar 2, 1854.

57. Information: Jorgensen, *Thermodynamic,* 49. Continuity: JR, Dec 29, 1853; WA, 294–298.

58. Leaf: Sattelmeyer, *Depopulation,* 442–443. Plants: JR, Dec 15 and 19, 1953.

59. Parasitic: WA, 298.

60. Earlier phrase: JR, Feb 5, 1854.

61. Flock: JR, Sep 2, 1850; Worster, *Economy,* 321.

62. Chemistry: Ilya Prigogine, *The End of Certainty* (London: Free Press, 1997).

63. Equilibrium: Jorgensen, *Thermodynamics.* Sphere music: JR, n.d. before Feb 5, 1840.

64. Touch: JR, Jan 13, 1854. Common sense: Joyce Carol Oates, "Introduction" to *Walden,* by Henry David Thoreau (Princeton: Princeton University Press, 1989).

10. Mythology

1. Harding and Bode, *Correspondence,* 200. They add on page 201: "Sanborn identifies the traveling professor as Agassiz."

2. Ibid.

3. My question presumes he wrote nothing only because I have been unable to find something on the subject. Publication: Kenneth Walter Cameron, *Thoreau's Fact Book: Annotated and Indexed, Volume II* (Hartford, CT: Transcendental Books, Drawer 1080, 1966).

4. Natural theology used as a proper noun refers only to the bible-based Christian version. Every religious tradition has its own natural theology. And philosophically, the term refers to a theology originating naturally in the human brain, meaning from within, rather than from without.

5. Assigned: Sattelmeyer, *Reading,* 16. Deist: Ibid. Exemplary: Harding and Bode, *Correspondence,* 26 (Letter of March 26, 1838); Borst, *Log,* 289 (May 27, 1854).

6. Attributes: Hovencamp, *Theology,* 44.

7. Man's needs: Ibid., 140. Bible: Ibid., 43. Religion: Ibid., 47. Seasons: Ibid., 42. Hail Mary: Hitchcock, *Final Report,* 403. Ararat: Ibid., 4a (Footnote), quotes a letter of November 6, 1840, from "Rev Justin Perkins . . . Missionary at Oroomiah in Persia."

8. Merrill, *First Hundred,* 404.

9. Diluvian: WA, 277.

10. Unknowable: Hovencamp, *Theology,* 51. Tea-party: WK, 406. Coal: Ibid., 81.

11. Humboldt: Walls, *Cosmos,* 215. Observer: Kuhn, *Autobiography,* 15. Pipkin, *Geographies,* 534.

12. Detheologization: Peck, *Morning,* 527. Christianity: Hovencamp, *Theology,* 43. This was an American version of the English Bridgewater Treatises, published in 1851. Connected: Ibid., 47.

13. Divorce: Hovencamp, *Theology,* 44.

14. Sacred: Cavell, *Senses,* 14. Renewal: Harding, *Days,* 94. Thought: WK, 166.

15. Eighteen: Cavell, *Senses,* 115. Bible: Lyon, *Symbol,* 294. Metaphor: Anderson, *Circle,* 88. Critique: Sattelmeyer, *Reading,* 67. Yogin: Ibid., 68. Library: Van Doren, *Annotated,* 58. Channing, *Poet,* 50. Krishna: Miller, *Consciousness,* 61. Trope: Friedrich, *Gita,* 148. Revisions: Ibid., 26. Absorbed: Sattelmeyer, *Reading,* 68.

16. Emerson: Richardson, *A Life,* 22–23. Society: WK, 154. West: WK, 154. Senses: Ibid., 188. Treasured: WK, 189. Fossils: WK, 192–194 (see also Walls, *Seeing,* 43–44).

17. *Gita* quotes: Mitchell, *Gita,* verses 2.17, 1.20, 8.17–18, and 9.4–8, in sequence.

18. Eliade, *Myth,* ix. Final: Ibid., 89. Rebirth: Van Doren, *Annotated,* 11 (citing Reginald Cook, 1965, 98).

19. Creation: Kuhn, *Autobiography,* 13. Intimate: Edward F. Mooney, *Lost Intimacy in American Thought* (London: Continuum International, 2009).

20. Engendered: Joel Porte, *Consciousness and Culture: Emerson and Thoreau Reviewed* (New Haven: Yale University Press, 2004). Divine: William James, *Varieties of Religious Experience* (New York: Random House, 1929, originally 1902) provides a backdrop for a review by John Gatta, *Making Nature Sacred* (New York: Oxford University Press, 2004), 129, 148. Exercise: WA, 86. Ganges: Ibid., 186. High Priest: Harding, *Days,* 196.

21. Theory: Sattelmeyer, *Depopulation,* 439. Correspondence: Ibid. Individuals: Walls, *Believing,* 18. Cosmology: Gatta, *Sacred,* 137. Sattelmeyer, *Depopulation,* 441. You: WA, 187.

22. Kouroo: WA, 318.

23. Editing out: Shanley, *Making,* 139. Forever: JR, Dec 29, 1853. Saxon: Ibid. At a time: Harding, *Days,* 465 (footnote 3, quoting Sanborn, *The Personality of Thoreau,* 68–69).

24. Forgot: Maynard, *History,* 20. Glaciology: Pipkin, *Geographies,* 531; McGregor, *Wider,* 11.

25. Publicity: Richardson, *A Life,* 252; WA, 41 (note 221 by Jeffrey Cramer, ed.) report that Thoreau avidly read newspapers, especially on topics like Arctic travel.

26. Unproven: Howarth, *Walking,* 13. Actinism: JR, Feb 18, 1851. Following Linnaeus and J. J. Audubon, Thoreau used the name *Colymbus glacialis.* The common loon is actually *Gavia immer.*

27. Fashionable: Rebecca Bedell, *The Anatomy of Nature: Geology & American Landscape Painting, 1825–1875* (Princeton, NJ: Princeton University Press, 2002), 2. Romantic: Wilson, *Consilience,* 61. The American Geological Society (1819), through two subsequent organizations, eventually gave rise to the AAAS in 1848, listing 461 "Men of Science." American Association for the Advancement of Science, "About AAAS: History and Archives," http://archives.aaas.org/. Pursued: Bedell, *Anatomy,* 2.

28. Ornithology: Harding, *Days,* 29. Emerson: Ralph W. Emerson, Journal entry for July 13, 1833, *Nature Writing: The Tradition in English* (New York: W. W. Norton & Company), 145 (excerpted from *The Journals of Ralph Waldo Emerson* (Boston: Houghton Mifflin, 1909–1914) Mad: Ibid., 3.

29. Queen: Laura Dassow Walls, "Science and Technology," *The Oxford Handbook of Transcendentalism,* ed. by Joel Myerson, Sandra Harbert Petrulionis, and Laura Dassow Walls (Oxford: Oxford University Press, 2010), 572. Astronomy: Herbert, *Darwin,* 31 (quoting Sir William Herschel). Society: Bedell, *Anatomy,* 7.

30. Chair: Lurie, *Agassiz,* 140.

31. Silliman: Lurie, *Agassiz,* 125. Audience: Ibid., 126. Brilliance: Merrill, *First Hundred,* 417. Interview: Harding and Bode, *Correspondence,* 244 (Letter of July 5, 1849). Zoology: Sattelmeyer, *Reading,* 83. Facsimiles: Harding, *Handbook,* 102. Letter: Harding and Bode, *Correspondence,* 296 (February 27, 1853, to H.G.O. Blake).

32. Scale: JR, Feb 26, 1854. Forbes: Elisha Kent Kane, *The U.S. Grinnell Expedition in Search of Sir John Franklin. A Personal Narrative* (New York: Harper & Brothers. [U.S.N. New Edition, Philadelphia: Childs & Peterson, 1854, E-text from 1856 edition, UMich http://quod.lib.umich.edu/m/moa/AJA5420.0001.001?rgn=main;view=fulltext]). Thoreau's fact book (Cameron, *Fact Book*) is indexed to Kane on page 217 and excerpts pages 48–143, mostly on botany. Thrust: Ibid., 455.

33. Works: Sattelmeyer, *Reading,* 87. Rival: Walls, *Oxford,* 580. Dissection: Sanborn, *Henry,* 245.

34. Proofs: Van Doren, *Annotated,* 37. Boston: JR, Mar 13, 1854. Indenture: Borst, *Log,* 283 (cites Ballou, 150–151, Mr. Fields aborted the trip).

35. Checkout: Sattelmeyer, *Reading,* 118, documents library records. Agassiz's book was widely known. Thoreau's fact book (Cameron, *Fact Book*, p. iv) contains

a reference to Agassiz and Gould's zoology text (indexed to page 8) but no citation to, or notes from, *Etudes sur les Glaciers.* Telescope: JR, Apr 13, 1854 (Borst, *Log,* 285, records for April 10 that Thoreau had long wanted a telescope).

36. First Proof: JR, Mar 27 1854. Voyagers: Ibid., Mar 28. Breaks the ice: Ibid., Mar 31.

37. By "glacial theory," I refer to Agassiz's catastrophic version. Thoreau may have been tempted to introduce Forbes's less catastrophic version, but chose not to.

38. Color: WA, 171 (footnote 12). Reflection: JR, Apr 27, 1854. Genius: Ibid.

39. Taunt: Harding and Bode, *Correspondence,* 326.

40. Moonlight: JR, Jun 28, 1852. Hitchcock, *Final Report,* 8a (Fig. 281 and 282). September: Howarth, *Walking,* 247–249. North and south: Hitchcock, *Final Report,* 389.

41. Grooved: JR, Jun 7, 1858 (from Howarth, *Walking,* 252–254). Thoreau added more on his last trip on August 2, 1860 (Howarth, *Walking,* 310).

42. Certain: WA, 177. Looked: JR, Feb 3, 1852.

43. Pow-wow: WA, 176–177. Lake: Van Doren, *Annotated,* 313 (footnote 31); WA, 177 (Note #30), referencing the Aug 11, 1821, Middlesex Gazette. The correct spelling is "Alexander." This post-publication "confession" is similar to his post-publication embrace of Monadnock's striations

44. Also possible is that native traditions derived from the visible (stratigraphic) evidence of collapse that is common wherever kettles are exposed by river or coastal erosion.

45. Mine: WA, 96; Shanley, *Making,* 143. Paver: WA, 176–177.

46. Glacier: WA, 177, annotation 32. Buried: WA, 133. Howls: Ibid., 133. Crept: JR, Feb 3, 1852. Dug: WA, 133. Keeping: Ibid. Note: this myth was present in *Walden,* Version I (Shanley, *Making,* 168).

47. Fertility: WA, 133. Expressed: WW, 123. Myself: WA, 134.

48. Father: WK, 467. Sister: Shanley, *Making,* 169.

49. Ax, snow, ice: WA, 39. Views: Ibid., 262. Snows: WA, 244.

50. Cat owl: WA, 124.

51. Unidentified: WA, 116 (note #50 by Jeffrey Cramer). Nevada: WA, 115–116. The plow/mouse connection may have derived from a poem by Robert Burns, *To a Mouse* (1785).

52. Existence: WA, 244. Victorian fear: Robert Macfarlane, *Mountains of the Mind* (New York: Pantheon Books, Random House, 2003), 128, describes it as their equivalent to "nuclear winter." Great Snow: Thoreau mentions it five times in *Walden.* By then the 1717 blizzard was old enough to have become mythical.

11. Simplicity

1. Laconic: Richardson, *A Life,* 65. Ascended: JR, Sep 10, 1839. Eagerness: Hodder, *Ecstatic,* 124. Positive: JR, Sep 6–10.

2. Factory: MW, 56. Nightmare: John Hanson Mitchell, *Living at the End of Time* (Boston: Houghton Mifflin, 1990), 166–167.

3. On lower summits like Wachusett, that "false" tree line is based on drainage and exposure, and typically reveals a small "bald" smoothed by the passage of glacial ice.

4. Cabin: Richardson, *A Life,* 147. Tower: Maynard, *History,* 82 (Letter from Emerson to Thomas Carlyle, May 14, 1846).

5. Conquer: Albert Boime, *The Magisterial Gaze* (Washington, DC: Smithsonian, 1991), 20. Lake District: Adam Black and Charles Black, *Black's Picturesque Guide to the English Lakes,* 2nd ed. (Edinburgh: Adam and Charles Black, 1844). Face of God: William Cronon, "The Trouble with Wilderness; or, Getting Back to the Wrong Nature," *Uncommon Ground: Rethinking the Human Place in Nature,* ed. William Cronon (New York: W. W. Norton & Co., 1995), 73.

6. Ambivalence: William Cronon, 73. Insolent: Boime, *Magisterial,* 19. No change: Harding and Bode, *Correspondence,* 296 (Feb 27, 1853, to H.G.O. Blake).

7. Massachusetts: WW, 90.

8. Observatory: Shanley, *Making,* 113. Thoreau's horizontal arc would have been somewhat larger, owing higher visibility due to more clearing. Walden Pond appears above the foreground at about –6° and the wooded horizon at +3°.

9. Alma natura: JR, Sep 15, 1838. First draft: Shanley, *Making,* 141. Essential facts: WA, 88. Resort: Ibid., 187. Heaven: Ibid., 274. Earth: JR, Oct 22, 1853. Burrowing: WA, 96. Thoreau's commitment to local, rather than global, travel is consistent with this approach.

10. Important: WA, 84.

11. Equation: Harding and Bode, *Correspondence,* 215 (March 27, 1848).

12. Pi or π: The integers after the decimal repeat endlessly and irregularly.

13. This fourth integration served as the title for the talk I presented to the Thoreau Society's 2010 annual gathering, which prompted my attendance on the field trips, which precipitated this book.

14. Noether: Natalie Angier, "The Mighty Mathematician You've Never Heard Of," *New York Times,* 27 March 2012, D4.

15. Simplicity: WA, 89. Note: Thoreau called all this sediment "aluvion," Darwin and Lyell called it "alluvium"; their term for all unconsolidated materials. "Contact!" is the spelling used in MW.

16. Regular: except for sunspot activity, magnetic storms, and other variables.

17. Troposphere is the technical term for "active atmosphere." It is about six miles thick over Walden. Ecologically, spring is the threshold moment when the system shifts from discharging its stored chemical power to recharging it.

18. Alive again: WA, 300.

19. The spring turnover is typically between late March and early April. The fall is greatly delayed, between mid-December and early January.

20. Robin: WA, 301.

21. Transcendentalism: Joel Myerson, Sandra Harbert Petrulionis, and Laura Dassow Walls, eds., *The Oxford Handbook of Transcendentalism* (Oxford: Oxford University Press, 2010). Transcend it: Gura, *Transcendentalism,* 269. Sensuousness: Friesen, *Huckleberry,* xiv.

22. Lowest: WA, 88. Fixate: Cameron, *Writing,* 51, focuses on the word "corner." Poets: Myerson, *Handbook*, 2010.

23. Simplicity: WA, 89. Father tongue: Ibid., 99. Patagonian: Ibid., 74. Wild animals: MW, 64. Cave: Ibid., 27. Agricultural: WK, 60. Huron: Ibid., 394. Truth: Ibid., 68. Skeletons: Ibid., 69. Particularity: H. Daniel Peck, "Lakes of Light: Modes of Representation in Walden," *Nineteenth Century Prose* 31, no. 2 (Fall 2004): 172. Intuition: Canby, *Thoreau,* 224. Pagan: Joseph Wood Krutch, Introduction to *Walden* (New York: Bantam, 1962), 12. Thebes: WA, 55.

24. Chaos: MW, 62. Lyre: WA, 119. Uncertain: WA, 170. Matchbox: James Russell Lowell, "Thoreau," *Walden and Resistance to Civil Government: Authoritative Texts, Thoreau's Journal, Reviews and Essays in Criticism, Second Edition,* ed. William Rossi (New York: W. W. Norton & Company, 1992), 337. Originally *The Writings of James Russell Lowell* (Boston: Riverside Edition, 1890) (written 1865).

25. Beginnings: JR, Jan 21, 1853. Slimy pools: Ibid., Jun 15, 1852. Sermons: JR, n.d., 1850; Torrey and Allen, *Journal of HDT,* V2, 14. Swamps: Walls, *Cosmos,* 579: "Down in the trenches—or rather, the ditches and swamps—was precisely where Thoreau wanted to be." Intriguing: WA, 170; Harding and Bode, *Correspondence,* 215 (March 27, 1848). Kith and kin: JR, Dec 15, 1841. Fox: Ibid.

26. Pond: Lyon, *Symbol,* 292. Well: Ibid., 291. City: James Carroll, *Jerusalem, Jerusalem: How the Ancient City Ignited Our Modern World* (Boston: Houghton Mifflin Harcourt, 2011).

27. Castles: WA, 315.

28. Kindred: WA, 128. Precisely: WA, 216.

29. Anthropic: Hawking, *Brief History,* 125 (pages 124–127 inclusive). More scientifically: "In a universe that is large or infinite in space and/or time, the conditions necessary for the development of intelligent life will be met only in certain regions that are limited in space and time." H. Daniel Peck (Peck, *Morning,* 117) considered this question from "Brute Neighbors" as "the central question that Walden seeks to answer and to which Walden itself is an answer." Evidence: Hawking, *Brief History,* 124.

30. Communion: JR, Jun 21, 1852. Fruits: Henry D. Thoreau, *Wild Fruits: Thoreau's Rediscovered Last Manuscript,* ed. Bradley P. Dean (New York: W. W. Norton & Company, 1999).

31. Variety: JR, Sep 7, 1851. Wilderness: Ibid., Jan 21, 1853.

32. Cowley: JR, Apr 18, 1852 [unidentified, but Sattelmeyer, *Reading,* 159, references Abraham Cowley's (seventeenth-century English poet and writer) *Prose Works* as cited in Thoreau's "Dated Reading List" (1840–1841) and "Literary Notebook" (1840–1848)]. Blackbirds: Ibid. Hieroglyphics: Ibid., Aug 23, 1852. Mysterious: Ibid., Apr 18, 1852. Ripple: Ibid., Nov 21, 1850. Fitted: Ibid., Aug 3, 1852. Rocks or earth: Ibid., Jan 17, 1854.

33. Who: JR, Feb 19, 1854. Why: WK, 195. Calyx: JR, Aug 6, 1852. Universe: Ibid., Feb 18, 1854.

34. Doubleness: JR, Aug 8, 1852. Designed: Ibid., Aug 6, 1852. Instincts: WG, 107. Birds: JR, Jun 1, 1853. Grand scheme: Birds and humans shared a common ancestor within the last 10 percent of geological time. Cockerels: WA, epigraph.

35. Feet: JR, Dec 16, 1850. Level best: Ibid., Apr 3, 1852. Helpless: Ibid., Aug 11, 1853. Unfitness: Ibid., Aug 31, 1852.

36. Public: JR, Aug 8, 1852. Frozen: Ibid., Dec 16, 1850. Silence: Ibid., Jun 11, 1851 (quoting Darwin). Religious: Ibid. God: Ibid.

37. Roger Lewin is a co-author.

Epilogue

1. Southwesterly: JR, Nov 7, 1851.

2. Cast up: JR, Nov 8, 1851. Dying: Ibid.

GLOSSARY

Note: Definitions as used in this book. Letters in parenthesis associate each entry with one of three main narratives: Geoscience (G), Thoreau (T), and History (H).

Acadian Orogeny Major mountain-building episode for central New England responsible for creating much of its metamorphic and igneous bedrock, which formed deep within the continental crust. (G)

Agassiz, Louis Harvard paleontologist, zoologist, and author of *Etudes sur les Glaciers*. One of Thoreau's two local intellectual antagonists, owing to his belief in progressive creationism through serial catastrophes. (H)

Alluvion Thoreau's version of the word "alluvium." In modern usage, it refers to unconsolidated river sediment. In Thoreau's era, following Darwin and Lyell, it referred to all unconsolidated sediments between the soil and bedrock thought to be deposited by agencies still active on the surface. (T)

Andover Granite Official bedrock unit mapped beneath Walden Pond by geologists. In Concord, it intruded preexisting strata of the Nashoba Terrane during peak metamorphism. (G)

Anthropic Principle Philosophical and cosmological principle stating that the world we know is the only one we could know, and explaining why the universe seems "just right" for humans. (G)

Anticyclone Persistent pattern of cold-dry glacial air flowing off of the lingering ice sheet and southwesterly over New England during the millennia of deglaciation. (G)

Aquifer Body of earth or fractured rock that is capable of holding and transmitting a significant quantity of water. At Walden, it is predominantly sand with minor gravel. (G)

Assabet River Northeast-flowing tributary of the Concord River that joins the Sudbury River at Egg Rock, Concord. Site of many excursions by Thoreau. (T)

Avalon Terrane Tectonic term for a discrete mass of Proterozoic continental crust present east of the Bloody Bluff Fault (also Avalonia). Near Walden, it created the topographic divide between the Concord River and the Charles River. Separated from the Nashoba Terrane to the west by the fault. (G)

Bhagavad Gita Sacred Hindu text that shaped Thoreau's lifelong thinking from his early transcendental phase through the Walden period. (T)

Bloody Bluff Fault Important geologic boundary separating the Avalon Terrane to the east-southeast from Nashoba Terrane to the west-northwest. (G)

Catastrophism Theoretical perspective on Earth history emphasizing periods of prolonged stasis separated by brief intervals of rapid change, usually accompanied by mass extinction. Counterpoint to uniformitarianism. (H)

Darwin, Charles Robert Nineteenth-century geologist and author of *Journal of Researches* (a.k.a. *Voyage of the Beagle*), a book that contributed greatly to Thoreau's decision in 1851 to be a sojourning field scientist. Thoreau read an 1846 edition. (H)

Debacle Catastrophic event invoked to explain what are now known as glacial deposits as the result of powerful surges of iceberg-laden water from the north. This explanation was favored by Massachusetts State Geologist Edward Hitchcock. (H)

Deep Cut Canyon-like excavation through a previously wooded ridge for the causeway of the Fitchburg Railroad. Thawing of fine-grained sediment on the exposed banks of the excavation produced intricate patterns used as metaphor in the climax of *Walden*. (T)

Delta-plain The flat upper surface of a local meltwater delta deposited by fluvial processes above the surface of a glacial lake. At Walden, the surface was more gravel-rich than the underlying sediment and was interrupted by "sink-holes" called kettles. The material from which the delta plain is composed is called the delta topset. (G)

Denudation Long-term removal of mass from the land caused by weathering and erosion (from the top down) averaged over long time intervals over large geographic areas. (G)

Descendentalism Practice or philosophy of descending toward meaning and truth as simplicity, rather than ascending or transcending toward complexity. Defined in this book to provide a counterpoint to the more familiar word "transcendentalism." (T)

Diluvium Official geological map unit used in nineteenth-century Massachusetts to identify and characterize what are now known as glacial sediments. The term is biblical in origin, traced back to the mythical deluge of Noah and Gilgamesh. (H)

Drift General term still in use today for what are now known as glacial deposits, but which were originally thought to have been deposited by debris-laden icebergs drifting in the sea. This was the prevailing interpretation for the origin of the unconsolidated sediment above rock during Thoreau's era. (H)

Emerson, Ralph Waldo Concord philosopher, essayist, and leader of the American transcendental movement. Thoreau's mentor, patron, editor, friend, and employer. (T)

Epilimnion Wind-stirred and sun-warmed surface layer of lake water floating above colder, more stable water at depth. (G)

Etudes sur les Glaciers "Studies of Glaciers" was the most important book responsible for promulgating the glacial theory in America. Authored by Louis Agassiz in 1840. (H)

Exhumation The passive uplift of rock from deep within the Earth's crust (usually a mountain root) caused by denudation at the surface. The principle mechanism is buoyant adjustment (isostasy) associated with the removal of mass at the top. (G)

Fairhaven Bay: Walden-sized kettle lake through which the Sudbury River flows from south to north. One of Thoreau's favorite sojourning destinations. (T)

Final Report on the Geology of Massachusetts Edward Hitchcock's major work as State Geologist and an important reference work for Thoreau. Published 1841. (H)

Focus For an earthquake, the site of original rupture on fault plane within rock. Technically known as the hypocenter, it lies directly below the epicenter. By analogy with Walden Pond, the site of maximum subsidence (gravitational ground zero). (G)

Fractal Short for fractional geometry. A self-similar pattern independent of scale. Technically, a mathematical set with a specific fractal dimension. (G)

Glacial Lake Sudbury Major glacial lake dammed up in the Sudbury Valley by the retreating Laurentide Ice Sheet to the north. It drained southward to the Charles River over a divide in Weston. The Walden kame delta was built into an arm of the lake, burying the ice blocks that later melted to create its kettle basins. (G)

Glacial Theory Explains the materials and landforms of presently non-glaciated regions as a consequence of greatly expanded glaciers of the past. For this book, an explanation for the New England landscape involving expansion of a continental ice sheet centered over eastern Canada during the recent geological past. (H)

Great Year A pedagogical device used in this book. One complete glacial cycle at Walden Pond treated as if it were a single year broken down into four seasons: Fall = ice advance; Winter = ice cover; Spring = recession after deglaciation; Summer = nonglacial conditions. (G)

Hitchcock, Edward State Geologist of Massachusetts, professor of natural theology, Congregational pastor, and President of Amherst College during the Walden era. One of Thoreau's two local intellectual antagonists for his ardent, lifelong belief that geology and Christian scripture must be concordant. (H)

Hutton, James Scottish gentleman-farmer considered the father of modern geology and the most famous of the plutonists. His *Theory of the Earth* (1795) emphasized cyclical creation and destruction of the crust by geothermal and aqueous mechanisms, respectively. (H)

Hypolimnion Mass of deep, cold, dense water in Walden that is nearly motionless between times of thermal mixing. It lies below the transitional layer (metalimnion), which lies below the warmer, less dense, and more active layer of surface water (epilimnion). (G)

Induction The creation of general ideas and/or fundamental principles by linking particular observations and more restricted ideas together. The goal is to create unifying laws and theories. (G)

Intransitive System A system with two or more stable states of equilibrium. Used to explain Thoreau's tendency to be alternately very scientific or very poetic, and Walden Pond's different limnological configurations during winter vs. summer. (T)

Isostasy Geological principle similar to buoyancy, but applied to the "flotation" of discrete and less-dense masses of the Earth's crust above and within its more dense, but ductile, upper mantle (asthenosphere). (G)

Kame delta Mass of glacial meltwater sediment built outward from the edge of stagnant ice into a glacial lake, usually during a local pause in retreat. Much of the chaotic terrain of Walden Woods formed by the collapse of a large kame delta. The remaining uncollapsed portion is the Walden deltaplain. (G)

Kane, Elisha Kent American Arctic explorer and author of *Personal Narratives* (1853), which greatly influenced Thoreau's thinking. Kane's descriptions of the Greenland Ice Sheet helped vindicate the glacial theory after the Walden period. (H)

Katabasis Greek word for descent used for the spiritual component of Thoreau's "down to the pond" experiences. The same word root is used for the persistent northeasterly "katabatic" winds that swept the area of Walden Pond during deglaciation. (T)

Katahdin, Mount Maine's tallest mountain. Site of Thoreau's 1846 epiphany regarding wildness, and the place where his powerful intuition brought him to a largely correct view of Earth's origin. Thoreau's spelling was "Ktaadn." (T)

Kettle Closed topographic depression caused by collapse and subsidence associated with stagnant glacial ice. Walden Pond fills a basin consisting of four related kettles that coalesced during meltdown. (G)

Laurentide Ice Sheet The physical mass created by the last major advance of glacial ice from Canada. It flowed south-southeast over Concord to reach a terminus somewhere near western Martha's Vineyard, and culminated between 25,000 and 20,000 years ago. (G)

Limnology Scientific study of lakes (physics/chemistry/biology/geology) and also extended to inland rivers. Similar to oceanology, but terrestrial. (G)

Lyell, Charles Scottish geologist and writer whose *Principles of Geology* gathered geology in to a true discipline that became the most popular science during Thoreau's era. Lyell was an ardent opponent of the glacial theory and promulgator of the drift theory. (H)

"Men of Science" Thoreau's generally pejorative term for the institutionalized American science of his day. A mix of both natural theologians guided by scripture and scientific secularists seeking to study nature objectively and dispassionately. Thoreau objected to both approaches. (T)

Monadnock, Mount Thoreau's favorite mountain in southwestern New Hampshire, which he climbed four times. The name was appropriated by geologists for a type of mountain isolated by differential erosion. The nineteenth-century spelling was Monadnoc. (T)

Muck General term for soft organic sediment at the bottom of Walden and other lakes. Synonymous with the word "detritus" of ecologists and "gyttja" of limnologists. (G)

Nashoba Terrane: Tectonic term for a discrete mass of Paleozoic crust beneath Concord, Walden Woods, Walden Pond, and Thoreau's western sojourning country, especially in the lower Concord River and its Assabet River tributary. Separated from the Avalon Terrane to the east by the Bloody Bluff Fault, and from the Merrimack Terrane to the west by the Clinton-Newberry Fault. (G)

Natural theology Historical movement during the eighteenth and nineteenth centuries seeking concordance between Christian scripture and natural science. Its American acme coincided with Thoreau's *Walden* revisions and generation of *Walden*-Part II. (H)

Neptunism Geological school of thought during the late eighteenth and early nineteenth centuries. Held that Earth's rocks were created by crystallization from aqueous brines during its early history. Associated with catastrophism and biblical scripturalism. Contrast with plutonism. (H)

Orogeny Discrete interval of mountain building, usually associated with the convergence and collision of crustal masses driven together by tectonism. Creation of New England was broadly the result of three orogenies: the Taconic, the Acadian, and the Alleghenian. (G)

Paleo-valley Former tributary valley of the Sudbury River created by long-term denudation of the bedrock grain. Filled during deglaciation with glacial deltaic sediment and residual blocks of stagnant ice, which melted to create Thoreau's "chain of ponds," one of which is Walden. (T)

Pangaea Earth's largest supercontinent created during the Alleghenian orogeny. Rifted apart to create the Atlantic Ocean. (G)

Plumbago Nineteenth-century term for graphite, a carbon mineral used to make pencil lead and for printing. Erroneously called "black lead." This commodity provided the main source of income for the Thoreau family during the Walden years. (T)

Plutonism Geological school of thought during the late eighteenth and early nineteenth centuries. Held that Earth's rock was created by crystallization from a molten state caused by geothermal heat. Associated with uniformitarianism and deism. Contrast with neptunism. (H)

Point d'appui Word of French origin used metaphorically by Thoreau to signify the solid point of support on rock where true reality could be found. In Concord, it is equivalent to the unconformity separating glacial "alluvion" from the underlying Andover Granite. (T)

Pond Survey Thoreau's mapping and bathymetric surveying (fathoming) of Walden Pond in the winter of 1846–1847, and his subsequent research project using that data. (T)

Sand foliage Inclusive term for a variety of transient features created by flowing slurries of saturated sand, silt, and clay at the Deep Cut during episodes of thaw. Use of the term "foliage" is metaphor. (T)

Stage Level or relative elevation of the surface of a body of water, usually a lake or a river, indicating its height above some reference datum. When Thoreau says "the pond rises and falls," he is describing changes in stage. (G)

Stagnant ice Glacial ice that was formerly part of an actively moving mass, but which has thinned to the point where it can no longer move. In terrain like Concord, it is usually rich in crushed rock debris and milled (subrounded by glacial shear) stones of all sizes. The voids that became Walden Pond were created by the melting of blocks of stagnant ice. (G)

Steady state Type of equilibrium in an open system in which negative feedbacks keep the system fluctuating around some average condition. Also called "homeostasis." The stage of Walden is a good example. (G)

Sudbury River North-flowing tributary of the Concord River that joins the Assabet River at Egg Rock, Concord. Near Thoreau's Walden, especially to the south, it was mostly agricultural country with broad fertile meadows. (T)

Surveying Land surveying for legal property description was Thoreau's primary vocation during the Walden years. (T)

Terrane Geological term for a discrete mass of rocks that share a common history, and which accrete with other terranes to form larger land masses. Not to be confused with "terrain," which is a topographic or descriptive surface term. (T)

Transcendentalism In Thoreau's America, a complex historical, spiritual, literary, and philosophical movement surrounding Ralph Waldo Emerson that sought correspondence between natural facts and spiritual facts. (T)

Turnover General term for the density-driven mixing of a lake volume. For Thoreau's Walden Pond, this happened twice per year: once during freeze-up; and more significantly during ice-melt. The lake mixes at a critical temperature of approximately 39°F and 4°C, when water reaches its maximum density. (G)

Unconformity Physical contact where the materials of a younger geological period overlie those of a previous period, and where a significant break of time is present. Locally, the contact between Nashoba bedrock and glacial sediment. Equivalent to Thoreau's *point d'appui*. (G)

Uniformitarianism Theoretical perspective on Earth history emphasizing the slow and uniform development of landscapes and sediments. Generally associated with the gradual summation of causes now in operation over long periods of time. Counterpoint to catastrophism. (H)

Walden Pond Lake in Concord, Massachusetts, created by melting of multiple residual blocks of stagnant ice and maintained by the filling of large voids with groundwater beneath a steady state water table. Its western basin was the site of Thoreau's famous experiment in deliberate living and the inspiration for his book *Walden*. (T)

Walden System Discrete, open, steady state system containing four realms of equal importance: Walden Pond, the air above it, the adjacent drained land, and the aquifer. Powered by sunlight and activated by life. Idealized as a radial system surrounding the nadir of its western basin. Thoreau lived within the system, not on it. (G)

Walden Woods Historic woodland surrounding Walden Pond, especially to the north. Generally characterized by irregular topography and sterile soils associated with meltdown collapse of the Walden kame delta. (T)

Werner, Abraham German mineralogist and father of the scientific neptunist school. Mentor and professor to Alexander Humboldt, Wolfgang Goethe, and Carl Linnaeus, all of whom influenced Thoreau. (H)

INDEX